EXPERIMENTAL PHYSICS
AND
ROCK MECHANICS
(Results of Laboratory Studies)

EXPERIMENTAL PHYSICS
AND
ROCK MECHANICS
(Results of Laboratory Studies)

A.N. Stavrogin
High Pressure Rock Physics Laboratory
Mining Institute (Technical University)
St. Petersburg
Russia

B.G. Tarasov
Department of Civil and Resource Engineering
The University of Western Australia
Nedlands, Western Australia
Australia

Edited by Charles Fairhurst

 A.A. BALKEMA Publishers

LISSE   ABINGDON   EXTON (PA)   TOKYO

Published by: A.A. Balkema, a member of Swets & Zeitlinger Publishers
www.balkema.n1 and www.szp.swets.n1

ISBN 90 5809 213 5

Printed in India

Foreword

This monograph by Professors Stavrogin and Tarasov describes a comprehensive series of laboratory tests to determine the complete load-deformation response of a variety of rock types over a wide range of testing conditions. Intrinsically high stiffness loading systems, specially designed by the authors, has been used for the tests. The information contained in the results is not limited to the laboratory, however, but is of more general value in a way that is of particular value in rock mechanics.

One of the main purposes of laboratory testing of materials is to determine mechanical properties such as deformability and strength for use directly in engineering design. Rock testing can and does serve this purpose but, since variability and uncertainty are intrinsic characteristics of rock behaviour, especially on the large scale, most laboratory studies on rock have been aimed at developing a better *qualitative* understanding of the physics and mechanics of rock deformation, disintegration and collapse. Valuable practical insights have been gained in this way, as illustrated below - and there is much more to be done!

Rock belongs to a class of materials known as *frictional-cohesive* including soils, cements, concretes, bitumens, ceramics, grains and powders - in which the strength is a function of the normal pressure applied to the material. The strength of a frictional material is considerably higher in compression than it is in tension; a consequence of the fact that the strength derives from the frictional resistance to shear deformation across 'point' contacts (particle-particle interfaces, or asperities on planar contact surfaces such as joints) within the material. Rocks and other granular materials with substantial cementation or cohesion between the grains or particles have considerably greater strength unconfined or under low confining stress than a more or less cohesionless soil or sand, but both rock and soil exhibit a rapid increase in strength (shear resistance) with increase in confining pressure. The (shear) strength, increases following Amonton's Second Law of Friction $\tau = \mu\sigma_n$ [where τ the resistance to shear along an interface, σ_n is the stress normal to the interface, and μ is the coefficient of friction between the surfaces, Scholz (1990.)] The *strength* is reached when frictional slip occurs. Where the cohesion (c) is significant, the relationship becomes $\tau = c + \mu\sigma_n$, corresponding to the linear Mohr-Coulomb failure criterion.

Laboratory test values can be applied in the classical manner (i.e. directly to design) for most frictional materials. By contrast, in many rock engineering problems - stability of slopes, tunnels and mine excavations, earthquakes, for example - the mechanical response of the large rock mass involved in design (and/or analysis) is dominated by the properties of interfaces and discontinuities (bedding planes; joints; faults) that are not represented in the typical laboratory specimen. While it is possible, in principle, to obtain and test samples of the interfaces and discontinuities, albeit on a smaller scale, and to incorporate these measured properties into numerical models that will predict a composite ('intact' specimen plus discontinuities) response, it is not possible, in general, to conduct direct tests to verify the predictions. It is certainly not possible to conduct enough such tests to establish the variability of the structure. There is also the difficulty of assessing the influence of the even larger-scale geological features - changes in lithology, folds, fault and other complicating heterogeneities that will influence the loading conditions on the 'outer boundary' of the local rock mass.

An additional complication with assessment of the mechanical behaviour of a rock mass is the fact that rock *in-situ* is pre-loaded, by forces due to gravity; tectonic effects; possibly residual stresses; and fluid pressure. The act of making an excavation, be it a borehole or a tunnel, causes these forces to be re-distributed. Stress concentrations in the immediate vicinity of the excavation may be sufficient to cause damage locally to the rock; fluid under pressure will tend to drain towards the excavation. Stress changes imposed during the coring process can also produce irreversible mechanical effects on and possible damage to the core. (This topic is discussed in Chapters 2 and 4 of this book). Where the pores contain fluids under pressure, the normal stress and, as a consequence, the frictional shear resistance or strength are effectively reduced, as defined by Terzaghi (1923) to the relationship $\tau_{eff} = \mu (\sigma_n - p)$ where τ_{eff} is the *effective shear stress*, and p is the pore fluid pressure. (The influence of pore fluids on rock deformation behaviour is discussed in Chapter 4 of this book.)

Given these complications, rock engineering design of large-scale structures tends to rely on estimates of the rock mass mechanical characteristics, based on a relatively large body of practical cases, codified into empirical *rock classification systems*. Although these systems usually include a laboratory compressive strength (or modulus of deformation, in the case of rock mass deformability estimates) as one of several inputs to the classification, precise values are not required.

What, then, is the significance of laboratory testing of *intact* specimens in rock mechanics?

First, it is important to recall that there are many rock engineering problems where laboratory - scale tests are directly relevant - borehole stability in petroleum engineering; rock drilling and tunnel boring; rock crushing for aggregates and mineral processing, for example. However, given the general impossibility of conducting large scale prototype tests in rock, engineers have also tried to use laboratory testing, so far as possible, to develop an understanding of full scale behaviour. The strength of a rock mass is typically much lower than the strength of an intact specimen, but both the rock mass and the specimen are frictional meterials and both exhibit a qualitatively similar (Mohr-Coulomb) response to applied loads. This appears to be true over the range of loading up to and beyond the peak resistance (strength), where disintegration and collapse take place. Thus, considerable general insight as to the ultimate behaviour of rock on the large scale can be gained by study of the corresponding behaviour of laboratory specimens.

An excellent example of the use of laboratory studies to gain this general insight is provided by the pioneering experiments by the late Professor Neville Cook (Cook 1965), to observe and control the progressive disintegration and loss of strength of rock specimens loaded beyond the peak strength i.e. to obtain the *complete load-deformation curve*, by use of specially designed stiff testing machines. At that time, in the early 1960's, the notions of *peak* and *residual strength* were well established for soils, but the *postpeak* response of rock was unknown. The reason was simple. Testing machines available then were designed primarily for testing ductile materials, or for determining the deformability and compressive strengths only of brittle materials such as concretes. Elastic design of structures was prevalent and little attention was given to behaviour beyond the peak load or *strength of the specimen*. (Barnard (1964) did design, independently, a stiff testing machine to determine the complete load-deformation response of concrete specimens). Cook recognized that the standard hydraulic ram machines were too compliant so that, once the specimen started to lose its load-carrying ability (i.e. to disintegrate), the loading system released more energy than could be absorbed by slow deformation of the specimen. The excess energy was used to accelerate the disintegration process - the specimen disintegrated violently.

Cook's research on stiff testing machines soon led to the introduction, by MTS Systems Corporation, Minneapolis, Minnesota, USA (Hudson et al 1972), of servocontrol systems—an alternative procedure for limiting the energy released into the specimen. These systems are now standard for the testing of rock and other brittle materials, especially when post-peak behavior is of interest.

The factors controlling energy exchange between the disintegrating specimen and the elastically unloading test machine were quickly seen by Cook et al (1966), Starfield and Fairhurst (1968), Salamon (1970) and others, to be analogous to collapse of mine pillars driven by the energy released by the surrounding elastically loaded rock-mass. Control of the energy release by control of the *local mine stiffness* was seen as a possible way to control rockbursts. The Energy Release Rate (ERR) method of mine design (Cook et at, 1966) was introduced based on this principle. More recently, the excess Shear Stress (ESS) approach by Ryder (1987) has been introduced, based on similar *energy release/energy dissipation* balance concepts but applied to slip on pre-existing joints. On a larger scale, the mechanics of energy release in earthquakes are explained in essentially the same manner (Scholz 1990).

Thus, major advances in understanding of the mechanics of unstable collapse in geotechnical engineering and valuable practical design principles have resulted from studies of the mechanics of damage and disintegration of laboratory specimens of rock. Comparable progress is possible today, albeit based on different technical developments.

The remarkable advances, over the past decade or so, in numerical modelling techniques designed for geotechnical applications provide exceptional opportunities, as yet virtually unexplored, for integration of laboratory studies and field applications in rock mechanics. This is especially so when the modelling is used in conjunction with the impressive developments in geophysical techniques, applicable also in the laboratory and in the field. [Young et al. (2000), Mendecki (1997)]. Two and three-dimensional discontinuum numerical modeling codes such as UDEC (1999) and 3DEC (1998), for example, provide a foundation on which to combine the results of laboratory tests on intact specimens and on joints to arrive at an estimate of rock mass properties.

Estimates of the strength of the rock mass in *unjointed* rock -- as in the granites of the Underground Research Laboratory, at Pinawa, Canada indicate that the strength is not very dependent on size. Some reduction in strength does occur with increased duration of loading, saturation of the rock, etc. that arises in field situations. A similar *size independence* appears to exist in other rocks e.g. salt, clays, where joints are absent. The greater unknown is how to conduct appropriate laboratory tests on joints and extend these results to the field scale and the field environment. This could be a fertile area of research. With data on rock joint properties available it would be posssible to compare the rock mass behaviour predictions from the numerical modeling studies with estimates suggested by the various empirical rock classifications systems. This would be informative as to the sources of the differences. If successful, it should be possible to identify the appropriate laboratory testing procedures for intact specimens and joints to allow good estimates to be made of the rock-mass behaviour, in both stable and unstable (dynamic) situations.

A fundamental limitation of empirical classification systems, such as now used widely in rock engineering, is that they should not be applied outside the range over which data has been gathered in the formulation of the empirical rule. Development of a more rationally based system, one that incorporates the empirical data base within a numerical modelling framework, should overcome this limitation. This is an increasingly important concern as practical projects seek to go beyond current experience.

Controlled laboratory tests, combined with geophysical (acoustic emission) monitoring e.g. to observe the development of damage around holes in rock specimens, can serve to verify the predictions of numerical codes, and help identify and resolve shortcomings before the codes are used to design and monitor costly in situ tests. In situ tests can also be instrumented to monitor the overall mechanical response of the rock, again as a check on the validity of the numerical prediction.

Long-term isolation of nuclear and hazardous waste deep in rock presents an especially complex challenge for rock mechanics [Fairhurst (1999)]. Since groundwater is the primary vehicle by which the toxic radionuclides or hazardous chemicals can be transported from the underground repository to the biosphere, the development of preferential pathways due to mechanical changes induced in the vicinity of the excavations must be either avoided or corrected. Heat generated by radioactive decay of nuclear waste generates increased temperatures, thermal stresses and displacements in the rock mass. Convective circulation of the groundwater away from the excavations may result, associated perhaps with chemical dissolution and, later, re-precipitation. These effects could in turn modify the flow pathways. Sin-fflarly, where flow occurs through fractures, thermal stresses will change apertures and hence the flow resistance.

These *coupled* hydrological-thermal-mechanical-chemical interactions require interdisciplinary collaboration in which rock mechanics has an important role. Questions of how the rock around the waste-filled drifts will degrade over the tens of thousands or more years that the waste must be isolated are not easily answered, and will require a deeper understanding of the fundamental physics of rock strength, Potyondy and Cundall (1998), than has been needed in rock engineering to date. Laboratory studies of rock deformation and degradation have an essential role in this task-as do larger scale in situ experiments.

Professors Stavrogin and Tarasov, of the (then) Leningrad Mining Institute, USSR, inspired by Cook's original work in the early 1960's, proceeded on an independent path to control and study the disintegration and collapse of rock specimens. Rather than rely on electronics and servocontrol systems, the Russian investigators devised ingenious novel designs to limit the energy stored in their testing machines. Carefully and painstakingly, working essentially out of contact with colleagues in other countries for over 30 years, Stavrogin and Tarasov have conducted many thousands of laboratory tests. They have explored the postpeak load response of many rock types over a very large range of loading rates and test conditions. Their work provides an unparalled wealth of data on the complete load deformation behaviour of rock specimens; data that should be of considerable interest to researchers elsewhere. Questions have been raised, for example, as to whether the use of the servo-control principle, in which a short (milliseconds) period of unstable deformation occurs before the servo-valve acts to correct the instability, will produce a *post-peak response* different from that which would have been observed if the same specimen had been tested in an intrinsically very stiff testing system without servo-valves. In particular, does the *snap-back* or "Class 11" behaviour, Wawersik (1968) [see Fig. 3.11 and associated discussion in this book] observed in some tests, represent the actual behaviour (stability locus) of the specimen or is it to some unknown extent an artifact of the specimen/servo-control system response? Certainly, Class II behaviour is not prohibited on fundamental grounds, Berry (1960), but it would be informative use the intrinsically stiff equipment to test specimens of rock types that have been found to exhibit a Class II response in servo-controlled systems.

Publication of this monograph comes at a time when important developments are taking place in the numerical modelling of the deformation response of frictional materials, based on micrqmechanics, i.e. the study of how the elementary particle-particle (cohesive-frictional) interactions combine to produce the macroscopic response of the material Cundall (1995), Potyondy and Cundall (1998). These developments offer new insights into the mechanics of deformation, damage and disintegration of rock specimens and perhaps the possibility of a better understanding of these phenomena also on the large scale. Analysis of the data presented in the monograph book by these numerical methods, supplemented by additional special tests as needed, could yield a better understanding on the critically important topic of unstable collapse of rock structures.

The publisher has done rock mechanics a valuable service by bringing the work of Stavrogin and Tarasov to the attention of the worldwide community.

Charles Fairhurst
Professor Emeritus of Mining Engineering and Rock Mechanics,
University of Minnesota.
Senior Consultant,
Itasca Consulting Group Minneapolis, Minnesota, USA
September 2000

REFERENCES

Barnard, P.R. (1964) Researches into the complete stressstrain curve for concrete, Mag. Concr. Res. Vol. 16 pp 203–210.

Berry J.P. (1960) Soine kinetic considerations of the Griffith criterion for fracture, J. Mech. Phys. Solids, vol 8.
I. Equations of motion at constant force pp. 194-206; II. Equations of motion at constant deformation pp. 207-16
Cook, N.G.W. (1965) The failure of rock. Int. J. Rock Mech. Min. SC. Vol. 2 p. 389.
Cook, N.G.W, E. Hoek, J.P.G. Pretorius, W. D. Ortlepp and M. D.G Salamon (1966) Rock mechanics applied to the study of rockbursts: J. So. Afr. Inst. Min. Metall. Vol. 66 pp 436–528
Cundall, P.A. (1995) Numerical experiments on rough joints in shear using a bonded particle model. Proc. Georg Mandel Symp.; Vienna, Austria.
Cundall, P.A. and D.O. Potyondy (1998) Modeling notchformation mechanisms in the URL mineby test tunnel using bonded assemblies of circular particles Proc. Third No. Amer. Rock Mech. Symp. (NARMS'98)
Fairhurst, C. (1999) Rock mechanics and nuclear waste repositories, First Int'l Workshop on Rock Mechanics of Nuclear Waste Reposaitories, 37th U.S. Rock Mech. Symp. June 5-6, Vail. Colorado, Balkema, Rotterdam.
Mendecki A.J. (Editor) (1997) Seismic Monitoring in Mines. Chapman and Hall.
Ryder, J. A. (1987) Excess shear stress (ESS) An engineering criterion for assessing unstable slip and associated rockburst hazards Proc. Sixth Int. Congr Rock Mech. pp 1211-1214 Balkema. (Rotterdam).
Starfield, A.M. and C. Fairhurst (1968) How high speed computers advance design of practical mine pillar systems. Eng'g. Min. J. vol. 169, May pp 78-84.
Scholz C.H. (1990) The Mechanics of Earthquakes and Faulting (Cambridge) New York 439p.
Salamon M.D.G. (1970) Stability, instability, and the design of pillar workings. Int. J. Rock Min Sci. Vol 7 pp 613-31.
Terzaghi K. Van (1923) Die Berectmung der Durchlassigkeitsziffer des Tones aus dem Verlauf der hydrodynamischen Spannungserscheinungen. Sher. Akad. Wiss. Wien. Vol. 132, p. 105.
UDEC (1999) Universal Distinct Element Code Version 3. 1. Itasca Consulting Group Inc, Minneapolis, Minnesota MN 55415, USA.
3DEC (1998) Three-Dimensional Distinct Element Code. Itasca.
Wawersik W.R. (1968) Experimental study of the fundamental mechanisms of rock failure in static uniaxial and triaxial compression, and uniaxial tension. Ph.D thesis, University of Minnesota.
Young R.P. J.F. Hazzard and W.S. Pettitt (2000) Seismic and micromechanical studies of rock fracture. Geophysical Res. Letters vol. 27, pp. 1667-1670.

Preface

This monograph summarises the results of experimental investigations on the mechanical behaviour of rock, conducted over a period of 40 years in the Laboratory for the Physics of Rocks at High Pressures.

Until 1986, the Laboratory was a part of the Vsesoyuznogo Instituta Gornoi Geomekhaniki i Marksheiderskogo Dela ((VNIMI) [All-Union Institute of Mine Geomechanics and Mine Surveying] in St. Petersburg. In 1986, the Laboratory was attached to the Mining Institute of St. Petersburg and remains an integral part of this Institute to date.

The Laboratory has participated in and continues to be involved in solving severe problems encountered in the mining industry of Russia and the former USSR. These problems include rockbursts, coal bumps, rock and gas outbursts, stability of mine workings at great depths, stability of deep and ultradeep boreholes, and problems associated with the construction of industrial underground structures for various purposes.

The fundamental investigations conducted in the Laboratory are directed towards the study of the mechanical behaviour of rock over a wide range of conditions. The principal characteristics of these conditions and types of investigations are listed below.

— Stress states: uniaxial compression and tension, tension-compression and triaxial compression, at confining pressures σ_2 up to 1000 MPa.

— Strain states, both pre-peak strength and post-peak strength regimes including the zone of residual strength, i.e., the 'complete load-deformation' response.

— Effect of loading path (history of variation in stress state) has been studied under conditions of proportional (simple) loading while changing the ratio of confining to axial stress from 0 to 1, and under conditions of complex loading, mainly according to Von Karman's scheme, while holding the confining pressure constant.

— Strain rates have been varied over 12 orders of magnitude, from $\dot{\varepsilon}_1 = 10^{-10}\,\mathrm{s}^{-1}$ to $\dot{\varepsilon}_1 = 10^{+2}\,\mathrm{s}^{-1}$. The highest strain rates ($\dot{\varepsilon}_1 = 10^{+2}\,\mathrm{s}^{-1}$) approach the speed of explosive loading while the lowest rates correspond to creep rates. The long-term strength of rocks was also studied in the creep loading regime.

— Pore pressure effects on mechanical properties of rock were studied for both liquid and gas for pore pressures varying from zero to equality with the confining pressure.

— Gas and liquid permeability of rock were investigated for various stress states and loading paths.

— The energy balance during brittle failure of rocks and brittle model materials was likewise studied under conditions of uniaxial and triaxial compression for a wide variation in rock brittleness, stiffness of the loading system (LS) (varied by a factor of more than 1000), inertia mass of the LS and volume undergoing fracture (VF), and acoustical impedance of the LS/VF boundary.

— Time-dependent deformation processes in rocks after release of applied loads, in which the rock was subjected to permanent strain under conditions of differential compression ($\sigma_1 \neq \sigma_2$) at various levels of stress states were studied.

— Fluid flow and strength properties of rock specimens deformed under the simultaneous effects of temperature and pressure of drill fluids of different composition were studied to evaluate borehole stability in deep and ultradeep holes.

Conduct of these series of complex investigations over such a wide range of test conditions required the development of several unique testing systems. All of the test equipment was developed by Laboratory personnel.

Two major stages of investigations in the Laboratory can be defined—pre-1974 and post-1974; 1974 marked the start of the development of a new generation of rock-testing machines.

During the decade 1964–1974 research was directed towards the pre-peak load deformation behaviour of rock. Investigation of dilatation phenomena under different stress states, loading paths and strain rates yielded very significant results. A sudden jump-like transition from an elastic Poisson ratio lateral strain behaviour to a 'coefficient of permanent lateral strain (μ)' behaviour was observed to occur immediately upon reaching the elastic limit of deformation. The lateral strain coefficient remained constant during the entire process of permanent strain, from the elastic limit until the ultimate strength was reached, under conditions of proportional loading at a constant value of the parameter C (= σ_2 / σ_1), as well as for conditions of complex loading at constant confining pressure σ_2. The coefficient of permanent lateral strain μ, as distinct from Poissons ratio, was observed to be a monotonic function of the parameter C and the confining pressure σ_2 and varied from 3 or more under uniaxial compression to 0.5 at high values of C and confining pressure σ_2, when dilatation (volumetric expansion) ceased. The invariability of μ for a given C and σ_2 leads to a linear relationship between the volumetric strain and the magnitude of principal permanent strain.

Significant results were obtained while studying the effect of loading path on the mechanical behaviour of rocks in the range between the elastic limit and the maximum load. It was found that the limit curves are independent on the loading path.

The notion that solids exhibit two inherent types of strength, viz. (tensile) fracture strength and shear strength, was adopted in research studies in the Laboratory. It was experimentally established that these two strengths and two resistance (fracture and shear) are, strictly speaking, bounding cases. The first occurs under tensile stresses only while the second results from shear stresses developed under high confining pressure, non-dilatant conditions. Intermediate between these two limits, both shear and fracture rupture can occur in the body during deformation and failure. The quantitative relationship between the two types of failure depends on the stress state. The mechanism of deformation and fracture between the limit conditions is explained by the authors in terms of a statistical model of a heterogeneous solid, which considers both failure modes and the quantitative ratio between them. The model provides a quantitative assessment of dilatation, strength, permanent lateral strain, orientation of the shear planes and their density as a function of stress state and confining pressure.

Analytical expressions for the elastic limit and strength are proposed in the form of exponential equations to describe the entire range of stress states between the above-mentioned two limiting conditions. As with the usual Mohr failure envelopes, the proposed analytical expressions do not consider the role of the intermediate principal stresss, but unlike the Mohr envelopes, the proposed conditions are quite universal. This has been verified in tests on more than 100 types of rocks and materials. Mohr envelopes cannot be described analytically by a unique expression.

A combination of the above-mentioned analytical expressious for the limiting conditions with results of the study of time-dependent behaviour allowed a more general limit-state criterion to be developed, one that includes time. The time factor is introduced in terms of the kinetic theory of strength of solids, the equation for which is solved together with the exponential limit-strength equations. The kinetic equation describes the experimental results quite satisfactorily in most cases.

The first results of the aforementioned studies were published in the 1960s [65, 69]. A large volume of additional experimental data has been collected for a wide range of rock types since that time. These data have been summarised in monographs published by the VNIMI Institute during the 1970s [71, 72, 75, 76, 77]. All these results concern the pre-failure strength zone since, during this period, the Laboratory was not equipped with stiff testing machines.

In 1974, the Laboratory began the development of stiff testing machines that would allow deformation studies to be made of the complete deformation curve including the post-peak strength regime. The machines developed are of original design with a very high stiffness (up to 2×10^4 MN/m). This allows extremely brittle rocks to be deformed stably in the post-peak zone without the use of servo-control devices. This set of stiff machines allowed materials to be tested under both static and dynamic conditions, and in uniaxial and confined compression. Fluid flow processes and energy balance during brittle failure under pore pressures generated by liquids or gas and by temperature could also be studied. Acoustical properties of the rocks were determined under loaded conditions. The machines used for these experiments are described in this monograph. The unique design of the machines and of individual components in them has been confirmed in patents awarded to the authors. Some of the patents are listed at the end of the Literature Cited.

Studies of rock deformation over the complete load-deformation cycle have allowed the growth of deformation in the post-peak zone to be observed and interpreted. A significant feature of this behaviour is the invariability of the coefficient of permanent lateral strain μ in the pre-peak and post-peak strength regimes. This indicates that deformation in the post-peak strength zone develops along shear planes that formed in the pre-peak zone. The number of shear planes, according to the statistical model, does not affect the index μ.

As in the case for strength and elasticity, the relationship between the residual strength of rocks and the stress state can be described by a similar exponential equation.

Investigations over a wide range of loading conditions and large permanent deformations continued to the residual strength state have revealed that the 'crushability' of rock under increasing deformation, and increasing confining pressure, has a finite limit. After reaching a minimum size distribution, increase in confining pressure σ_2 does not result in further size reduction. The minimum size of the crushed fragments is determined by the dimensions of the structural elements of the rock, i.e., the grains and crystals. This discovery provided valuable insight into the mechanics of development of deformation under high pressures and also resulted in major changes in the deformation model. Since there was no limit to the size reduction, this earlier model allowed crushing to occur continuously and contained no upper limit to hardening and to increase in the elastic limit of the material with increase in σ_2. This limit results in a gradual transition of the failure envelope towards the horizontal. This principle allowed the critical conditions for the horizontal transition of limit curves and conditions for the possible growth of disintegration in rocks under high pressure to be defined and also provided an explanation for the swift rise in plasticity under these conditions.

Investigations into the mechanics of rock hardening with increase in confining pressure showed that up to 70% of the total hardening could be accounted for by strain hardening, the proportion increasing with increase in plasticity of the material.

The role of the rock structure factor was also investigated—in particular the percentage distribution of structural elements in a body with respect to shear resistance in the formation of strength and strain properties of rocks.

Investigations into the processes of dynamic uncontrolled fracture of rocks in the post-peak zone under widely varying degrees of brittleness of rocks, stiffness indices of the LS, inertial masses of the LS and VF, stress state etc., allowed the authors to establish the mechanics by which potential elastic energy of compression stored in the system before failure was converted into dynamic energy. A complete system of equations was developed to describe the energy balance during the process of dynamic uncontrolled fracture. These tests were directed towards solving the problem of rockbursts.

Complex comparative tests were conducted on two types of Donbass sandstones—burst-prone (BP) and non-burst-prone (NBP)—to study the problem of sudden rock and gas bumps in mine workings. The two types of sandstone differ basically in that the first type is hazardous with respect to sudden bumps of rock and gas while the second is not. The study searched for criteria whereby these sandstones could be reliably distinguished and sought to establish the mechanics of initiation of sudden outbursts. Though identical in mechanical properties, the sandstones exhibited a strong difference (by 2–3 orders of magnitude) in initial permeability and the degree to which fluid flow varied under pre-peak and post-peak deformations. The permeability of BP sandstones at different levels of confining pressure changed by one order of magnitude while that for NBP sandstone changed by 3 or more orders of magnitude. Such a difference in properties of fluid flow is attributable to the difference in rock structure, in particular the difference in the cement filling of the intergranular spaces.

The mechanics of initiation of sudden outbursts of gas and rock in mine workings was analysed on the basis of the above study of fluid flow properties of BP and NBP sandstones.

With respect to the stability of deep and ultradeep boreholes, studies were conducted on the effect of the pore pressure developed by various media and drilling fluids at different temperatures on the mechanical properties of rocks. These studies led to the development of a methodology for testing the effectiveness of drilling fluids in increasing borehole stability.

Tests were conducted under a wide variety of test conditions to study the consequences of decompression and subsequent deformation processes on the rock structure. These tests were valuable in solving problems of recovery of intact cores from great depths.

Results of the foregoing investigations, undertaken in the second stage of the Laboratory's scientific programme, were first published in 1981 [79, 80, 84, 85, 86, 87, 88, 89, 91, 92, 94, 97, 98, 99, 100, 102]. Summaries of the results have been published periodically since that time [83, 92, 101].

The Authors

Andrei Nikolaevich Stavrogin, Professor, Doctor of Engineering sciences, Honored scientist of Russia, founder of the Laboratory of Physics of Rocks and High Pressures, was Head of the Laboratory for over 40 years. Currently he is Professor of Saint-Petersburg Mining Institute (Technical University), Russia.

Boris Grigorievich Tarasov, Professor, Doctor of Engineering sciences, leading scientist in the Laboratory for over 20 years, was a Director of the Research Center "Geotest" established on the base of the Laboratory (1990–1998). Currently he is a Professor at the University of Western Australia.

ACKNOWLEDGEMENT

Several scientists have participated in research activities at different periods over the 40 years of operation of the Laboratory. Special mention must be made of the services rendered by the most senior and current leader of research in the Laboratory, Oleg Aleksandrovich Shirkes, who has been a leading scientist at the Laboratory over the entire 40 years of its existence. Credit for the development and fabrication of the first prototypes of the Laboratory's high-pressure testing machines goes to Mr. Shirkes.

Former research scholars working at the Laboratory, who have since become leading scientists at VNIMI, made significant contributions to various investigations: E.D. Pevzner A.T. Karmanskiy, E.V. Lodus and G.N. Tanov. G.B. Mikheev participated in the design of stiff machine components. Testing and processing of experimental results over the years were mostly handled first by V.S. Georgievskiy, then respectively by Yu.P. Korolev, N.V. Fokeev, M.S. Ognev, V.P. Sapunova, E.Yu. Semenova and N.M. Polkovsky. Petrographic studies were conducted by V.F. Avksent'eva and G.N. Yurel.

The authors express heartfelt gratitude to all the persons involved in the laborious and painstaking projects at the Laboratory.

In recent years, difficult times for Russian scientific research, the Laboratory has been supported by the Russian Fund for Fundamental Research through grant No. 93-05-9850 (Stavrogin) and No. 96-05-66180 (Tarasov) for the period 1993–1998.

The authors appreciate the constant support of Prof. Charles Fairhurst. It was he who encouraged us to start the book and he has maintained interest in it's progress to the end. His voluntary effort in editing the initial translation was immense. His advice and corrections transformed the initial draft into this book. We hope that it will prove useful to readers interested in the experimental mechanics of rock deformation and stability.

Andrei N. Stavrogin
Boris G. Tarasov

Contents

1

Pre-failure and Post-failure Strength and Deformation Properties of Rocks under a Wide Range of Stress States and Loading Paths

1.1 INTRODUCTION

Procedures for the experimental testing of rocks have been developed by the authors that allow study of the pre-failure and post-failure strength and deformation behaviour over a wide range of stress states and loading paths. The testing systems have been designed to have very high stiffness (up to 2×10^4 MN/m). These testing machines allow even very brittle rocks to be deformed stably through the post-failure regime, without the use of servo-controlled devices. The machines are of compact design and reliable in operation.

The entire range of experiments developed by the authors is discussed briefly. The wide range of stress states includes tensile and 'true triaxial' (i.e., non-equal confining stresses) compression and a variety of loading paths. The results are outlined in general terms in this chapter, together with a statistical model of the deformation behaviour of solids, which considers the development of irreversible deformation (damage) in rocks, before and after peak strength.

1.2 METHODS USED TO STUDY STRENGTH AND DEFORMATION BEHAVIOUR OF ROCKS IN THE PRE-FAILURE AND POST-FAILURE REGIMES

1.2.1 Sample Preparation for Testing and Methods of Recording Strains and Forces

This section outlines general principles for the preparation of test specimens for the machines designed by the authors. Specific details are given in the

a)

b)

Fig. 1.1: Two types of specimens readied for testing.

discussion of each experiment. Most tests were conducted on cylindrical specimens of 30 mm diameter and 60–80 mm length. Fig. 1.1 shows two types of specimens readied for testing.

The sample (1) (Fig. 1.1a) is placed between two spherical steel thrust bearings (2) and centred with reference to the loading ram in the machine. Extensometers for recording axial (3) and lateral (4) deformations are attached to the sample. In many of the experiments, deformation was measured by electric resistance strain gauges (5) affixed directly to the specimen. A thermocouple (6) was also attached directly to the specimen to measure temperature.

The method of isolating the specimen from the fluid required to develop hydrostatic confining pressure in the test chamber and construction of the extensometers are described later. The specimen shown in Fig. 1.1(b) was prepared for testing in a stiff triaxial compression machine; it is positioned between the loading piston of the high-pressure chamber (bottom) and the load cell (top).

It is very important that care be taken in preparing the surfaces of specimens for testing. A parallelism of 0.02 mm between the upper and lower surfaces is imposed for a 30 mm diameter specimen. Such a stringent requirement is necessary since spherical bearings are not used—in order to achieve maximum stiffness of the testing system. To achieve this high precision, specimens are first fixed in special mandrels (see Fig. 1.2). The special mandrels also ensure that all specimens are of the same length. The mandrel consists of a heavy gauge cylinder (2) with slots (3) to provide flexibility. The inner dimension of the mandrel can be altered using the bolt (4). The ends of the clamped specimen (1) are then ground parallel on a grinding wheel.

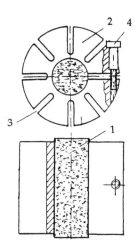

Fig. 1.2: Device for preparing specimens.

The precise dimensions and smoothness of specimens were obtained using a non-centring grinding machine. Such precision was particularly important when conducting experiments to detect low-level effects.

Specimens for confined pressure tests (Fig. 1.3) were sealed along the vertical sides with a thin-walled (0.5 to 0.2 mm thick) polythene tube (2). Hermetic sealing of the sides was accomplished either (a) by extending the

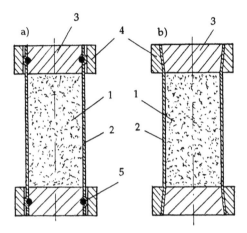

Fig. 1.3: Hydraulic sealing of specimens.

tube over the thrust bearing (3) using the rings (4) and rubber seals (5) or (b) by compressing the tube between the conical surfaces of the thrust bearings (3) and rings (4). In the second case, the taper angle of the bearing was 3°. This method of sealing specimens proved very reliable both for tests at high pressures (up to 1000 MPa) and tests involving large permanent deformations (lateral deformations up to 50%, see Fig. 1.44). For tests at elevated temperatures, the polyethylene was replaced by heat-resistant rubber or copper.

Specimen preparation also includes attaching gauges to the specimens to detect axial and lateral deformations. Different methods of deformation measurement were used, depending on the type of test and accuracy required.

In some cases electrical resistance strain gauges were affixed directly to the specimen surface. Although relatively expensive, since the gauges can be used once only, this method is sometimes the only option available. This is so, for example, in the determination of elastic properties during tests involving low-magnitude deformations and when studying strain waves under dynamic loading conditions.

In other cases, electrical resistance extensometers were used to measure both axial (or longitudinal) and lateral (or transverse) deformation. These transducers are depicted schematically in Fig. 1.4 [70].

Extensometers for sensing the lateral deformation of a specimen (Fig. 1.4a) consist of a steel clip (1) with a small steel ball (2) attached at each end, and a regulator screw (3) to ensure that the clip (clamp ring) is in good contact with the specimen. When the specimen deforms, the clip

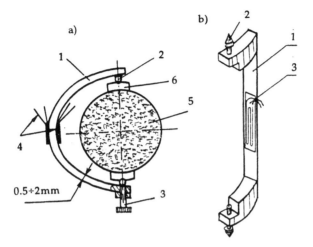

Fig. 1.4: Extensometers for measuring (a) lateral and (b) axial strains in specimens.

deflects, generating a signal in the gauges (4) affixed directly on the clip. The resistance gauges were connected in a Wheatstone bridge network for accurate reading of the electrical output. The gauges were calibrated directly using special devices. This allowed the measurement error to be held to no more than approximately 1%. Lateral deformation of the specimen was measured at mid-height with the clip attached to the specimen (5) through the steel bearings (6).

The number of gauges attached to the specimen ranged from 1 to 4 depending on the purpose of the experiment. Clips of different thickness were used, depending on the expected magnitude of deformation in the particular series of experiments. Up to 30% lateral deformation was recorded in some tests.

The extensometer for recording longitudinal deformation of the specimen (Fig. 1.4b) consisted of a steel plate (1) (beam), to the end of which were attached supporting elements in the form of pivots (2), placed off-centre from the plate axis. The gauge length of the extensometer could be adjusted by means of threaded screws. The longitudinal displacement of the pivots produced deflection in the beam, which was recorded by strain gauges (3) bonded to the beam. Each extensometer was calibrated using a special calibrating device. The extensometer measurements were accurate to within 1%.

The error in recording the axial deformation of specimens could also be determined by attaching the extensometer to the specimen. The method of attachment was dictated by the type of experiment to be conducted. Two methods are shown in Fig. 1.5. In the first method (a), the beam is located between the machine platens. In the second (b), the beam is placed between special steel rings (1) attached to the side of the specimen, using a device (2) similar in construction to the clamp ring illustrated in Fig. 1.4 (a). The advantages and disadvantages of each method are described below.

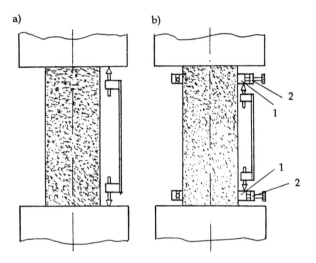

Fig. 1.5: Two methods of locating axial strain gauges on a specimen.

In method (a), the measured deformation includes deformations due to crushing or machining irregularities on the end surfaces. This introduces large errors in the initial stage of loading. For tests involving large permanent deformations (especially at high confining pressures), lateral deformation along the length of the specimen was very non-uniform, and largest in the centre, resulting in a 'barrel-shaped' specimen. In such cases, the extensometer recorded the mean deformation over the full length of the specimen.

These disadvantages can be avoided by using the second method (b). In this case, however, cracks that develop in the post-failure period may displace the steel rings bonded to the specimen surface, thereby introducing error into the observed deformations.

For certain tests, the two methods are used simultaneously to avoid loss of valuable information.

The load on the specimen was measured by load cells placed in contact with the specimen inside the high-pressure cell. The load cell consists of a steel cylinder with electrical resistance strain gauges attached to its surface. For a 2:1 (height (h) to diameter (d)) cylinder the error in the measured force did not exceed 1%. In some cases, the h/d ratio was reduced to increase the stiffness of the loading system.

Special stiff dynamometers with a h/d ratio of 0.25 [92] were developed for the stiff testing machines. Here, the error in force measurements was larger—but did not exceed 3%. Relatively high precision in recording the load on the specimen was achieved by attaching the strain gauges (1) (see Fig. 1.6) to the surface of a narrow groove machined along the dynamometer axis. This eliminated almost all of the load cell friction effect that often arises from contact of this cell with the machine during loading. Stiff dynamometers of other configurations were also developed.

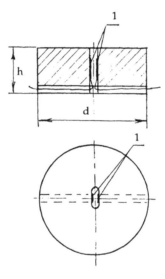

Fig. 1.6: Schematic diagram of a stiff dynamometer.

1.2.2 Stiff Loading and Stiff Testing Machines for Uniaxial Compression

It is possible for a specimen to undergo stable deformation in the post-failure range [14, 15, 6, 60, 32, 44], provided the energy is supplied in strictly controlled amounts, each not exceeding the amount required for the increment in deformation and associated crack development processes. In addition to the specimen, elements of the loading system also undergo elastic deformation during loading, thereby storing elastic energy. It is the release of this stored energy that leads to uncontrolled dynamic failure of the specimen as the deformation transits to the post-failure regime. Principles of (a) stable and (b) unstable post-failure deformation are shown in Fig. 1.7.

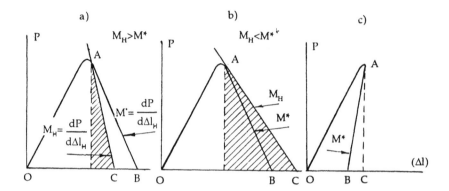

Fig. 1.7: Stable and unstable post-failure strength deformation of a specimen.

The 'load (P)—deformation Δl' diagram of specimen OAB is shown, together with the stiffness characteristic of the loading system AC. The shaded triangular portion indicates the reserve of elastic energy in the loading system when the load is applied at point A. If the post-failure branch in the diagram is denoted by coefficient $M^* = dP/d\Delta l$, and the loading system characterised by $M_H = dP/d\Delta l_H$ (where Δl is the deformation in the specimen and Δl_H is the deformation in the loading system), then the condition for stable deformation in the post-failure region may be stated as the requirement that $M_H > M^*$ (or $M_H/M^* > 1$) while unstable deformation corresponds to the condition $M_H < M^*$ (or $M_H/M^* < 1$).

Historically, two general approaches have evolved for testing brittle rocks beyond their peak strength:

1) Modification of relatively low stiffness machines by introduction of servo-control monitoring systems, which allow any excess energy of deformation to be removed from the loading system as the load-bearing capacity of the specimen decreases.

2) Development of stiff loading machines which minimise the storage of elastic energy in the system during loading of the specimens, such that the post-failure deformation process becomes stable and, consequently, servo-control is not necessary.

The testing machines developed by the authors and described in this book belong to the second category. Stiff machines of this type cannot be used to develop stable deformation of rocks and tests for which the post-failure deformability coefficient M^* is positive (this condition is illustrated in Fig. 1.7c). In this case, uncontrolled failure of the specimen occurs due to release of excess elastic energy stored in the specimen itself, even though the loading system may be absolutely rigid. However, rocks having such

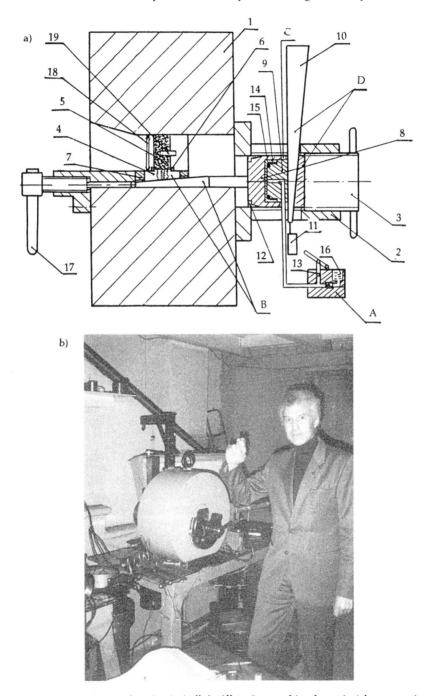

Fig. 1.8 (a): Basic scheme of an 'intrinsically' stiff testing machine for uniaxial compression; (b) testing machine, with Professor Boris Tarasov standing alongside.

properties are rarely encountered and it is possible, in such cases, to en-
hance the performance of the intrinsically very stiff loading systems by
introduction of servo-control. The authors have developed certain devices
of this nature which are discussed later. Several stiff testing machines have
been designed for study of the post-failure deformation properties. Two
such machines are described below.

The stiffness index of the machine illustrated in Fig. 1.8 is 2×10^4 MN/m
[104, 116, 78, 92]. This system allows very brittle rocks to be deformed
stably in the post-failure strength regime.

The high stiffness of the machine is achieved by a unique design of the
hydromechanical loading drive whereby (i) essentially no energy is stored
in the compressed fluid in the hydraulic jack and (ii) the number and longitudinal dimensions of the various components of the structure subject to
elastic deformation during loading are reduced to a minimum.

The hydromechanical drive consists of the 'self-braking' wedge pair B,
the hydraulic ram C, the source A from which fluid is fed in pulses to the
hydraulic ram, and the slippage mechanism for the ram piston. The slippage mechanism consists of the self-braking wedge pair D and the regulating screw (3).

The loading wedge pair B is located within the working space of the
monolithic stiff frame of the press (1). Wedge (4) serves as the base on
which the test specimen (5) and a load cell are mounted. The specimen is
placed in the press, which contains no spherical seatings. The stringent
requirements for maintaining parallelism of the specimen end faces are
usually met by preparing them in the mandrel (described earlier). The
specimen is centred on the platen/load cell by means of the clamp ring (6).
Wedge (4) can be moved vertically within a guiding frame (7) fixed between
the platens of the machine.

Hydraulic ram C applies load to the specimen through the wedge pair
B. To minimise storage of elastic energy of the pressurised fluid in the
ram—which would reduce the stiffness of the loading system—the working
fluid is fed into the loading system in metred doses. A pulsating pressure
is created by a plunger-type pump A, which is permanently connected
hydraulically to the working chamber of the hydraulic ram. When the
plunger (13) of the pump is moved downwards, pressure in the hydraulic
ram increases, causing movement of the cylinder (8) and wedge pair B, and
thus in loading and deforming the specimen. When the plunger (13) displaces upwards, the pressure in the ram drops to zero. The wedge surfaces
are inclined at an angle lower than the angle of friction, so that they are
'self-braking', i.e., they do not slip when the pressure is released. The load
on the specimen and the deformation are maintained on the down stroke
by downward sliding of the wedge (10) under the effect of the weight (11).
The fluid drains completely from the hydraulic ram and is returned to
pump A. On the next loading stroke of the pump, the cylinder (8) of the
ram starts its forward motion from the new (advanced) position.

In this way, the small volume (1 cu. cm) of fluid below the plunger (13) is able to generate a total deformation sufficient to produce complete failure of the specimen and to develop the corresponding forces on the specimen.

The mass of the weight (11) is chosen such that, for the given frequency of loading pulses, it is sufficient to move the piston (9) into the new position between pulses. A mass of 10–15 kg has been found sufficient to accomplish this at a frequency of 300 strokes per minute.

A collet (12) is fixed on the cylinder (8) of the hydraulic ram in order to ensure that the force developed by the wedge (10) and the weight (11) is not transferred to the specimen. The frictional force developed between the housing (2) and cylinder (8) is sufficient to counteract this force.

An elastic membrane (14) secured by a rubber ring (15) is used to increase the rate of flow of the fluid out of the cylinder ram after each working pulse. The membrane deforms when the fluid is fed into the ram, transmitting the force to the bottom of the ram. When the fluid pressure drops, the fluid drains through the membrane, which returns to its undeformed shape.

The deformation produced per pulse can be varied from 0.001 mm to 0.05 mm by varying the volume of fluid fed into the hydraulic ram. The rate of deformation of the specimen can be changed over several orders of magnitude by adjustment of the volume of fluid and pulse frequency. The minimum duration of an experiment is 5 seconds, which corresponds to a deformation rate not exceeding 10^{-3} s^{-1}.

Screw (3) is used to set the gap when the specimen is mounted, while screw (17) allows the specimen to be unloaded and wedges B to be returned to their initial position.

The axial and lateral deformations of the specimen are recorded by extensometers (18) and (19). The error in the measured quantities is of the order of 1%. The load on the specimen is recorded by the load cell (4). The stiffness of the load cell is 10^5 MN/m and the error in the measured forces is approximately 3%. The design of the extensometers, their attachment to the specimen and the design of the stiff load cell have been described in the previous section.

The loading press has a capacity of 500 kN and dimensions 350 × 500 × 500 mm. The test specimen has a—diameter 30 mm, and length 60 mm.

The second version of a stiff press is illustrated in Fig. 1.9 [111, 101]. In contrast to the previous design, the hydraulic drive of the ram consists of a variable volume receiver (1) connected by the pressure tube to a hydraulic ram (2). The lower wedge (3) of the pair of loading wedges is positioned between the ram and stop screw (5) with a thrust beating (4). The specimen (7) is tested as follows:

The pressure source (6) is activated, developing an initial pressure in the receiver (1) such that the energy stored in it is adequate to deform and

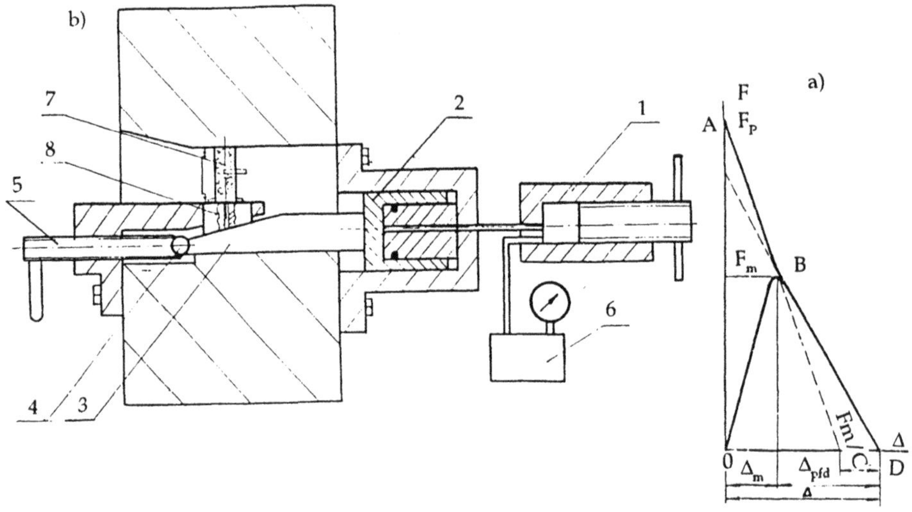

Fig. 1.9: Schematic diagram of a stiff press for uniaxial compression (alternative design 2).

fracture the specimen. Pressure is applied by fluid from chamber (1) to the inside face of the piston (2). This exerts a force through the wedge (3) which is taken by the top screw (5). The wedge does not displace and no force is transmitted onto the specimen (7). The screw (5) is then loosened, the frame (2) and the wedges (3) and (8) move, there applying the load and deforming the specimen. Movement of the ram results in a pressure drop in the receiver due to the increase in working area on the hydraulic ram. At this point fluid flows from the receiver into the piston chamber. When the applied load exceeds the peak strength on the specimen, the screw (5) prevents the wedge (3) from moving, thereby preventing release of the energy of the pressurised fluid in the receiver and hydraulic ram, as well as failure of the specimen. The volume of the receiver and pressure of the working fluid are so selected as to ensure a minimum reserve of elastic energy in the hydraulic system during the post-failure deformation stage of the specimen. Then, when the frame of the hydraulic ram moves, the resultant pressure drop applies a force on the specimen corresponding to its decreased load-bearing capacity.

A typical situation is illustrated in Fig. 1.9 (a). OBD represents the complete load-deformation curve for the specimen, while curve ABD corresponds to the force-deformation characteristic of the press resulting from the pressure of the fluid flowing from the receiver to the ram. In this case, the characteristic ABD is observed to be non-linear since the elastic energy stored in the loaded elements of the press at the peak strength of the specimen is also taken into account. This energy is determined by the stiffness characteristic of the loading system. Initial pressure (P) in the receiver and its volume (V) can be computed from the expressions:

$$V = nS_d \, (\Delta - F_m \, /C\,)/\beta \tag{1.1}$$

$$P = F_m \, (\Delta - F_m/C\,)/S_d \, (\Delta_{pfd} - F_m/C\,) \, n, \tag{1.2}$$

where C is the machine stiffness; S_d is the cross-sectional area of the hydraulic ram; β the compressibility of the working fluid; Δ the total (absolute) deformation of the specimen; Δ_{pfd} the post-failure deformation of the specimen; F_m the maximum load on the specimen; and n the transmission ratio of the loading wedge pair.

Since the compressibility coefficient of the fluid (β) is a function of pressure, the average value of β is used to provide an estimate for the first experiment while the deformation and strength parameters of the specimen, essential for computation, are assumed. The corrected values are used in subsequent experiments.

This press design has advantages over the machines described earlier, especially when studying acoustic and electromagnetic emission of rocks during deformation and failure, since the noise (disturbance) associated with operation of a hydraulic pump is eliminated. With this machine it is simple to automate the process of loading the specimen at a given (constant) rate of deformation. This press, designed and used by the authors, has a force capacity of 150 kN. It's stiffness is quite low compared to the version described earlier. Other parameters and dimensions, as well as the methods of recording strains and forces, are similar to those of the first version.

1.2.3 Stiff Machine for Testing Materials under Triaxial Compression

Using the principles described above, the authors have designed stiff machines to test specimens under confined lateral compression, i.e. $\sigma_1 > \sigma_2 = \sigma_3$ [106, 112, 92]. A schematic diagram and general photograph of one of their stiff compression testing machines is shown in Fig. 1.10.

The machine consists of a stiff monolithic frame (1), high-pressure cell (working cell) (4), pressure source (3) and a hydromechanical drive. The specimen (5), covered in an impermeable jacket, is placed between the piston (2) and the load cell (8). Axial and lateral deformation transducers are attached to the jacket. Low-viscosity oil is used to pressurise the cell. The working area of the cell is connected to a compensating cell (15) by a tube (13). This connection allows the pressure in the cell to be held constant when the piston (2) is displaced into the cell during loading. The compensating cell is located in the frame (10).

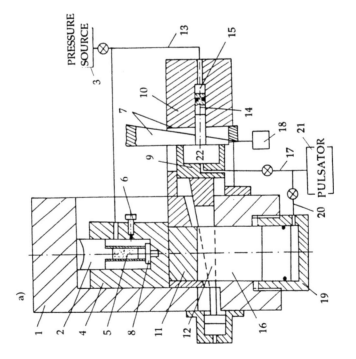

Fig. 1.10: (a) Schematic diagram and (b) general photograph of a stiff testing machine.

The hydromechanical drive allows two methods of loading to be used. The first method is for tests at high stiffness (10^{10} N/m) and is used for testing brittle materials under conditions close to uniaxial compression, i.e., confining pressure σ_2 between 0 and 10 MPa. The second loading method is used for confining pressures up to 300 MPa. In this method, the system can develop loads up to 2000 kN and axial deformations up to 20 mm, while the stiffness of the machine can be reduced due to the increasingly 'ductile' behaviour of the test specimens.

The hydromechanical drive designed for the first method of loading consists of a loading and 'self-braking' wedge pair (11) and (12), and a hydraulic ram (9) connected to a fluid pressure source (Pulsator) (21) through a tube (17). Fluid is fed in fixed increments into the hydraulic ram. The slippage mechanism for the piston (22) consists of the wedge pair (7), the weight (18) and the plunger (14). The specimen is loaded as follows: The ram (9) applies a load, which is transmitted to the specimen through the wedge pairs (11) and (12). As mentioned earlier, build-up of energy in the compression fluid (which would decrease the stiffness of the loading system) is avoided by delivering the fluid to the ram in metred portions using a pulsator. After each (measured) input of fluid, the ram (9) and wedge pairs (11) and (12) are displaced and the pressure in the ram then drops to zero. Friction in the 'self-braking' wedge pairs (11) and (12) ensures that the applied load and associated deformation of the specimen are maintained while the wedge pair (7) and the plunger (14) completely drain the fluid from the ram cylinder (9), moving the piston (22) to the bottom of the frame of the ram. Each new pulse is transmitted into the empty ram cylinder, i.e., the volume of the cylinder is reduced to zero between pulses. Thus a minimum volume of fluid (equal to the volume of each loading pulse) suffices to deform the specimen to any extent desired. Each withdrawal of fluid from the ram is accompanied by a movement of the plunger (14) so that the volume of the compensating cell (15) increases, becoming filled by the fluid drained from the working chamber (4) by the piston (2). During this period the pressure in the working cell is maintained constant since the ratio of internal diameters of the working and compensating cells is equal to the transmission ratio of the loading wedge pair.

The second method of loading is provided by the ram (19), fixed to the frame of the press and connected to the pulsator (21) through the tube (20). The ram piston (16) rests against wedge (11) through a slot in wedge (12). Loading starts with wedge pair (7) in the lifted position. The tube (17) is closed. The pulsator (21) feeds working fluid into the ram (19) in individual pulses. Each injection of fluid into the ram transmits a force to the cell (4) through the piston (16) and wedge (11), thereby deforming the specimen. Simultaneously, wedge (12), under the action of the plunger (14), moves forward while maintaining contact with wedge (11) along the wedge surface. When pressure in the hydraulic ram drops to zero, the force applied on the specimen is maintained, because wedge (11) 'locks' against wedge (12). When the ram is not pressurised, an additional drive (not shown in Fig. 1.10) turns a cylinder (19) of the ram in the housing (1), thus completely draining the fluid from the ram and reducing the working volume of the cell to zero. This is repeated after every loading pulse so that, even for large specimen deformation, the volume of fluid in the hydraulic ram remains at a constant minimum value, i.e., the volume of fluid injected in one loading pulse. Each successive loading pulse injects the same small increment of fluid under pressure into the empty ram cylinder.

In both methods, the rate of loading and the rate of deformation can be controlled smoothly by selecting the volume of fluid appropriate to a given working pulse amplitude and pulse frequency. The maximum deformation rate achieved in this set-up is roughly 10^{-3} s^{-1}. The process can be changed to reduce the deformation rate several orders of magnitude. Errors in measuring axial and lateral deformations of a specimen are less than 1% and errors in measuring forces are about 3%.

The stiffness of the press for different methods of loading ranges from 0.5×10^4 MN/m to 10^4 MN/m. The maximum hydrostatic pressure in the cell is 300 MPa. The maximum axial force developed by the press is 2000 kN. The dimensions of the press are $800 \times 400 \times 400$ mm.

The machine is capable of studying acoustic and permeability properties of rocks under different levels of confining pressure σ_2; recording changes in the volume of crack and pore space in a specimen during deformation, both in the pre-peak and post-peak strength ranges; and investigating the effect of pore pressure on the mechanical properties. A heating element is located in the high-pressure cell for conducting experiments at high temperature.

A mentioned earlier, it is necessary to use servo-control systems when investigating rocks with a positive coefficient for the post-peak branch of the diagram $M^* = dP/d\Delta l$, as well as under absolutely stiff loading systems. Mechanical attachments which minimise response time and decrease load on the specimen when a signal is received from the monitoring system have been described in earlier publications [105, 107].

1.2.4 High-pressure Testing Machine for Proportional and Complex Paths of Loading

This testing machine is capable of generating pressure in the hydraulic cell up to $\sigma_2 = \sigma_3 = 1000$ MPa and conducting investigations for stress states $\sigma_1 > \sigma_2 = \sigma_3$. The maximum axial force developed by the press is 1000 tons. Two paths of loading are possible with this machine:

1) Proportional (or 'simple', according to A.B. Iluyshin's classification) loading when the relationship (C) between the confining component of pressure on the specimen $\sigma_2 = \sigma_3$ and the axial component σ_1 is held constant throughout the test ($C = \sigma_2/\sigma_1 = $ const) [67].

2) Complex loading, in which the specimen is subjected to confining pressure, $\sigma_1 = \sigma_2 = \sigma_3$ and axial compression is applied later at a constant pressure selected at the beginning of the experiment, $\sigma_2 = \sigma_3$ (Karman's loading scheme) [67, 101].

A schematic diagram and general photograph of this testing machine are shown in Fig. 1.11.

Experiments involving simple loading of a specimen are conducted as follows. The specimen (1) is placed in a hydraulic cell (2) where the pressure $\sigma_2 = \sigma_3$ is generated by a pressure source H_1. Axial stress σ_1 is imposed automatically on the specimen by the hydraulic ram (3) since the cell (2) and cavity (13) are interconnected by a groove (14). A constant ratio $C = \sigma_2/\sigma_1$ is thus obtained throughout the range of loads. The value of C can be changed in other experiments by using a set of replaceable rams (3) of different piston (5) diameters (d_{vario}). $C = \sigma_2/\sigma_1$ can be varied from $C = 1$ under hydrostatic pressure up to $C = 0$ under uniaxial compression when $\sigma_2 = \sigma_3 = 0$. The magnitude of C depends on the ratio of diameter d_1 of the upper piston (4), diameter of the specimen d_2 and piston diameter d_{vario} in the lower loading hydraulic ram. C can be computed from the expression:

$$\frac{\sigma_2}{\sigma_1} = \frac{\sigma_3}{\sigma_1} = C = \frac{F_2}{F_1 - F_3} \qquad (1.3)$$

where $F_2 = \dfrac{\pi d_2^2}{4}$; $F_1 = \dfrac{\pi d_{vario}^2}{4}$; $F_3 = \dfrac{\pi(d_1^2 - d_2^2)}{4}$.

Here C in Equation 1.3 does not take into account frictional forces introduced by the piston seals (4) and (5). This error is handled by the elastic tensometric load cell (6) positioned in the high-pressure cell (2), which measures the axial stress component σ_1. Pressure $\sigma_2 = \sigma_3$ is recorded by a manometer. After placing it in the machine, the specimen is pushed into the stiff load frame, which takes up the external force developed by the hydraulic ram (3). The loading frame is indicated by the cross-hatching on the upper and lower parts of the base in Fig. 1.11. Experiments in simple loading are conducted using a single pressure source. For the case in which $d_1 = d_{vario}$, the pressure σ_2 developed in the cell will exert a hydrostatic pressure on the specimen since the value of C is equal to unity.

When diameter $d_1 = d_{vario}$, complex loading is achieved using the hydraulic ram (3) fed from a second pressure source H_2. The frame of the ram (3) in the cylinder (7) serves as the piston. The cavity (13) above the

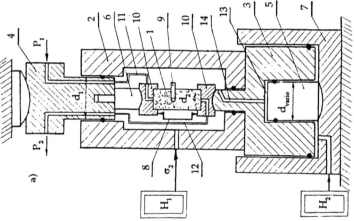

Fig. 1.11: (a) Schematic diagram and (b) general photograph of high-pressure testing machine for simple and complex loading of a specimen.

piston (5) serves as a compensating chamber in this case, allowing the pressure $\sigma_2 = \sigma_3$ in the cell (2) to be held constant. When the plunger (4) is pushed into the cell as the specimen deforms, the fluid is directed into the compensating chamber; the volume of the latter is equal to the volume of fluid displaced from the cell (2).

Axial and lateral deformations of the specimen are measured by gauges (8) and (9). This machine can also be used to study the fluid flow properties of specimens subjected to strain and the effect of pore pressure. Fluid is directed into the specimen through a hole in the thrust bearing (10), using capillary steel tubes (11) and (12). These tubes slide out of the frame through grooves in the piston (4). The machine also facilitates investigation of the processes of acoustic emission, measuring the propagation velocity of ultrasonic waves and the electrical resistivity of the specimen.

Electrical connections are introduced into the cell (2) through grooves in the piston rod (4) (grooves not shown in the diagram).

The hydraulic pump is used to develop fluid pressure in the cell up to 200 MPa. This can be extended to 1000 MPa using a intensifier assembled in a separate frame. Inert gas (nitrogen) is used to create pressure in the cell for experiments at elevated temperatures.

1.3 MECHANICAL PROPERTIES OF ROCKS IN THE PRE- AND POST-FAILURE STRENGTH ZONE FOR A WIDE RANGE OF STRESS STATES AND LOADING PATHS

1.3.1 Effect of Stress State on the Mechanical Characteristics of Rocks

'Stress-strain' curves are the basic experimental data from which most of the constitutive behaviour of rock is derived. A typical 'stress-strain' curve obtained by testing a rock specimen in the machine described earlier, is shown in Fig. 1.12. The diagram is plotted in the $\Delta\sigma_1 - \varepsilon_1 - \varepsilon_2$ coordinate system, where $\Delta\sigma_1 = \sigma_1 - \sigma_2$ denotes the axial stress deviator; ε_1 and ε_2 are the axial and lateral strains respectively. The locations of important parts of the curve are indicated in the diagram.

The notations given here are used in subsequent discussions:
$\Delta\sigma_1^{el}$ is the axial stress deviator at the elastic limit;
$\Delta\sigma_1^{us}$ the axial stress deviator at the peak strength;
$\Delta\sigma_1^{res}$ the axial stress deviator at the ultimate residual strength;
ε_1^{el} and ε_2^{el} the axial and lateral strains at the elastic limit;
ε_1^{us} and ε_2^{us} the axial and lateral strains at peak strength;

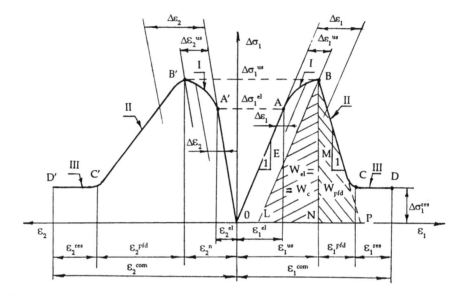

Fig. 1.12: Typical 'stress-strain' curve for rock specimens.

ε_1^{pfd} and ε_2^{pfd} the axial and lateral strains on the descending branch in the post-failure state;

ε_1^{res} and ε_2^{res} the axial and lateral strains in the residual strength zone;

ε_1^{com} and ε_2^{com} the complete (total) axial and lateral strains in the specimen;

$\Delta\varepsilon_1^{us}$ and $\Delta\varepsilon_2^{us}$ the permanent axial and lateral strains at peak strength;

$\Delta\varepsilon_1$ and $\Delta\varepsilon_2$ the permanent axial and lateral strains at any arbitrary point in the curve;

$E = d\Delta\sigma_1/d\varepsilon_1$ is the (Young) modulus of elasticity;

$M = |d\Delta\sigma_1/d\varepsilon_1|$ is the modulus of the descending (post-failure) branch of the curve;

$\nu = \varepsilon_2^{el}/\varepsilon_1^{el}$ the Poisson ratio;

$\mu = \Delta\varepsilon_2/\Delta\varepsilon_1$ the coefficient of permanent lateral deformation.

Segments AB, BC and CD in the curve represent permanent deformation processes and are indicated by Roman numerals I, II and III. Segment I indicates the segment of the curve from the elastic limit to the ultimate strength, segment II defines the region where the strength lies between the ultimate and residual strengths, and segment III indicates the residual strength zone. The deformation processes occurring in these segments are analysed below.

In the initial loading phase, the test specimens are deformed elastically, according to Hooke's law. The gradient of the straight-line portion of the curve is characterised by Young's modulus (E). Beyond the elastic limit $\Delta\sigma_1^{el}$, permanent strains are developed within the material. Beyond the ultimate strength limit $\Delta\sigma_1^{us}$, the zone of post-failure deformation is encountered. This zone continues until the ultimate residual strength $\Delta\sigma_1^{res}$ is attained. The post-peak branch of the curve is often approximated as a straight line, with the gradient of this zone being characterised by the 'drop modulus' M.

Figure 1.13 shows actual curves $\sigma_1 - \varepsilon_1 - \varepsilon_2$ obtained for a series of rocks by careful testing of specimens under uniaxial compression ($\sigma_2 = 0$; $\Delta\sigma_1 = \sigma_1$) [78, 92]. Results for the following rocks are shown in the Figure: 1—marble (Koelga); 2—biotitic granite (Kareliya); 3—biotitic plagiogranite (Yuzhuralzoloto); 4—sandstone 1 (Donbass); 5—plagiogranite (Yuzhuralzoloto); 6—diabase (Bratsk Hydroelectric Plant); 7—talc chlorite (Seg Lake); 8—sandstone 2 (Donbass); 9—magnetite-haematitic hornfels (the Urals); 10—magnetite-haematitic ore (the Urals); 11—albitite (the Urals).

The various rocks tested exhibit sharp differences in mechanical properties. For example, ultimate strengths (σ_1^{us}) range from 76 MPa (marble) to 620 MPa (albitite), while Young's modulus (E) varies from 0.25×10^5 MPa (sandstone 2) to 1.1×10^5 MPa (ore) and the drop modulus (M) varies from 0.17×10^5 MPa (marble) to 13.2×10^5 MPa (diabase). The magnitudes of permanent deformation $\Delta\varepsilon_1^{us}$ at the ultimate strength limit, Poisson ratio ν and the brittleness coefficient K_{br} are given in Table 1.1, together with the indices mentioned above.

Except for marble, all the rocks exhibit a fairly steep curve beyond the peak strength. The complete diagram (i.e., one that includes the post-failure branch), allows us to assess the brittleness of rocks. This index is important in assessing the susceptibility of rocks to dynamic uncontrolled failure, such as may occur, for example, with rockbursts in underground workings. As already mentioned, the condition for a stable deformation process in the post-failure strength zone is determined by the following ratio between the characteristics of the post-failure branch of the curve (M^*) and the stiffness characteristics of the loading system (M_H): $M_H/M^* > 1$. The drop modulus M and characteristic M^* are related as follows: $M = M^*l/F$ (where l is specimen length and F the cross-sectional area of the specimen). For specimens having the same l/F ratio, the drop modulus M serves as a measure of the relative brittleness of the material. Based on this index, diabase is the most brittle of the rocks listed in Table 1.1.

The degree of brittleness of rock can also be characterised in terms of the susceptibility of a particular rock to self-disintegration. Self-disintegration is possible when the amount of specific elastic energy stored in the material at peak strength load is equal to or exceeds the specific energy absorbed by the rock material while undergoing deformation and rupture during the post-failure phase. These energy types are indicated in Fig. 1.12 as follows:

a)

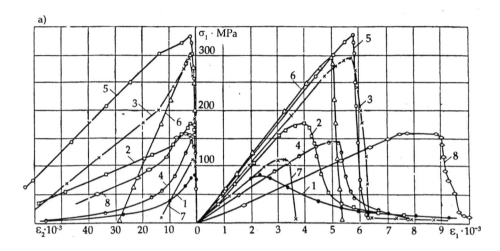

b)

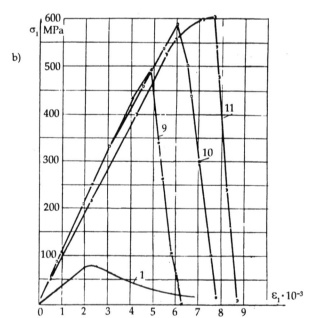

Fig. 1.13: Complete 'stress-strain' curves for a series of rocks tested under uniaxial compression. The legend is given in the text.

—Elastic energy stored in the rock at peak strength (at point B) is indicated as W_{el} (area of triangle LBN).

—Work done in the permanent deformation and failure of the material beyond point B, due to the elastic energy W_{el}, indicated as W_c (in this case,

Table 1.1

No.	Rock	$E \times 10^5$, MPa	$M \times 10^5$, MPa	σ_1^{us}, MPa	$\Delta\varepsilon_1^{us}$, 10^{-3}	ν	K_{br}
1.	Marble	0.4	0.17	76	0.6	0.18	0.30
2.	Biotitic granite	0.55	1.85	175	0.63	0.17	0.77
3.	Biotitic plagiogranite	0.57	4.17	293	0.36	0.24	0.88
4.	Sandstone 1	0.3	3	142	0.5	0.1	0.91
5.	Plagiogranite	0.6	8	355	0.2	0.18	0.93
6.	Diabase	0.67	13.2	295	0.28	0.22	0.95
7.	Talc chlorite	0.41	10.3	110	0.8	0.25	0.96
8.	Sandstone 2	0.25	8	157	1.1	0.33	0.97
9.	Magnetite haematitic hornfels	1.1	4	480	0.2	0.1	0.78
10.	Magnetite haematitic ore	1.1	3.5	580	0.5	0.1	0.76
11.	Albitite (soda feldspar)	0.94	6	620	1.2	0.1	0.86

the work done is equal in magnitude to the work of elastic deformation W_{el}, done on the specimen prior to point B).

—Work done in deformation and failure—which requires an additional supply of energy from outside the loading system—is indicated as W_{pfd} (area of the triangle NBP, with no consideration of the deformation at the ultimate residual strength limit). This energy is termed 'energy consumed in post-failure deformation' in subsequent discussions in this book.

The brittleness coefficient, characterising the susceptibility of rock to self-disintegration, is determined from the following relationships:

$$K_{br} = W_{el} / (W_c + W_{pfd}) = W_{el} / (W_{el} + W_{pfd}) = M/(M + E). \quad (1.4)$$

The brittleness coefficient computed from formula (1.4) may vary from zero to unity, depending on the properties of the material (under the condition $W_{pfd} > 0$). The nearer the value of the brittleness coefficient to unity, the higher the brittleness of a given material. Values of brittleness coefficients for all the rocks listed in Table 1.1 were found to be close to unity except for marble from the Urals. Such high values of brittleness index are not accidental since the rock specimens tested were taken from those fields and zones in deposits most prone to rockbursts and sudden outbursts of rocks and gas. Despite the highly brittle nature of the rocks, the deformation process in the post-failure strength zone was stable when tested in the stiff machines (Figs. 1.8 and 1.9) described earlier. The process can be stopped at any point in the post-failure curve and, if necessary, the specimen unloaded.

Quantitative evaluation of the brittleness index makes it possible to classify rocks more precisely according to their mechanical properties.

Based on the results shown in Fig. 1.13 (a), the relationship between the triaxial (volumetric) strain θ and the stress level σ_1 in the material in the pre-failure and post-failure strength zones, is plotted in Fig. 1.14. The volumetric strains were determined from the expression

$$\theta = \varepsilon_1 + 2\varepsilon_2. \tag{1.5}$$

Complete curves are shown in Fig. 1.14 (a) while curves up to the pre-failure strength are plotted in Fig. 1.14 (b) on a larger scale. In the post-failure zone the specimen tends to dilate with respect to the initial specimen volume. The total magnitude of disintegration of the rock was determined at the end of the experiment from the nature of crack growth during the crack formation process. Large-scale disintegration obtained in (1) marble, (2) biotitic granite, (3) biotitic plagiogranite and (5) plagiogranite resulted from an intensive crack formation process that developed fairly evenly over the entire volume of the specimen. In contrast to these rocks, the failure of diabase (6) and talc chlorite (7) occurred due to the formation of one or a small number of major cracks, accompanied by relatively less dilatation.

We shall now examine the results of experiments conducted to determine the properties of rocks under triaxial stress states. Numerous experiments were conducted under triaxial compression in which $\sigma_1 > \sigma_2 = \sigma_3$ on the

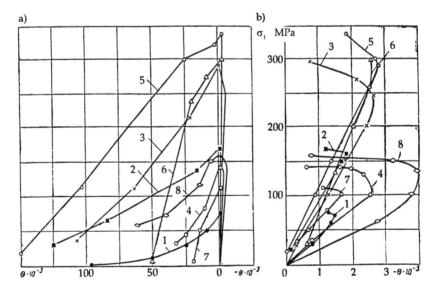

Fig. 1.14: Relationship between volumetric strain and axial stress intensity for eight rock types: (a) complete curves; (b) pre-failure curves. For legend, see text circa Fig. 1.12.

machines illustrated in Figs. 1.10 and 1.11. A few tests were conducted under biaxial tension, biaxial tension-compression, and tension under confining pressure; these are detailed later.

Experiments in which $\sigma_1 > \sigma_2 = \sigma_3$ were conducted on cylindrical specimens of 30 mm diameter and 80 mm length. The complete curves obtained at different confining pressures $\sigma_2 = \sigma_3$ for a series of rocks are shown in Fig. 1.15 (a—marble from the Urals (Koelga), b—lignite (Shurabsk), c—granite (Kareliya), d—sandstone NBP (Donbass), e—sandstone BP (Donbass), f—sulphidic ore (Noril'sk)) [79, 92]. The abbreviation NBP after sandstone stands for 'non-burst-prone' (non-hazardous) with respect to dynamic events in the form of outbursts of rock and gas into a mine working, while BP stands for 'burst-prone' (hazardous). Axial strains ε_1 and lateral strains ε_2 are plotted on the horizontal-axis while values of axial stress $\Delta\sigma_1$, the stress deviator between the principal stress σ_1 and the confining pressure $\sigma_2 = \sigma_3$, i.e., $\Delta\sigma_1 = \sigma_1 - \sigma_2$, are shown on the vertical-axis. The magnitudes of confining pressures σ_2 at which the curves were obtained are indicated on the respective diagrams.

The complete curves obtained for marble specimens are shown in Fig. 1.15 (a). At low values of σ_2, a sharply distinguishable stress maximum is observed, followed by a drop which eventually levels out in a horizontal position. The load-bearing capacity of the material in this section is termed *residual strength*. Here, the material has entirely lost its cohesiveness and further deformation occurs due to the relative slippage of two or more fragments of the specimen along rough surfaces caused by shearing. Friction over these surfaces gives rise to residual strength. With increase in σ_2, the post-failure part of the curve flattens, permanent deformations at the ultimate strength limit increase (reaching 30% and 40%), stress maxima are less distinct and the residual strength rises sharply. The value of this strength at high confining pressures can become equal to the ultimate strength.

The curves for lignite specimens from the Shurabsk coal basin in Tadzhikistan appear to be qualitatively similar to the results obtained for marble, as shown in Fig. 1.15 (b). The distinct stress maxima seen at low values of confining pressure σ_2, disappear at pressures of 50 and 100 MPa. Axial deformations under these pressures can exceed 30%.

When rock is subjected to high confining pressures, it ceases to behave like a brittle material. The modulus of post-failure deformation M decreases. Under such conditions it is not necessary to have a stiff loading frame in order to obtain complete curves. From a practical point of view, the results obtained at relatively low levels of confining pressure σ_2 are interesting because rocks at the exposed surfaces in mine workings are subject to such loading conditions. Under these conditions data on the behaviour of rocks in the post-failure zone are essential in making assessments of the stability

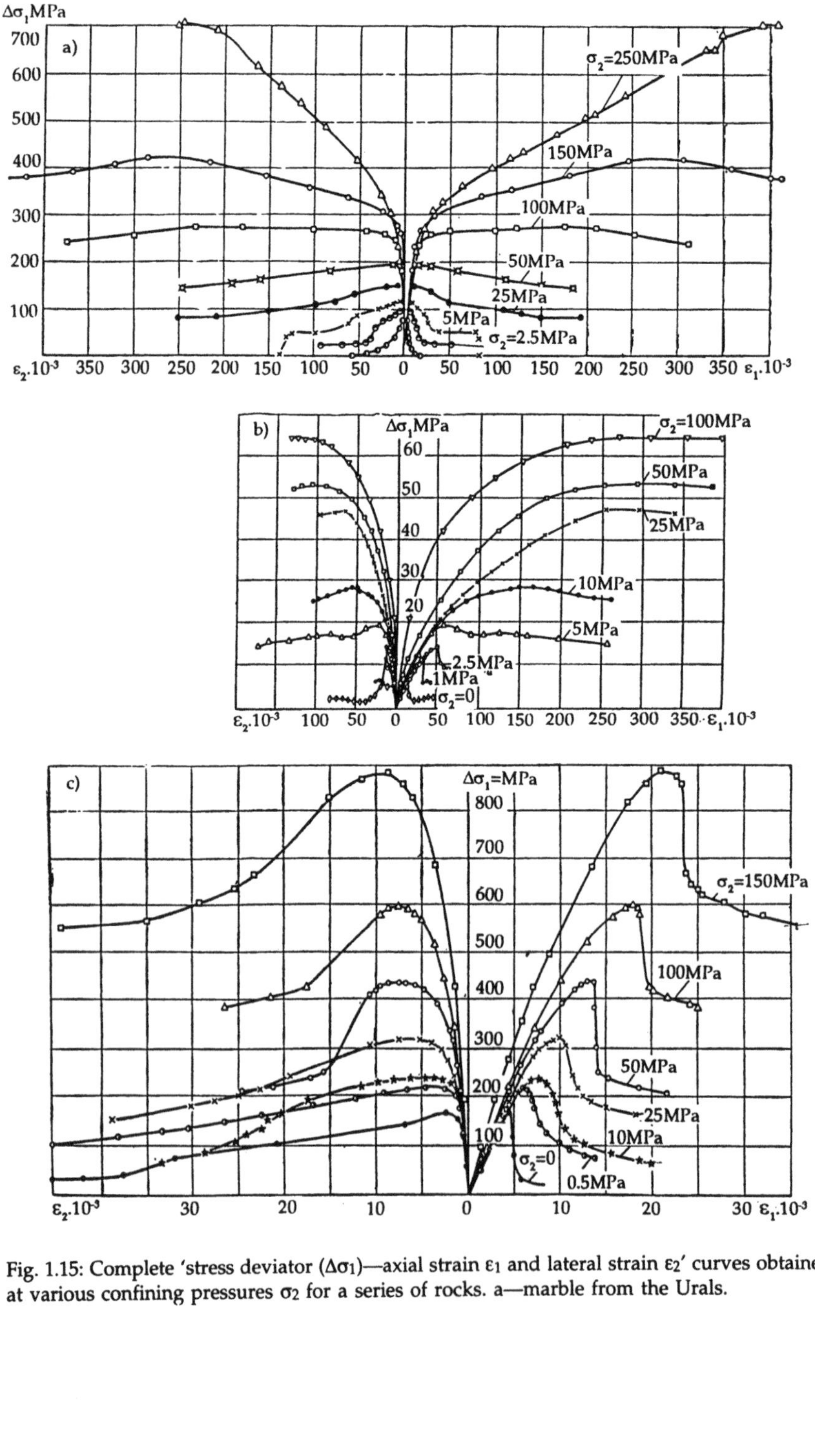

Fig. 1.15: Complete 'stress deviator ($\Delta\sigma_1$)—axial strain ε_1 and lateral strain ε_2' curves obtained at various confining pressures σ_2 for a series of rocks. a—marble from the Urals.

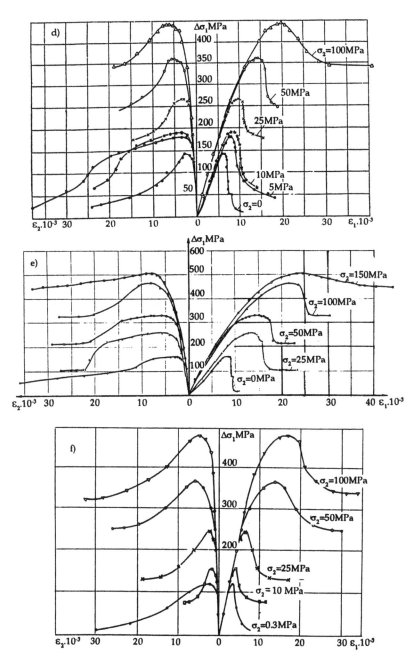

Fig. 1.15 (*Contd.*): b—lignite (Shurabsk); c—granite (Kareliya); d—sandstone NBP (Donbass); e—sandstone BP (Donbass); f—sulphidic ore (Noril'sk).

of mine workings and in forecasting the probability of dynamic phenomena (rockbursts, sudden outbursts etc.).

In the curves shown in Fig. 1.15 (c–f) for granite, sandstone (NBP and BP) and sulphidic ores, all the relationships exhibit a distinct maximum; only at higher values of σ_2 does the post-failure part of the curve flatten out.

The 'stress-strain' curves described earlier, in addition to providing elastic constants and post-failure characteristics, enable all conditions to be established for the three limit states: elastic limit, peak strength limit and residual strength limit. Customarily, these limit states (or values) are represented in the form of Mohr envelopes. There is no corresponding universal analytical form in which the envelopes for a wide range of stress states can be shown graphically. This constitutes a serious drawback. The authors have developed a new analytical representation [69, 74, 79] for all three types of limit states—in the form of exponential equations:

$$\tau_{el} = \tau_{el}^0 \exp{(BC)} \qquad\qquad (1.6)$$

$$\tau_{us} = \tau_{us}^0 \exp{(AC)} \qquad\qquad (1.7)$$

$$\tau_{res} = \tau_{res}^0 \exp{(OC)} \qquad\qquad (1.8)$$

The conditions for the elastic limit are given by equation (1.6), while equation (1.7) describes the conditions for peak strength. Equation (1.8) describes the conditions of residual strength limits. In these equations:

$$\tau_{el} = \left(\sigma_1^{el} - \sigma_2\right)/2 \text{ is the elastic shear limit,}$$

τ_{el}^0 is a constant representing the elastic shear limit under uniaxial compression;

$$\tau_{us} = \left(\sigma_1^{us} - \sigma_2\right)/2 \text{ is the shear peak strength;}$$

τ_{us}^0 is a constant to represent the peak shear strength under uniaxial compression;

$$\tau_{res} = \left(\sigma_1^{res} - \sigma_2\right)/2 \text{ is the ultimate residual shear strength;}$$

τ_{res}^0 is a constant that represents the ultimate shear residual strength under uniaxial compression.

B, A, O are constants (to be precise, parameters) reflecting rock hardening with increase in hydrostatic pressure;

$C = \sigma_2/\sigma_1$ is a parameter used to indicate stress state.

The constants B, A, O can be interpreted, in the terminology of Mohr envelopes, as angles of internal friction, although in this case they do not depend on the magnitude of σ_2. The constants τ_{el}^0, τ_{us}^0 and τ_{res}^0 are analogues of cohesion coefficients. The analytical expressions mentioned above, and Mohr envelopes, do not consider the influence of the intermediate principal normal stress.

Experimental curves for the elastic limit and the peak strength states plotted in (ln τ vs C) coordinates for marbles from (a) Kararsk and (b) the Urals, (c) talc chlorite (Karel'sk) and (d) diabase (Siberia) [69] are shown in Fig. 1.16. The petrographic composition of these rocks is given in Appendices I and II.

The experimental points lie satisfactorily on straight lines. This helps in approximating the relationships obtained by equations of the type (1.6) and (1.7). At $C = 0$, we have uniaxial compression. To the right of the origin, parameter C is positive and all components of principal stresses are compressive. To the left of the origin, parameter C is negative, which explains the tensile sign for the principal normal stresses.

Additional studies were undertaken of talc chlorite under loading conditions in the zone of negative values of parameter C (Fig. 1.16c) [65, 69]. Cylindrical specimens 30 mm in diameter with metal grips attached at the ends were tested in order to develop tensile stress σ_1 along the specimen axis. The specimen was subjected to compressive stress $\sigma_2 = \sigma_3$ applied to its sides by means of pressurised fluid in the cell. In this case parameter C was computed as the ratio, $C = \sigma_1/\sigma_2$.

The specimen and layout of the experiment are shown in Fig. 1.17 (a). The specimen (1) with a working length $l = 90$ mm was tested. A thin layer of glue (4) was applied to each end in order to affix metal grips (2) and (3). The specimen assembly was placed in a hydrostatically pressurised cell (5). The grip material was selected such that its modulus of elasticity was comparable to the modulus of elasticity of the test specimen. This was done to reduce stress concentrations at the *specimen-grip* interface. The upper grip (2) is supported by a spherical thrust bearing (6). The shank of the lower grip (3), diameter d_{vario}, is extended outwards. The shank diameter was varied to facilitate change of the value of C over a wide range and to implement only 'simple loading' of the specimen by hydrostatic pressure $\sigma_2 = \sigma_3$ in the cell (5). Pressure was developed by a pump (H). The parameter C was calculated on the basis of the specimen diameter d_1 and shank d_{vario} according to the following relation:

$$- C = \sigma_1/\sigma_2 = [(d_{vario}^2/d_1^2)-1].$$

If the two diameters are equal, then $C = 0$. This indicates equal biaxial components of compression. Under this compression, the peak strength and elastic limit for isotropic material coincide with the results obtained under

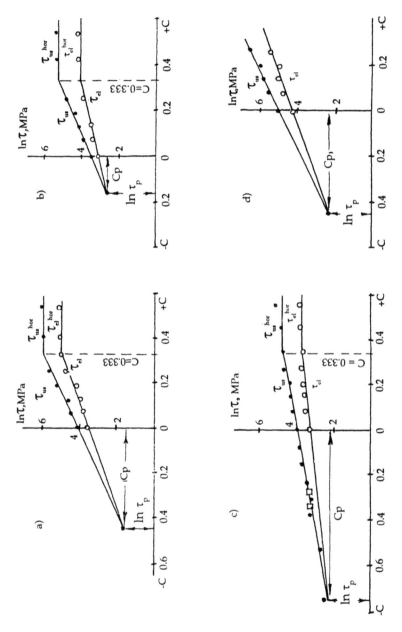

Fig. 1.16: Dependence of peak strength τ_{us} and elastic limit τ_{el} on parameter C for: a—white marble (Kararsk), b—white marble (the Urals), c—talc chlorite (Kareliya) and d—diabase from environs of Bratsk hydroelectric power station.

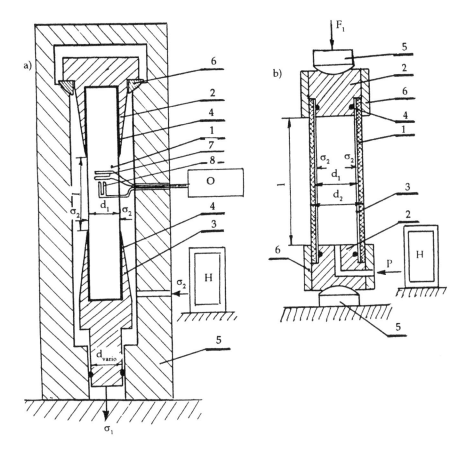

Fig. 1.17: Design of specimens and loading scheme for testing in the negative value zone for parameter C.

uniaxial compression, as will be demonstrated later. Lateral and axial deformations were measured by electrical resistance strain gauges (7) and (8) glued onto the surface of the specimen. Signals from these gauges were recorded by an oscillograph (O). The specimen surface was coated with an impermeable layer of epoxy resin to preclude entry of cell fluid into the specimen.

A second series of experiments was conducted on specimens of talc chlorite in the form of thin-walled tubes in order to study the zone of negative values of parameter C. The design of the experiment and configuration of the specimen are illustrated in Fig. 1.17(b). The specimen (1) dimensions were $d_1 = 40$ mm, $d_2 = 46$ mm and working length $l = 95$ mm. Metal thrust bearings (2) were affixed to the ends of the specimen. The inner surface of the specimen was protected by a thin polymer film (3) to preclude fluid entry. The cavity of the tubular specimen was sealed with rubber rings (4). An axial compressive stress σ_1 was applied to the specimen through spherical seatings (5). The edges of the specimen were firmly supported by hard polymer bushings (6)—used in order to strengthen the specimen grips to reduce stress concentration at these points. Tensile stresses $\sigma_2 = \sigma_3$ were produced in the specimen walls by the hydrostatic pressure P—generated by the pump (H). The axial compressive stress σ_1 was developed independent of $\sigma_2 = \sigma_3$. Parameter C was quantified by the ratio:

$$C = -\sigma_2/\sigma_1,$$

where $\sigma_2 = d_1 P/2\delta$ and $\sigma_1 = F/\pi d_1 \delta_1$ while $\delta = (d_2 - d_1)/2$. The force F was determined from the relationship which considers the hydrostatic pressure inside the tubular specimen:

$$F = [F_1 - \pi d_1^2 P/4],$$

where F_1 is the total axial force applied along the axis of the specimen while it was subjected to hydrostatic pressure P.

Axial and lateral deformations in the specimen walls were measured by electrical resistance strain gauges glued to the outer surface. These gauges are not shown in the Figure.

The results of the two series of experiments on specimens of talc chlorite are shown in Fig. 1.16(c). The black dots indicate hollow cylinder specimens and the white squares solid cylindrical specimens. The two results are seen to be in reasonably good agreement.

The method described above is more accurate and more reliable than conventional tests. However, the increased complexity and labour involved in special preparation of the specimens is a significant disadvantage. In subsequent studies the tensile strength was determined by the Brazilian method, i.e., the diametrical splitting of a circular specimen along the axis between the loading points.

Test results can be represented as inclined straight lines for τ_{us} and τ_{el}, which intersect in the region of negative values of C at the coordinates $-C_p$ and $\ln \tau_p$. The peak strength and elastic limit coincide, i.e., there is no macroscopic permanent deformation, fracture is brittle, and the fracture surfaces coincide with the surfaces of maximum tensile stress. Thus the strength is determined by the maximum tensile normal stress. This criterion is known as the classical strength theory *I*, i.e., failure by rupture under normal tensile stresses.

In Russia, the various classical strength theories are designated by Roman numerals I–IV. These designations are followed in this book.

I—theory of maximum normal stress;

II—theory of maximum strain
III—theory of maximum tangential stress;
IV—theory of maximum octahedral shear stress.
With $\tau_{us} = \tau_{el}$, the value of C_p is found to be

$$C_p = \ln (\tau_{us}^0/\tau_{el}^0) \, [1/B - A] \qquad (1.9)$$

The value of $\ln \tau_p$ for the point of intersection of the failure curves is obtained by substituting for C_p in equation (1.6) or (1.7). The rupture strength σ_p can then be determined from the expression

$$\sigma_p = 2\tau_p/[(1/C_p)-1]. \qquad (1.10)$$

On the right side of the relation under consideration, the horizontal section of the failure envelopes starts at $C = 0.333$, corresponding to the condition $\tau = \sigma_2$. Proof that the limit values are independent of σ_2 in this region has been presented in [3] and is discussed in detail below.

Over the horizontal sections of the failure envelopes τ_{el}^{hor} and τ_{us}^{hor}, the elastic limit and peak strengths are governed by either the maximum tangential or octahedral stress, i.e., by strength theory III or IV. Quantitatively, the difference between these two theories is insignificant. In future discussion, the horizontal section of the envelopes will be referred to simply as the *zone of simple shear*.

Equations (1.6) and (1.7) describe the conditions of ultimate stress in the transition region between the zone of simple rupture and the zone of simple shear.

A wide range of rocks were also tested for conditions of positive values of C [71]. Some of the results are given in Appendix I. The values of C corresponding to the strength and elastic limits, number of specimens tested, and values of the peak strength τ_{us} and elastic limits τ_{el} are listed together with a short petrographic description of each rock, modulus of elasticity E, Poisson ratio ν, coefficients of permanent lateral deformation μ and scatter of the strength results.

Experimental relationships between the peak (ultimate) strength τ_{us}, residual strength τ_{res} and C are plotted for a series of rocks in Fig. 1.18 [79] in the ($\ln \tau$ vs C) coordinate system. These relationships were obtained from the experimental results shown in Fig. 1.15. Experimental points are approximated by straight lines in accordance with equations (1.7) and (1.8). Parameter O, indicating the slope of the residual strength line, is always greater than parameter A obtained for the lines of peak strength. The horizontal part of the peak strength lines intersects at the point where parameter $C \approx 0.333$. The concept of *residual strength* becomes meaningless when the peak strength lines become horizontal. The point at which the peak (or ultimate) strength curves become horizontal (i.e., parallel) to the

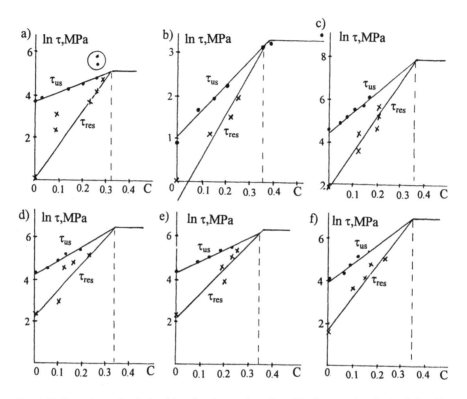

Fig. 1.18: Experimental relationships for the peak and residual strengths obtained for (a) marble, (b) lignite, (c) granite, (d) sandstone NBP, (e) sandstone BP and (f) sulphidic ore.

C-axis depends on the rate of loading. With increase in rate of loading, this point moves towards values less than 0.333. This is discussed in greater detail in Chapters 2 and 3.

The scatter of experimental points on line τ_{res} is considerably greater than that for points on the line τ_{us}. This is probably due to the difference in conditions on the (sliding) surfaces of rupture as well as to low accuracy in determining the values of ultimate residual strength (see Figs. 1.6–1.10).

Some rock types under conditions of high pressures and large plastic deformations exhibit a greater tendency towards hardening than indicated by equation (1.7). For example, the strength characteristics of marble obtained in the experiments shown in Fig. 1.15 (a) at $\sigma_2 = 150$ and 250 MPa (shown as black circles in Fig. 1.18a) do not follow the general trend. Increased hardness under the experimental conditions is probably the result of structural rearrangement of the material, due to the large plastic deformation. This aspect is being investigated further in our laboratory.

Before discussing the next series of experimental results, it is appropriate to recall the concept first proposed by N. N. Davidenkov in 1936 [17], of the dual nature of the strength of solids, viz. brittle strength—with reference to rupture, and ductile/viscous strength—with reference to shear failure preceding permanent deformation in shear. This idea was further explored by Ya. B. Freedman in 1941–1943 [25] who proposed a diagram to represent the mechanical state, which is a synthesis of the two theories of strength. The zone of failure by rupture in Freedman's diagram is said to be described by either strength theory I or II while the zone corresponding to failure by shear is described by either theory III or IV.

The mechanical state diagram is plotted in Fig. 1.19 in the coordinate system 'fracture strength $\sigma_{p\,max}$ and shear strength τ_{max}' in which the universal constants of material are initially reflected. Lines are drawn at different angles β from the origin. The point at which these intersect the peak strength lines is considered to characterise the *stiffness* of the stress state according to Freedman. Tan β is the ratio of maximum tangential stress to maximum normal stress at the moment of failure.

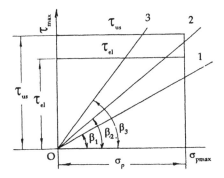

Fig. 1.19: Mechanical state diagram of materials proposed by Davidenkov [17] and Freedman [25].

Thus the nature of failure is determined by that branch of the diagram which is intersected first by the straight-lines from the origin. For example, line 1 first intersects the (vertical) line of tensile fracture hence, in this case rupture should take place without prior plastic (residual) deformation. Line 2 also intersects the (vertical) line of rupture first; however, it also crosses the elastic limit (horizontal) line and consequently undergoes some irreversible strain prior to rupture. Finally, line 3 first intersects the (horizontal) elastic limit and shear strength lines; in this case, it is concluded that the material failed due to shear produced by tangential stress.

The applicability of strength theory I or II in the tensile (rupture) zone was a subject of debate until numerous direct experimental results were accumulated in favour of strength theory I, i.e., failure due to the maximum normal tensile stresses. Some of these investigations, namely those more closely concerned with the problem of rock strength and the strength of brittle materials similar to rocks, are discussed below.

Experimental results obtained by several investigators to determine the strength of various materials loaded in plane-stress in the 'tension-tension' and 'tension' compression' quadrants are shown in Figs. 1.20 and 1.21. The authors are listed in the legend. Curves are plotted in dimensionless coordinates (σ_2/σ_p) and (σ_1/σ_p) (where σ_p is the fracture strength obtained under uniaxial tension) for ease of comparison between test results.

In the 'tension-tension' quadrant the strength conditions for all materials tested obey strength theory I. Theory I is only partially satisfied in the quadrant 'tension-compression'. In Fig. 1.20, talc chlorite satisfies theory I up to that level of compression at which $|\sigma_1| \approx |\sigma_p|$; for cast iron this ratio is roughly equal to 2, i.e., $|\sigma_1| \approx |2\sigma_p|$, while for gypsum $|\sigma_1| \approx |3\sigma_p|$. The maximum ratio of σ_1 to σ_2 is 7—for soda lime glass. Specially designed experiments for uniaxial compression of glass tubes showed, however, that the compressive strength of glass exceeded the fracture strength by a factor of 75–80. Thus strength theory I can be applied to a range considerably wider than $|\sigma_1| \approx |7\sigma_p|$.

Since the results illustrated in Fig. 1.21 coincide qualitatively with those described in Fig. 1.20, they are not discussed further here.

The results of strength tests conducted on specimens of talc chlorite in coordinates $(\sigma_1 - \sigma_2)$ are shown in Fig. 1.22. Results for hollow cylinders are indicated by squares. In the 'tension-compression' quadrant, two data points (open circles) were obtained for solid cylindrical specimens tested under axial tension with confining pressure. In the 'compression-compression' quadrant, the experimental points obtained for hollow cylinder or tubular specimens are again indicated by squares. Unlike the tests carried out in the 'tension-compression' quadrant, these specimens were subjected to lateral loading with no internal pressure. Instead, external confining pressure was applied to induce compressive tangential stresses in the walls of the hollow cylinders. The axial compressive stress was applied independently. Each hollow cylinder was tested in a hydraulic seal in the same manner as for the solid cylindrical specimens. The outer surface of the hollow cylinder specimen was coated with an impermeable layer of epoxy resin. The design of the experiment anad configuration of the specimen are illustrated in Fig. 1.17.

It can be seen from Fig. 1.22 that the strength under biaxial compression is equal to the uniaxial compressive strength. This situation, important for mining practice, agrees fully with the criterion of strength theory III for the maximum tangential stress attained under uniaxial compression. This plane-stress 'compression-compression' condition is often encountered at the surface of exposed faces in mine workings.

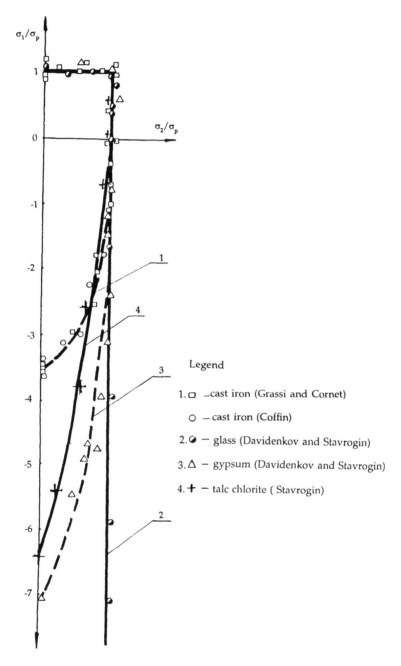

Fig. 1.20: Strength of (1) cast iron [13, 29], (2) glass, (3) gypsum and (4) talc chlorite [18, 65] under plane stress conditions plotted in dimensionless coordinates.

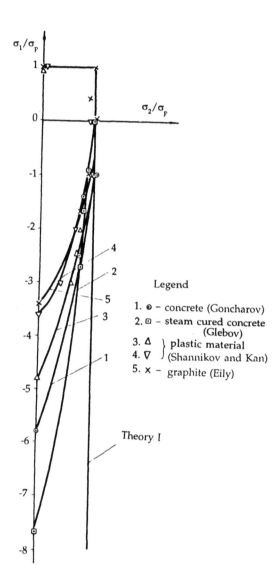

Fig. 1.21: Strength of (1) concrete [28], (2) steam-cured concrete [27], (3, 4) plastic material [55] and (5) graphite [20] under conditions of plane-stress.

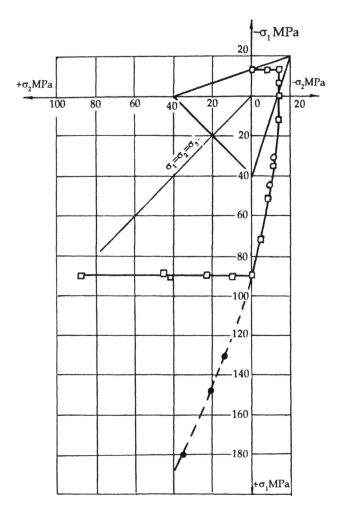

Fig. 1.22: Strength of talc chlorite under plane- and volumetric-stress.

The dashed line branch in the limit curve shown in Fig. 1.22, for which the experimental points are shown by solid black circles is an extension of this curve beyond the 'tension-compression' quadrant. This section was obtained under conditions of deviatorie triaxial compression of the type $\sigma_1 > \sigma_2 = \sigma_3$, when solid cylindrical specimens were subjected to confined pressure tests. With increase in hydrostatic pressure, the slope of the limit curve approaches the slope of the hydrostatic curve $\sigma_1 = \sigma_2 = \sigma_3$, which is the axis of symmetry for Coulomb's prism and Huber's, Mise's and Jenkin's cylinder. If the limit curve is parallel to the hydrostatic curve, then the limit conditions may be described by either strength theory III for maximum tangential stress, or strength theory IV for the octahedral stress criterion. Failure is due to simple shear preceded by a permanent shear strain. Strength theory II for talc chlorite is represented by the triangular region in Fig. 1.22. It is obvious that the experimental results do not conform to theory II.

Results obtained by various authors for quite strong and brittle rocks are shown in Fig. 1.23. Consider the 'tension-compression' and 'compression-compression' quadrants. Specimens were subjected to confining pressures with $\sigma_1 > \sigma_2 = \sigma_3$. The curve for soda lime glass is largely hypothetical, extended by analogy with the behaviour of other materials. The uniaxial compressive strength of glass was obtained experimentally, as shown in the 'tension-compression' quadrant, for the condition $\sigma_3/\sigma_p = 75$–80, while data points were obtained over the range $\sigma_3/\sigma_p = 1$ to $\sigma_3/\sigma_p = 7$.

The results given in Fig. 1.23 as well as those discussed earlier, provide justification for strength theory I. At the points where experimental curves deviate from conditions of the classical strength theory I, the limit conditions appear to be best described qualitatively by a Mohr envelope. However, such a curve does not have a unique analytical expression, which makes it difficult to use in engineering computations. Hence, as mentioned above, a simplified and rather rough approximation, in the form of a linear Mohr-Coulomb envelope, is used for practical purposes, to cover a narrow range of stress states.

With increase in hydrostatic pressure, the limit curves in the 'compression-compression' quadrant tend to become parallel to the line $\sigma_1 = \sigma_2 = \sigma_3$, i.e., in accordance with strength theory III.

Figures 1.24 (a) and (b) show the same data as in Figs. 1.20, 1.21 and 1.23 but redrawn by the authors in the (ln τ vs C) coordinate system. The legend for Fig. 1.24 is: 1—limestone; 2—marble [34]; 3, 7—limestone; 4, 6—marble; 5—dolomite; 8—sandstone [48]; 9—cement [28]; 10, 13—dolomite; 11—diabase; 12—quartzite [8]; 14—gypsum [18]; 15—graphite [20]; 16—steam-cured concrete [27]; 17, 18—plastic material [55]; 19—cast iron [29]; 20—cast iron [13].

All the above relationships, when plotted in the (ln τ vs C) coordinate system, were linear and can be described analytically by exponential equations (1.6) and (1.7).

1.3.2 Permanent Deformations in Rocks

The range of stress states in the interval between the zone of simple rupture at $C = -C_p$ and the zone of simple shear at $C \geq 0.333$ is characterised by the fact that two processes take place during permanent deformation: the

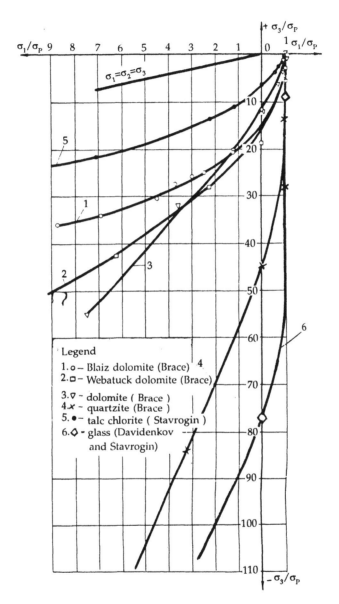

Fig. 1.23: Strength of a series of rocks and brittle materials under plane- and triaxial-stress conditions—data obtained by various authors.
1—Blaiz dolomite; 2—Webatuck dolomite; 3—diabase; 4—quartzite [8]; 5—talc chlorite [65]; 6—glass [18].

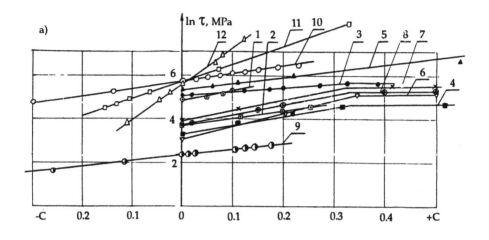

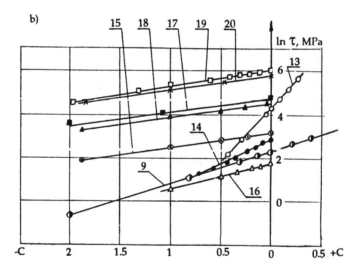

Fig. 1.24: Dependence of the strength of various materials on the parameter C, based on data obtained by various authors in both (a) compression and (b) tension.

process of microcracking, producing dilatant volumetric strain and the process of shear, resulting in a change in body shape, ostensibly with no overall volume change.

The complete 'stress-strain' type diagrams, such as shown in Fig. 1.15, represent an important compilation of experimental data for study of permanent deformation in rocks. This problem was investigated rigorously on specimens of white marble from the Urals. Test results obtained with this material are typical of a wide range of rock types; hence the relationships obtained with marble specimens can be extended qualitatively to other rocks, in particular those for which it is difficult to obtain the required number of specimens for statistically reliable interpretation of results.

The relationship between the permanent dilatant volumetric strain $\Delta\theta$ of marble from the Urals and the confining pressure σ_2 is shown in Fig. 1.25 [79]. Curve 1 shows the volumetric strains in the material at peak strength; curve 2 was obtained at the stress level $\Delta\tau$ at 66% of the difference between the peak strength and the elastic limit; curve 3 was obtained at the stress level $\Delta\tau$ at 33% of the above difference. Line 4 shows the volumetric strains at the peak residual strength. As can be seen from these curves, dilatant strains reach maximum values at specific levels of confining pressure. Curve 1, for example, reaches a maximum at $\sigma_2 = 100$ MPa. Volume increase under these conditions is about 20%. At pressures exceeding 100 MPa, volumetric strains reduce, while at pressure $\sigma_2 = 250$ MPa, they become negative (contractant). Two maxima are seen in curve 4, which indicates complete (crushing) deformation of the material. One maximum appears at low values of σ_2, while the second appears at those values of σ_2 at which maximum strain at the peak strength is observed. The presence of a maximum at low pressures is explained by a change in the mechanics of development of the

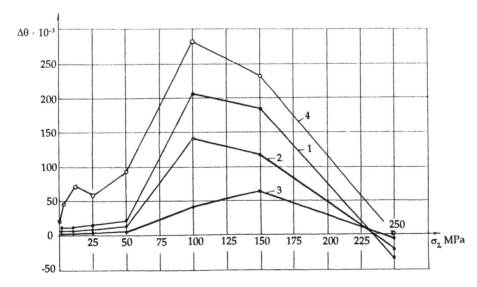

Fig. 1.25: Relationship between permanent volumetric strain $\Delta\theta$ of marble from the Urals and the magnitude of confining pressure σ_2.

deformation process in the post-failure regime beyond a certain level of confining pressure. This effect is manifested in several rocks. The essential features are discussed below.

Under conditions close to uniaxial compression, deformation of these rocks in the post-failure strength zone is accompanied by a considerable uniform disintegration of the rock specimen throughout its volume. With increase in confining pressure σ_2, the degree of disintegration increases—up to a specific limit. At a particular pressure level σ_2, uniform deformation over the entire volume of the specimen in the post-failure strength zone ends and deformation localises along one or few shear planes. At this point, volumetric deformations and the energy consumed in the rupturing process decrease considerably. The question of change in deformation mechanics at a specific level of confining pressure and the resultant effects are discussed in Chapter 3 in terms of the energy balance during brittle uncontrolled rupture.

Volumetric strain ($\Delta\theta$) is a function of the principal axial permanent strain $\Delta\varepsilon_1$. Curves of $\Delta\theta$ versus $\Delta\varepsilon_1$ obtained on specimens of marble from the Urals under different pressure levels σ_2 are shown in Fig. 1.26. The straight-line relationships (until these flatten horizontally) are plotted, based on the 'stress-strain' curves for the two sections (I) from the elastic limit to the peak strength and (II) from the peak (ultimate) strength to the residual strength, i.e., sections I and II—using the notation used in Fig. 1.12.

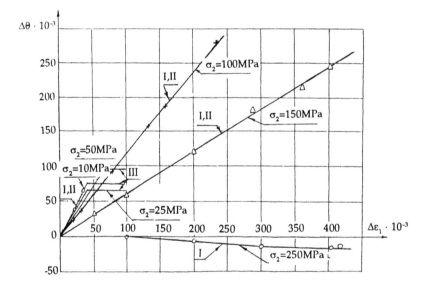

Fig. 1.26: Relationship between volumetric strain $\Delta\theta$ and principal axial permanent deformation $\Delta\varepsilon_1$ for marble.

The horizontal region III defines the zone of residual strength, where volumetric strains are essentially zero. Relationships between the volumetric and the magnitude of linear residual strain in the sloping sections are described appropriately by an equation of the type

$$\Delta\theta = \Delta\varepsilon_1 (1 - 2\mu), \tag{1.11}$$

where $\mu = \Delta\varepsilon_2/\Delta\varepsilon_1$ is the coefficient of lateral permanent deformation, which depends on the pressure σ_2; at a constant value of pressure σ_2, μ remains constant in sections I and II of the 'stress-strain' curve [69, 74, 79], as indicated by the experimental curves shown in Fig. 1.27.

The relationship between $\Delta\varepsilon_1$ and $\Delta\varepsilon_2$ for marble from the Urals is plotted in Fig. 1.27 (a). (The method for determining permanent strains based on $\Delta\sigma_1 - \Delta\varepsilon_1 - \Delta\varepsilon_2$ curves at various stages of the deformation process has been demonstrated in Fig. 1.12). The confining pressures σ_2 for the tests are indicated in Figure 1.27. The relationships for sections I and II were plotted separately. Solid black circles in the curves correspond to section I and open circles to section II. It can be seen that relationships I and II are well described by straight lines that practically coincide. Different scales were used on the graphs because of the large difference in deformations in sections I, II and III. The scales are labelled I, II and III.

The tangents of the angles of inclination of relationships I and II are equal to the corresponding values of the coefficient μ. At a pressure $\sigma_2 = 250$ MPa, the coefficient μ has the value 0.5, which indicates the absence of dilatation. At the level of residual strength, the strain is characterised by line III and in all cases the coefficient μ is close to 0.5, which is also indicative of the absence of volumetric strain. The residual strength level was not attained for the high confining pressures $\sigma_2 = 100, 150$ and 250 MPa.

Figure 1.27 (b) shows the relationships between the coefficient μ and the principal axial permanent strain $\Delta\varepsilon_1$, plotted from data of curves (a). Each line corresponds to a specific level of confining pressure σ_2 and indicates the independence of the coefficient μ from the magnitude of axial deformation. The scale chosen for $\Delta\varepsilon_1$ varies at different levels of σ_2. The 'crossed' circles define the boundary between sections I and II, i.e., the values of $\Delta\varepsilon_1$ at the ultimate strength.

Figure 1.28 shows similar relationships obtained for specimens of (a) sandstone and (b) granite. The confining pressure levels at which the experiment was conducted are indicated on the curves. Unlike the preceding curves, sections I, II and III here were plotted sequentially. The boundaries between the sections are marked by crossed circles. In each experiment the gradient of the curves in sections I and II is constant. As in the case of marble, the volumetric strain is described by equation (1.11). Curves in

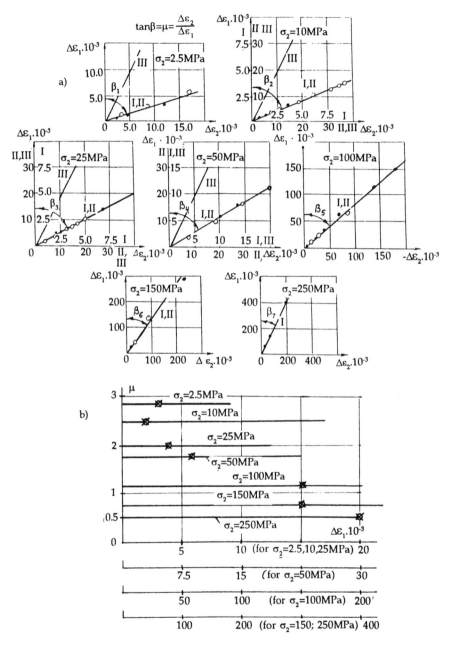

Fig. 1.27: (a) Relationship between the permanent lateral strain $\Delta\varepsilon_2$ and the permanent axial strain $\Delta\varepsilon_1$ for marble at different confining pressures σ_2. (b) Relationship between the coefficient of permanent lateral deformation μ and the principal axial permanent strain $\Delta\varepsilon_1$ plotted from data of the curves shown in (a).

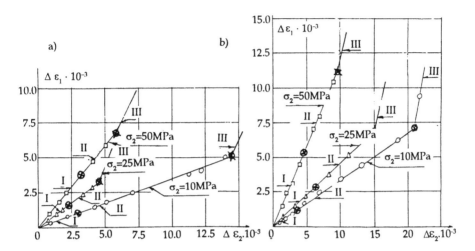

Fig. 1.28: Relationship between permanent lateral strain $\Delta\varepsilon_2$ and permanent axial strain $\Delta\varepsilon_1$ for different levels of confining pressure σ_2, obtained on samples of (a) sandstone and (b) granite.

Fig. 1.29 show the linear relationships between the volumetric strain $\Delta\theta$ and the principal permanent strain $\Delta\varepsilon_1$, obtained from data of Fig. 1.28.

It should be noted that in order to obtain a linear relationship between $\Delta\varepsilon_1$ and $\Delta\varepsilon_2$, the non-uniform growth of permanent lateral strain in various directions should be taken into consideration. In experiments to study this question, lateral strain gauges were positioned on the specimen in four directions at 45° (see Fig. 1.30). Lines of slip and the orientation of strain gauges with respect to their slope are also shown. In the given case, gauge 1 recorded the largest lateral strain in the specimen and gauge 2 the least. In a similar situation of uneven lateral growth of the deformation process, which often occurs during a transition through the peak strength, the average value of lateral strain can be determined from readings of all the gauges. Some experimental results concerning the anisotropic properties of rocks in relation to the development of lateral strain [90], are shown in Fig. 1.31. These results were obtained by testing specimens of (a) sandstone NBP, (b) sandstone BP and (c) granite. The levels of confining pressure σ_2 at which the tests were conducted are shown on the curves. The point corresponding to the peak (ultimate) strength is indicated as *max*. Readings of all lateral strain gauges were the same in section I, while the relationship between $\Delta\varepsilon_1$ and $\Delta\varepsilon_2$ is almost linear. After passing through the peak strength level, the deformation became non-uniform. Gauge readings giving the largest and least lateral strain are shown in the curves in sections

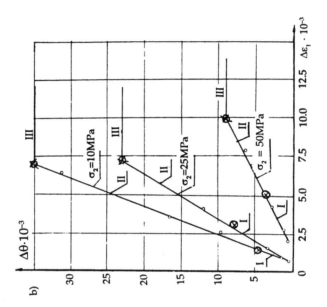

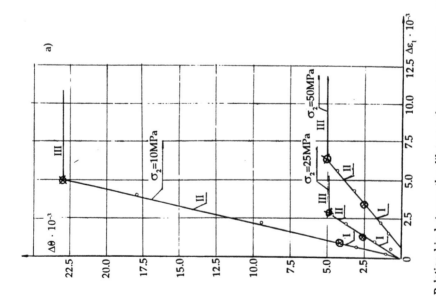

Fig. 1.29: Relationship between the dilatant volumetric strain $\Delta\theta$ and the principal permanent strain $\Delta\varepsilon_1$ for (a) sandstone and (b) granite under different levels of σ_2.

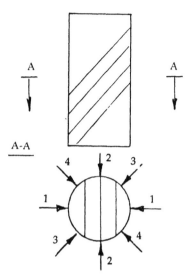

Fig. 1.30: Arrangement of lateral strain gauges on a specimen to study anisotropy in the growth of lateral deformation.

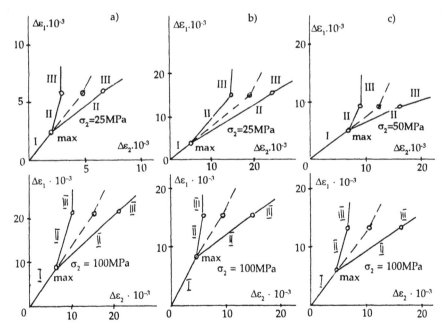

Fig. 1.31: $\Delta\varepsilon_1$ versus $\Delta\varepsilon_2$ for (a) sandstone NBP, (b) sandstone BP and (c) granite, taking into account anisotropy in growth of lateral deformation.

II and III. The dashed line, indicating average values of lateral deformation in section II, has the same slope as line I, which indicates no change in the lateral deformation coefficient μ even after passing through the peak strength. The average value of the coefficient μ in section III is close to 0.5.

The constancy of the coefficient of permanent lateral deformation μ in the pre-failure strength (section I) and post-failure strength regions (section II) of a heterogeneous solid using the model developed by the authors, indicates the invariability of the growth mechanics of the deformation process in these two sections. Essential features of the model are discussed below.

The linear relationship between $\Delta\varepsilon_1$ and $\Delta\varepsilon_2$ is an important feature of development of the deformation process in the zone of permanent deformations. This feature is observed clearly during testing of rocks under triaxial compression along different loading paths, as shown below. However, under uniaxial compression and compression at low levels of confining pressure, the linear relationship often deviates towards an increase in lateral deformation. The causes of such behaviour are not discussed in this monograph. The relationship between $\Delta\varepsilon_1$ and $\Delta\varepsilon_2$ determines the coefficient μ, dependent on the level of confining pressure σ_2. Relationships between the coefficient of permanent lateral deformation μ and the level of confining pressure σ_2 for (1) marble, (2) sandstone and (3) granite are shown in Fig. 1.32. These curves were plotted from data taken from Figs. 1.27 and 1.28; their shape is typical for most rocks. The value of the coefficient μ decreases with increase in level of confining pressure, tending in the limit to a value of 0.5.

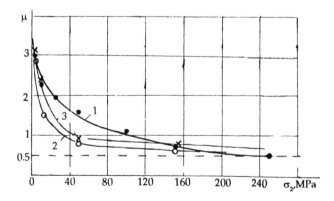

Fig. 1.32: Relationship between the coefficient of permanent lateral deformation μ and level of confining pressure σ_2 for (1) marble, (2) sandstone and (3) granite.

1.3.3 Effect of Loading Path on the Mechanical Properties of Rocks

The mechanical properties of rocks are affected significantly not only by the current stress state, but also the stress history, or loading path by which the current stress state was reached. The loading path may be represented visually by the trajectory of the load changes in a coordinate system of principal normal stresses. This is shown pictorially in Fig. 1.33 (a). The principal stresses σ_1 and $\sigma_2 = \sigma_3$ in the quadrant are all compressive. Any point in the given coordinate field corresponds to a specific stress state. This point can be reached through various loading paths. One such loading

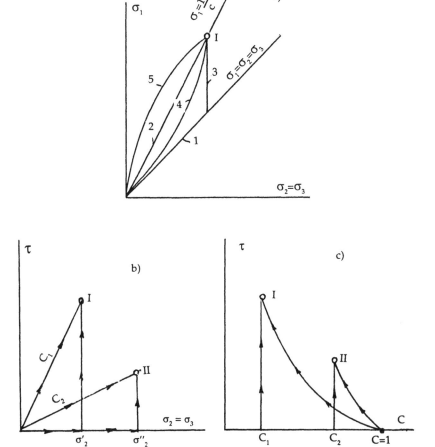

Fig. 1.33: Various loading paths in three different coordinate systems.

path is the straight line ray from the origin. Such a loading path is characterised by the parameter $C = \sigma_2/\sigma_1$, which is equal to the tangent of the angle between the ray and the vertical axis. The ray coinciding with the vertical axis corresponds to $C = 0$; the ray at an angle of 45° corresponds to $C = 1$. Loading applied at a given constant value of parameter C is termed *proportional loading* or, according to the terminology adopted by A.A. Iluyshin, *simple loading*. All loading paths other than simple loading are termed *complex paths*. Various complex loading paths are illustrated in Fig. 1.33(a) by trajectories 1–3, 4 and 5, all of which lead to the same end point I. The study of the effect of loading path on properties of materials is based on the principle of comparing the properties when the same final stress state is reached (i.e., all arriving at one point in the coordinate field) through various paths. In the present study, each final stress state was reached by two paths: simple but for different values of C and complex, reached by way of the path 1–3 (Fig. 1.33) for different levels of confining pressure σ_2. Under complex loading, the specimen in the test chamber was first subjected to hydrostatic compression $\sigma_1 = \sigma_2 = \sigma_3$, up to the level $\sigma_2 = \sigma_3$, corresponding to the horizontal coordinate of point I; additional axial stress $\sigma_1 > \sigma_2 = \sigma_3$ was then applied separately to reach point I (i.e., along the vertical path 3 in Fig. 1.33)—which is the loading principle of T. Karman.

The experimental investigations discussed in this monograph were conducted under stress states lying above the ray $C = 1$, i.e., in the range $O < C < 1$. For these stress states, it is convenient to work with the coordinates $\tau - \sigma_2$ or $\tau - C$ in order to represent results clearly. Figures 1.33 (b) and (c) show the simple and complex paths of loading in these coordinate axes. In the $\tau - \sigma_2$ coordinates, simple loading paths plot as straight rays from the origin. The ray coinciding with the vertical axis represents $C = 0$ while the ray along the horizontal axis corresponds to $C = 1$. Rays passing through points I and II [Figs. 1.33 (b)] are characterised by values of C between O and 1, i.e. $O < C < 1$. The complex path 1–3 is a discontinuous line reaching the same points I and II: it initially lies on the horzontal axis up to the point corresponding to the required level of hydrostatic pressure $\sigma_1 = \sigma_2 = \sigma_3$, and then becomes vertical. In the $\tau - C$ coordinate system, the simple loading path is represented by a vertical line while the path of complex loading reflects only that part of the path answering to the condition $\sigma_1 > \sigma_2 = \sigma_3$, and is indicated by a curve starting from the end point $C = 1$, where $\sigma_1 = \sigma_2 = \sigma_3$. Arrows on all the lines indicate the direction of stress change during the loading process.

Studies of the effect of loading path on rock properties were concentrated initially on the peak elastic and peak strength states [66, 77]. These studies established that the limit curves for both simple and complex loading coincide. This is demonstrated in the experimental curves shown in Fig. 1.34,

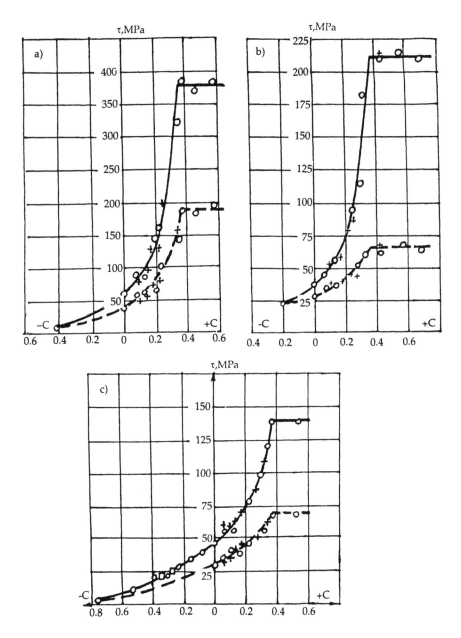

Fig. 1.34: Experimental curves of limit states for (a) marble from Kararsk, (b) marble from the Urals and (c) talc chlorite, under simple and complex paths of loading.

obtained by testing specimens of (a) marble from Kararsk, (b) marble from the Urals and (c) talc chlorite. Points obtained in simple loading are shown by circles and those obtained in complex loading by crosses. This experimental proof of the coincidence of limit curves for different loading paths is an important result, which served as the basis for developing mechanical state diagrams for rocks. These diagrams facilitate analysis of the mechanical properties of rocks for a wide range of stress states and loading paths. The procedure for plotting mechanical state diagrams is discussed later.

Figure 1.35 shows the relationships between permanent strain $\Delta\varepsilon_1$ and parameter C for simple (curves 1) and complex (curves 2) loading paths for (a) marble from Kararsk, (b) marble from the Urals and (c) talc chlorite. The strains generated in these rocks when the same stress level equal to peak strength is applied via various loading paths are compared. The stress state is characterised by C. It can be seen that under simple loading, the magnitudes of permanent deformations at peak strength are three times (or more) higher than those attained under complex loading. As mentioned earlier, since permanent volumetric strains are proportional to the axial strain $\Delta\varepsilon_1$, the dependence of rock disintegration on C for similar states of stress arrived at by different loading paths, will be similar to the curve shown in Fig. 1.35. The reasons for such behaviour will be discussed in the section dealing with the diagrams of mechanical state for rocks.

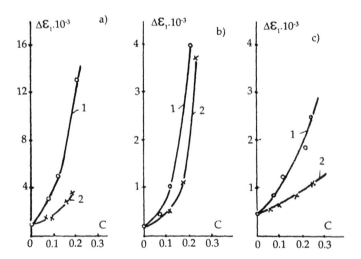

Fig. 1.35: Relationships between permanent axial strain $\Delta\varepsilon_1$ and parameter C for (a) marble from Kararsk, (b) marble from the Urals and (c) talc chlorite for the same stress state, arrived at by (1) simple and (2) complex loading paths.

Figure 1.36 shows the relationship between the coefficient of permanent lateral deformation μ and the magnitude of permanent axial strain $\Delta\varepsilon_1$, obtained under simple loading for different values of C for (a) marble from Kararsk, (b) marble from the Urals and (c) talc chlorite. The values for C are indicated on the lines. As these curves show clearly, for the deformation produced by loading at a constant value of C, the value of coefficient μ remains constant. (Some deviation from this relationship is observed for small values of C and $\Delta\varepsilon_1$.) At the same time, coefficient μ is a function of C and was found to have a value varying from 4 to 0.5 when C was varied from 0 to 0.5. The relationship between μ and C for simple loading is illustrated in Fig. 1.37.

Results of other experimental studies on the effect of loading paths on properties of rocks will be analysed later, when discussing the mechanical state diagrams, model of deformation processes and rock permeability.

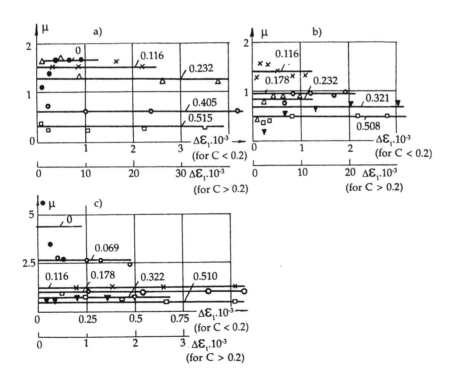

Fig. 1.36: Relationship between the coefficient of permanent lateral deformation μ and magnitude of permanent axial strain $\Delta\varepsilon_1$ obtained under conditions of simple loading and different values of C for (a) marble from Kararsk, (b) marble from the Urals and (c) talc chlorite.

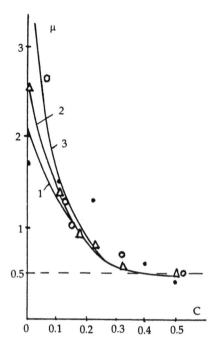

Fig. 1.37: Relationship between the coefficient of permanent lateral deformation μ and C determined for conditions of simple loading for (1) marble from Kararsk, (2) marble from the Urals and (3) talc chlorite.

1.4 MECHANICS OF PERMANENT DEFORMATION IN ROCKS

1.4.1 Some Features of Permanent Deformation in Rocks

In order to understand the mechanics of permanent deformation in rocks, it is necessary to first examine the processes by which microcracks and microshears develop in the rock. Experimental results that help in understanding these micromechanisms and the development of permanent deformation in rocks are outlined below.

As shown earlier, rock strength is a function of the stress state. Thus, marble under a confining pressure $\sigma_2 = 100$ MPa exhibits strength hardening by 3.5 times compared to that under uniaxial compression (see Fig. 1.15a). Under such conditions, the peak strength is attained after the development of large permanent strain, i.e., of the order 20%. If such a specimen, having high strength under triaxial loading is unloaded and its resistance to loading under uniaxial compression is determined, it will be

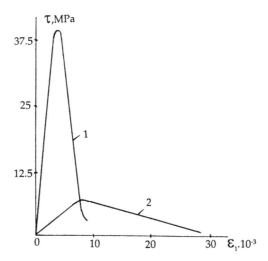

Fig. 1.38: 'Stress-strain' curves obtained in testing (1) priorly undeformed and (2) pre-deformed specimens of marble under uniaxial compression.

seen that the specimen is much weaker than the initial material tested under uniaxial compression only. Fig. 1.38 shows two diagrams: diagram 1 was obtained by subjecting a previous by unloaded specimen of marble to uniaxial compression; diagram 2 shows the results for a specimen previously deformed by 20% at a confining pressure of $\sigma_2 = 100$ MPa. The strength of the pre-deformed specimen was about 6 MPa and that of the other (undeformed) 40 MPa, i.e., pre-loading reduced the strength by a factor of 6.7. The modulus of elasticity of the deformed specimen was reduced more than 10-fold and the drop modulus M more than 100 times.

A second series of confining pressure tests with pre-deformed and undeformed specimens was conducted for different values of σ_2, starting from zero and increasing to $\sigma_2 = 100$ MPa. The results are shown in Fig. 1.39. The pre-deformed specimens had a lower strength compared with the original strength and only at a confining pressure of $\sigma_2 = 100$ MPa were the strengths of the two specimens equal. At $\sigma_2 = 100$ MPa the pre-deformed specimen acquired its original load-bearing capacity (possessed also by the initially undeformed specimen). This large reduction in compressive strength of the pre-deformed specimen in diapason of σ_2 between 0 and 100 MPa is explained by the extensive damage and growth of microcracks and cavities sustained during the initial loading cycle. Further, tensile strength of the deformed specimens was reduced to such an extent that they could be disintegrated into small pieces by hand. After that the fractured particles of the specimens were sieved, sorted according to size and weighted. The size distribution of the fractured particles was determined

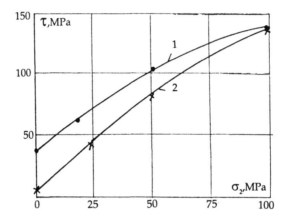

Fig. 1.39: Peak shear strength as a function of confining pressures σ_2 for specimens of marble (1) previously undeformed and (2) pre-deformed at $\sigma_2 = 100$ MPa.

and is shown in Fig. 1.40 [87, 101]. The size distribution obtained for specimens deformed under confining pressures $\sigma_2 = 1$, 10 and 50 MPa is given in Fig. 1.40 (a). The maximum linear dimension of the fragments is shown on the *d*-axis. The size distribution changed with change in confining pressure. As the confinement increased the specimen sizes became progressively finer. However, from confining pressure $\sigma_2 = 50$ MPa up to $\sigma_2 = 800$ MPa (the maximum confining pressure used in the tests) there was no increase in crushing of the fragments. Size distributions obtained at confining pressures $\sigma_2 = 50$, 150, 200, 400, 500 and 800 MPa are shown in Fig. 40 (b). It can be seen that there is only little change due to confining pressure. The maxima of all size distributions correspond to the same size fraction, approximately 0.3 mm. Notations are given in the diagram. The size distribution and percentage composition of marble grains, obtained through petrographic analysis of thin sections and polished microsections from an undeformed specimen, are indicated by crosses in the Figure. The initial size distribution, obtained using the petrographic method (see Fig. 2 in Appendix II), was reconstructed by taking into account the possibility of screening out grains used in the experiment during the sieving process. As can be seen, the distribution obtained petrographically coincides well with the results from the sieve analysis. This is a direct proof that deformation and rupture of marble occur mainly along the boundaries of grains and crystalline grains in the rock.

Petrographic studies of microsections of deformed marble under pressure $\sigma_2 = 0$ to 400 MPa were conducted and the results compared with thin sections of undeformed specimens. These studies are briefly discussed below.

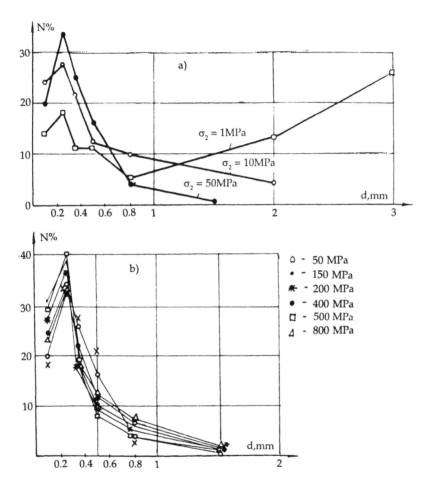

Fig. 1.40: Fractional size distribution functions (per cent) of rocks grains generated in the crushing process as a result of subjecting marble specimens to deformation at different levels of confining pressure σ_2.

1) Permanent deformation is followed by significant rearrangement of the rock microstructure. This occurs due to translational slippage along grain boundaries almost without effect on the continuity of the calcite grains. Only in certain grains are short 'blind' (i.e., not traversing the crystalline mass entirely) rupture cracks observed. The phenomenon of grain cataclasis is observed in narrow localised microscopic zones of crushing, attributed to grain boundaries; this phenomenon results from large relative movements in these zones.

2) The structural changes begin during the initial stage of loading a few microvolumes of low strength. With increase in applied load, the density of the microvolumes increases, eventually encompassing the entire volume under study.

3) Ruptures along grain boundaries, the weak links in the system, exert major damage on rock cohesion. At high hydrostatic pressures the dimensions of rupture cracks are limited by grain size.

Figure 1.41 shows a photomicrograph of marble deformed under a confining pressure $\sigma_2 = 400$ MPa. This demonstrates that the grains remain largely intact with small amounts of damage along the boundaries. Thus petrographic studies of marble match well qualitatively with the results obtained by sieve analysis of the crushed marble specimens.

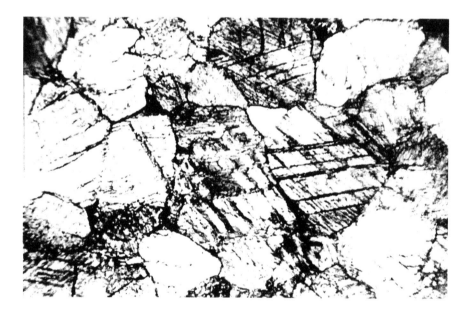

Fig. 1.41: Photomicrograph of marble deformed by 15% at a confining pressure of $\sigma_2 = 400$ MPa.

Analysis of crack formation and growth in sandstones [85] has shown that the residual (permanent) deformation of BP sandstones is caused mainly by intergranular cracking and displacements along grain boundaries. Permanent deformation of BP sandstones is accompanied by reorientation of the clayey minerals in the cementing material and disintegrated material produced by bending and sometimes folding of individual films of mica. Reorientation facilitates the formation and growth of intergranular cracks that separate grains and cement; opening of such cracks leads to failure.

Permanent deformation of NBP sandstones commences with twinning of calcite crystals contained in the cementing material of these rocks and growth of intracrystalline cracks. Grooves and pores are formed. Rupture of calcite grains occurs due to twinning of cement grains. Quartzite grains remain intact.

Thus in the case examined, the deformation process affects very weak structures, i.e., calcite grains and grain boundaries. Additional results of petrographic investigations of these two types of sandstones are discussed in Chapter 4, in which studies on the flow and permeability properties of rocks are considered.

In summarising the entire data from petrographic studies of the microstructural changes that occur in rock during the deformation process, the following features important to understanding the mechanics of development of permanent deformation in heterogeneous solids should be emphasised.

1) The deformation process, manifested through growth of the crack network (jointing) and relative displacement of microstructural elements, occurs along the weakest links in the structure, especially along grain boundaries and grains of low-strength minerals in polymineral rocks.

2) Notwithstanding the effects of high stresses and deformations, the rock preserves its structural base; grains do not lose their integrity as microstructural elements, even though they are subjected to various disturbances.

3) At high confining pressures, rupture cracks generated in the microstructural elements are usually limited to intergranular contacts and their dimensions are limited by the grain size.

4) In the initial stages of the development of permanent deformation, structural changes occur in those discrete volumes of the rock that are weakest. As the deformations and stresses increase, the structural changes propagate throughout the entire volume of the test specimen.

5) As the confining pressure is increased, the degree of crushing of the rock material increases as an increasing number of microstructural elements become involved in the permanent deformation process. However, there exists an upper limit to the intensity of crushing of the material, beyond which further increase in confining pressure has virtually no effect. This limit is reached when all the microstructural elments have been brought into the process of permanent deformation.

The phenomena of disintegration and dilatation of rock during permanent deformation are caused mainly by the simultaneous initiation of microruptures (microcracks) and microshears along grain boundaries. Microcracks due to rupture form predominantly in the direction parallel to that of the maximum principal stress. Results of experimental studies by various authors on formation of rupture cracks are analysed later in this monograph.

The density of microruptures and microshears depends on the magnitude of the hydrostatic pressure σ_2 and the stress state, characterised by the parameter C. Macroscopic shear planes form when these micro defects merge. The process of permanent deformation occurs along these planes. Slip lines, often seen on the surfaces of deformed specimens, are traces of such macroscopic planes of shear. In physical metallurgy, these planes are referred to as 'Chernov-Luders' lines, after the scientists who first observed them. Two systems of intersecting slip lines have been defined, whose angles of orientation with reference to the σ_1-axis and the density of lines lying within a unit surface depend on the stress state. Photographs of deformed and failed specimens indicating distinct planes of slip and failure are shown in Figs. 1.42–1.44.

When the limit state curves become horizontal (i.e., when simple shear occurs), the slip lines are oriented at 45° to the σ_1-axis, i.e., they coincide with the planes of maximum tangential stress. For stress states between simple shear and uniaxial compression, the angle of inclination of the slip planes varies from roughly 45° to 18–20°. In the tensile stress zone, the angle of fracture planes varies from approximately 18° under uniaxial compression to zero under uniaxial tension. Specimens of rocks deformed under high confining pressures are shown in Fig. 1.42. Traces of slip planes are clearly visible on the specimen surfaces. Since the confining pressures for the two bottom specimens were quite high, it can be seen that the inclination of the slip planes to the specimen axis and their density over a unit surface area are also quite high.

The dynamics of initiation and growth of these planes is as follows. The first plane appears on the surface of the specimen at stresses close to the elastic limit. Later, as the stress is increased, the number of shear planes increases continuously, until the peak strength is reached. Although the deformations become large—as indicated by the barrel-shape of the specimens (two bottom photographs)—the deformation process is distributed uniformly over all planes. The resistance of the rocks in such cases is either increased or stays constant at the peak strength. Specimens in which a failure plane has formed due to the development of permanent deformations along a number of slip planes, are shown in Fig. 1.43. Localisation of the deformation on one or several slip planes occurred in the post-peak strength region of deformation. In such cases, the deformation which had been taking place along other planes during the pre-peak permanent deformation ceased. Localisation of the deformation process in this way can occur under low and medium confining pressures. In such cases, the 'stress-strain' curve exhibits a distinct descending branch in the post-failure region of deformation.

A previously untested marble specimen that was subsequently subjected to high permanent deformations is shown in Fig. 1.44. The effect of the large deformation that occurs in such a distorted specimen introduces

Fig. 1.42: Photographs of rock specimens showing clear traces of slip (shear) planes.

Fig. 1.43: Photographs of rock specimens showing failure planes.

Fig. 1.44: Photographs of an untested marble specimen and the same specimen after being subjected to large permanent deformation under high pressures.

unacceptable errors in the standard procedure for calculating strains and stresses. A laboratory method was therefore developed for study of large plastic deformations in rocks whereby a change in shape which could distort the normal evolution of the deformation process is excluded. This method is discussed in Chapter 4.

Figure 1.45 shows the relationships between the orientation angles α of the shear planes relative to the σ_1-axis and (a) the parameter C defining the stress state and (b) magnitude of confining pressure σ_2 for (1) marble from Kararsk, (2) marble from the Urals and (3) talc chlorite from Karel'sk. A different scale of pressure σ_2 was selected for different rocks (Fig. 1.45b). It can be seen from these curves that the angle α increases with increase in C and σ_2, tending ultimately to 45°. This angle was attained by different rocks under various confining pressures but all at the same value of C, approximately 0.33.

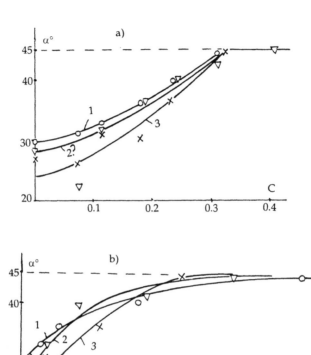

Fig. 1.45: Relationship between the angle of orientation α of shear planes relative to the σ_1-axis and (a) stress state C and (b) magnitude of confining pressure σ_2 for (1) marble from Kararsk, (2) marble from the Urals and (3) talc chlorite from Karel'sk.

1.4.2 Processes of Failure and Formation of Rupture Cracks

The authors of this monograph have not been directly involved in studies of the initiation and formation of rupture cracks in solids (except for petrographic studies on the development of deformation processes in rocks such as those described above). However, fundamental concepts of this process, studied by other researchers, were used to build a model of a deformed heterogeneous solid, the essential features of which are discussed later. Here, we shall confine the discussion to a brief review of the experimental results to establish the validity of certain concepts in the model.

Numerous studies have been conducted in several countries, including Russia, on the mechanics and processes of brittle failure. Several studies of the micromechanics of failure were published in the Proceedings of the International Conference on Failure Studies held in April 1959 in Swampscotte (USA), under the general title 'Atomic mechanics of failure' [2].

Results of investigations into the micromechanics of brittle failure conducted in the former Soviet Union were also published in the Republican Interdepartmental Bulletin *Physical Nature of Brittle Failure of Metals*, Metallofizika series, Naukova Dumka, Kiev (1965) [23].

Numerous experimental and theoretical investigations into brittle failure processes were published in the collected papers *Failure* edited by G. Libovitz, vol. 7, pt. I, *Inorganic Materials* (translation), Mir Publishers, Moscow (1976) [22].

A short overview of the literature pertaining to the study of brittle failure processes cited in the aforesaid bulletins and publications [2, 23, 22] is given below.

The foundation for study of the effect of microcracks on strength and failure of brittle materials was established in two publications by Griffith in the early 1920s. Griffith explained that the large difference between technical strength of brittle materials and computed theoretical strength was due to the existence of submicroscopic flaws and defects in all real materials. He outlined the mechanics of the process of initiation and growth of rupture in brittle materials.

Griffith's hypothesis holds that crack growth leads to a continuous reduction in the free energy of a body [22]. With increase in crack length, the surface energy of the body increases, while the stored elastic energy decreases. The energy U_s required to form two new surfaces in a crack is determined by the expression

$$U_s = 2U_1\, l, \tag{1.12}$$

where U_1 is the surface energy per unit area and l is the crack length.

The net reduction in potential energy of the solid due to the introduction of a crack of length l is given by the expression

$$U_{el} = \pi\, l^2 \sigma^2 / E, \tag{1.13}$$

where σ is the average stress in the body prior to crack initiation and E is the modulus of elasticity of the material.

The condition for possible crack growth is determined by the expression

$$\partial U_{el}/\partial l > \partial U_s/\partial l. \tag{1.14}$$

From this, the condition for the strength of a brittle solid is obtained

$$\sigma = \sqrt{EU_1/\pi l}. \tag{1.15}$$

Thus the stress required for propagation of a crack is inversely proportional to the square root cf the crack length. This is Griffith's criterion for brittle failure, which requires no information about the shape of the crack tip.

The Griffith criterion describes very well the failure of glass and brittle materials similar to it, in which there is no evidence of plastic deformation prior to failure. In experiments studying the failure of glass tubes and spherical shells by internal pressure, Griffith [cited from 22] established an equivalence between the tensile stress generated and the fracture strength of the glass.

But Griffith's criterion does not explain the failure process for materials exhibiting plastic deformation during failure. It is thus necessary to increase the value of the surface energy by several orders of magnitude in order to achieve reasonable agreement between his criterion and the observed strengths. In 1934, Orowan suggested that the quantity of energy U_1 in expression (1.15) should be modified to include the work done on plastic deformation U_{pl}, which is essential for crack initiation in such materials. Total energy U in (1.15) could then be represented as a sum:

$$U = (U_1 + U_{pl}). \tag{1.16}$$

Using the data available for the surface energy of brittle rocks, Brace [9] computed the strength of several types of rocks and obtained reasonable agreement with measured results.

Subsequent modifications to Griffith's theory have prompted its application to a wide range of materials subjected to complex states of triaxial compression as well as the simpler conditions originally considered.

Under uniaxial compression, Griffith's criterion gives a uniaxial compression strength that is 8 times greater than the strength under uniaxial tension.

One such modification considered frictional forces between crack faces compressed against each other under conditions of general triaxial compression. This led to the prediction that the uniaxial compressive strength should be between 10–17 times greater (in absolute terms) than the uniaxial tensile strength. Moreover, this additional condition resulted in a linear envelope in the form of a Coulomb-Mohr failure criterion.

Experiments to determine the strength of rock under tension and plane stress conditions, obtained by various authors (and shown in Figs. 1.20–1.23), indicate that the uniaxial compressive strength usually exceeds the uniaxial tensile strength by a factor of 3.5 to 80 for various materials. Further, the highest uniaxial compressive strength values were observed in highly brittle materials (glass, quartzite, dolomite and diabase) while less brittle materials (gypsum, talc chlorite, pressure-cured concrete, concrete) under conditions of uniaxial tension and uniaxial compression are in good agreement with Griffith's original criterion.

Griffith's theory provides a mechanism for the initiation and growth of rupture, while the mathematics of the theory includes physical parameters and properties of solids that can (to some extent) be measured. This aspect of Griffith's theory gives it a unique scientific value. Griffith's theory is most appropriately applied, both in qualitative and quantitative respects, to the case of simple tensile rupture when the maximum tensile stress is directed normal to the length of the crack. In this zone the Griffith strength criterion and the phenomenological strength theory I of maximum tensile stresses coincide. For the case of heterogeneous solids subject to complex stresses, the parameters contained in the mathematical model are difficult to determine and for rocks in particular, almost impossible to obtain. Thus, for the practical application of Griffith's theory to complex stress states, one has to resort to the concepts of compressive strength and tensile strength as applied in the phenomenological strength theories of Mohr and Coulomb-Mohr.

As already noted, the authors support the concept of two strength criteria in the same solid body: a fracture strength criterion and a shear strength criterion. In compressive stress regimes, when experimental points in the tension-compression quadrant of Figs. 1.20–1.23 cease to obey simple rupture criteria (in Fig. 1.16, this situation appears on the right side of the point of intersection between the lines of peak strength and elastic limit), we invoke a mechanical statistical model to explain the deformation and failure process. In this model, processes of rupture and shear are presumed to occur simultaneously. The extent to which each mechanism participates in the failure is determined by the stress state. When the conditions for tensile rupture disappear, the zone of simple shear 'takes over'.

We shall now turn to a review of published experimental studies on the mechanics of crack initiation and propagation in heterogeneous solids (materials having various defects).

The growth of rupture cracks in epoxy resin plates has been studied by Filatov and Belyakov [24]. A system of initial cracks of various orientations was chosen for study, as shown in Fig. 1.46 (a). The angle of inclination of the cracks relative to the direction of applied load is seen to be variable. Figure 1.46 indicates that the cracks merged to form a macroscopic failure plane. All the cracks that extended are oriented along the line of action of the principal compressive stress. Given the different orientation of the initial cracks, the stress required for propagation of the rupture cracks also varied. The general failure picture, in the case of initial cracks being oriented across the plate and along the line of action of the compressive stress, is illustrated in Fig. 1.46 (b). In both cases, the effect of loading was to induce the initiation

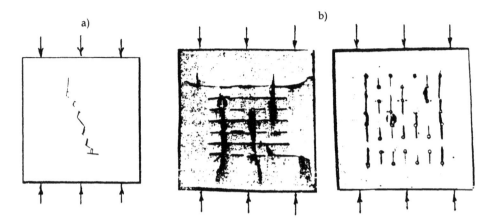

Fig. 1.46: Orientation of rupture cracks along the direction of compressive stress when the initial cracks differ in orientation in the plates [24].

and propagation of cracks oriented along the direction of compression. Some cracks were initiated from holes made in the plate.

Experiments conducted on epoxy resin plates [24] containing artificially introduced cracks allowed a relationship to be established between the uniaxial compressive strength and uniaxial tensile strength and the orientation of the initial cracks. Results are shown in Fig. 1.47. At an angle of inclination 30° to the line of action of the load, the minimum strength was found to occur under uniaxial compression. In the case of tension, the strength remained essentially constant at the minimum tensile failure value even though the angles of the cracks varied from 90° to 30°.

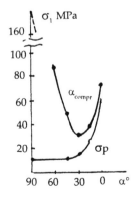

Fig. 1.47: Variation in breaking stresses in testing plates with singular initial cracks oriented at different slope angles to the direction of action of load [24].

Jaeger [33] established that the most critical angle of orientation of the initial crack depends on the shape of the crack itself. In the case of an elliptical crack, the most favourable angle for initiation of a new rupture crack seems to be 30°, while in the case of an ideally narrow crack (slit) the angle is 45°. In the latter case, the original crack is oriented along the line of action of the maximum tangential stress, while the rupture crack induced by it extends in the direction of action of the maximum compressive stress.

Photographs that record the failure of models with cracks initially inclined and bounded or not bounded by holes, are shown in Fig. 1.48 [24]. Rupture cracks propagate along the axis of the maximum compressive stress because this direction requires the least amount of energy for crack propagation. Minimal strength was seen for model (a) with no holes at the crack tips. Model (b), with holes of 3 mm diameter, was very strong, while model (c) exhibited a medium strength. Crack propagation started from the hole-free end and only after further increase in load did the crack begin to propagate from the holed end. Increase in diameter of the hole reduced the stress concentration and resulted in strength hardening. This method (i.e., introducing a circular hole at the tip of a sharp crack) is widely used in practice to prevent further growth of an extending crack.

The process of natural initiation and growth of rupture cracks in materials, without artificially creating stress concentrators, has evoked wide interest. Stokes et al. [22] obtained photomicrographs of crystals of MgO

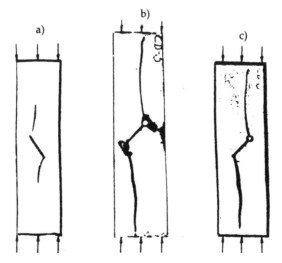

Fig. 1.48: Failure of models with initially inclined cracks including cracks bounded by holes [24].

and LiF deformed under conditions of uniaxial compression. Development of a microcrack in an MgO crystal from the point of intersection of slip planes is shown in Fig. 1.49 (a). The authors inferred that crack initiation was associated with a large non-uniform plastic deformation. Where the propagation of slip bands was hindered by a microstructural barrier, such as a grain boundary, stress concentrations developed, resulting in the formation of rupture cracks. With increase in plastic deformation, the number of cracks at the intersection of the slip bands increased and merged, thereby forming extended cracks. Photomicrographs of deformed crystals of LiF and MgO are shown in Fig. 1.49 (b) and (c) respectively. The crack system is seen to be quite extensive. The direction of loading is indicated in the

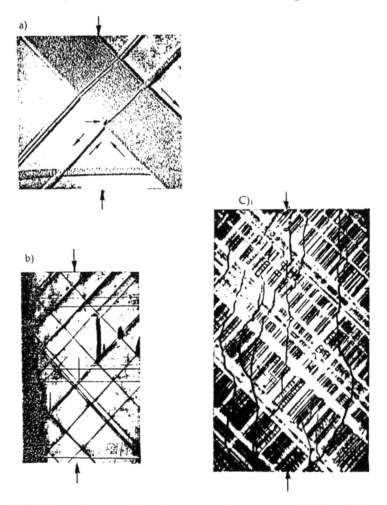

Fig. 1.49: Initiation and propagation of rupture cracks in crystals of MgO and LiF at the intersection of slip bands [57, 58].

Figure. Cracks so formed are oriented parallel to the line of action of the compressive stress.

Aragon and Orowan [1 cited in 22] have demonstrated that cracks in MgO are initiated as a result of the applied macrostresses and not to stresses generated at the dislocation level. They showed that internal macrostresses are generated by the resistance of the dislocating band to induction of a slip band propagating towards the dislocating band from the opposite direction.

Propagation of a crack under confining pressure is more difficult. The greater the confining pressure, the more difficult crack growth. Fairhurst and Cook [21] established the effect of confining pressure σ_2 on crack propagation in the direction of axial stress σ_1 and proved that the stress σ_1 required for initiation and propagation of a crack increases linearly with increase in σ_2 (see Fig. 1.50 in which C_0 denotes the uniaxial compressive strength). The linear relationship between σ_1 and σ_2 agrees qualitatively with the straight-line envelope obtained using the Mohr-Coulomb criterion and agrees well also for real materials over a narrow range of confining pressures σ_2. For the same range of variation in σ_2, a linear approximation of the peak strength in τ versus C coordinates is possible, as shown in Fig. 1.34.

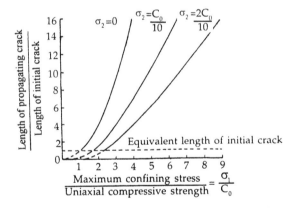

Fig. 1.50: Effect of lateral stress on crack growth [21].

Based on the studies discussed above, the following conclusions can be drawn regarding the mechanics of initiation and growth of rupture cracks in heterogeneous solids under tension and compression.

1) Initiation of rupture cracks in solids results from the existence of various types of heterogeneities in the material: defects, microcracks, pores, grain boundaries etc.

2) Concentration of stresses at these defects, due to non-uniform elastic or plastic deformations, leads to initiation of rupture cracks.

3) Under compressive loading conditions, rupture cracks propagate parallel to the direction of the maximum principal compressive stress, but under tensile conditions they propagate perpendicular to the direction of maximum tensile stress.

4) The orientations of initial cracks relative to the applied load favourable for the initiation of rupture cracks are: for compression loading 30–45° (depending on the configuration of the initial crack) and for tension loading, 90–45°.

5) Under conditions of triaxial stress, a confining pressure σ_2 inhibits the initiation and propagation of rupture cracks, resulting in an increased strength of the material and reduction in length of rupture cracks that develop compared to those at the same stress under low confinement.

1.4.3 Mechanical State Diagrams and Model Simulating Development of Permanent Deformation in Rocks

The mechanics of development of permanent (irreversible) deformation in rocks can be illustrated graphically by mechanical state diagrams. These are useful in analysis of the mechanical properties of rocks over a wide range of stresses and loading paths. Experimental results demonstrating the independence of the peak strength and elastic limit of rocks on the loading path, as shown in Fig. 1.34, serves as the basis for developing the mechanical state diagrams. Typical diagrams for rocks, plotted in the coordinate systems $(\tau - C)$ and $(\tau - \sigma_2)$, are shown in Fig. 1.51 [65, 96, 100, 101].

The main elements in these diagrams are the limit curves, corresponding to the peak strength τ_{us} and the elastic limit τ_{el}; experiments have shown them to be similar for simple and complex paths of loading. A section for tensile stress σ_2 is also shown in the diagrams. Three characteristic points can be distinguished in these diagrams: point K, at which the limit curves merge, and points M and N at which the limit curves become horizontal, i.e., they become independent of C and of confining pressure σ_2. Points M and N lie at the intersection of the limit curves τ_{us} and τ_{el} with the simple loading ray defined by $C = 0.333$. Equivalence of the tangential (shear) stress and confining pressure $(\tau = \sigma_2)$ is typical for these points. Critical levels of confining pressure, at which the elastic limit and peak strength curves become horizontal (points M and N), are indicated in diagram (b) by σ_{2cr}^{el} and σ_{2cr}^{us} respectively. The region of stresses corresponding to the development of permanent deformations in the material, lies between the limit curves τ_{us} and τ_{el}. Part of this zone, in which permanent deformation is accompanied by growth of crack and pore space in the material and overall increase in volume of the material, is shaded and indicated by the letter B (the theoretical and experimental basis for the existence of this region is discussed later

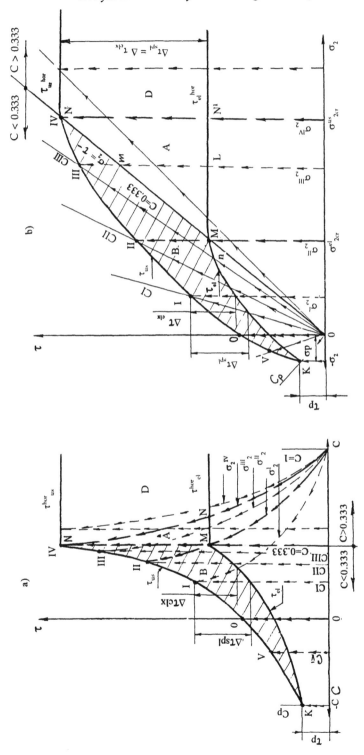

Fig. 1.51: Typical mechanical state diagrams for rock, plotted in the coordinates (a) $\tau - C$ and (b) $\tau - \sigma_2$.

in this monograph). In the remaining parts of the zone (A and D), permanent deformation is purely plastic in nature and occurs with no increase in volume. At the boundary between zones B and A (inclusive of points M and N), the state of stress is characterised by $C = 0.333$, which is equivalent to the condition $\tau = \sigma_2$. This follows from the relations:

$$C = \sigma_2/\sigma_1 = \sigma_2/(\Delta\sigma_1 + \sigma_2) = \sigma_2/(2\tau + \sigma_2)$$

$$= \sigma_2/(2\sigma_2 + \sigma_2) = 1/3 \approx 0.333.$$

Reasons for the transition of limit curves to the horizontal at $\tau_{el} = \sigma_{2cr}^{el}$ and $\tau_{us} = \sigma_{2cr}^{us}$ (or $C = 0.333$), and possible increase in dilatation when $\tau > \sigma_2$ (or $C < 0.333$) are discussed later. The experimental results will then be explained and generalised in terms of a statistical model of non-uniform deformation of a solid that has been formulated and developed by the authors [69, 74, 79, 97, 98, 101].

Eight points are marked on the peak strength curve, corresponding to different stress states. We shall discuss features of the mechanics of the development of permanent deformation in rock and its failure when these points on the limit curve are reached through simple and complex paths of loading. Discussion of complex loading paths will be restricted to those defined by Karman only. Loading paths are indicated by arrows in the diagrams. In diagram (a), the paths of simple loading for different values of C are shown as vertical straight lines, while complex loading paths for various levels of σ_2 are indicated by curved rays emanating from a single point, defined by the condition $\sigma_1 = \sigma_2 = \sigma_3$. In diagram (b), simple loading paths under different values of C are indicated by straight line rays emanating from the origin while complex loading paths for various values of σ_2 are shown by vertical straight lines.

Before studying the mechanics of formation of strength and deformation properties of rocks for various states of stress and loading paths, we re-emphasise that rock is structurally complex. Rocks consist of various crystals, grains, cementing material, pores and cracks differing in type and dimensions, inclusions and other defects. All these structural elements possess different mechanical properties. When such a rock is loaded, each structural element is subjected to shear and tensile stresses. Microfailure commence in the weakest structural elements but with increments in load, stronger elements become involved in the failure process. The strength and deformation properties of rocks depend largely on the loading conditions. This relationship is applicable to a wide range of rocks. Secondary indices, such as the angle of development of deformation planes, coefficient of permanent lateral deformation, degree of disintegration, degree of crushing etc., also vary in a specific manner depending on the loading conditions. The mechanism of formulation of these indices when rocks are subjected to deformation under various loading conditions is taken into account by the authors in the model developed by them for deformed heterogeneous solids.

We shall first consider the mechanics of deformation under uniaxial compression. In this case, the loading path is the same for simple and complex loading, and is represented in the diagrams as a vertical straight line $0 - 0$ up to the point on the limit curve. Here $C = 0$ and $\sigma_2 = 0$. A model of the growth in deformation in a heterogeneous solid under uniaxial compression is shown in Fig. 1.52. The various stages of deformation are marked by point A in the curves τ vs ε_1 typical for the conditions indicated in successive stages of the model ($0_{(1)}, 0_{(2)}...$). Structural elements in the modes are bounded by a grid inclined at $45°$ to the specimen axis (i.e., in the direction of line of maximum tangential stress). The term 'structural elements' implies grains and parts of grains, solid inclusions etc. in the rock, which remain intact even under high stresses, corresponding to the horizontal section of the ultimate strength curve τ_{us}^{hor}. For simplicity, in the model the dimensions of all structural elements are considered to be identical. Irreversible shear deformation occurs only along boundaries of the structural elements. Structural elements may have very different values of fracture strength and shear strength along their boundaries. For such a heterogeneous body, the statistical distribution of concentration of the various elements versus their shear toughness and fracture strength is a key characteristic. This function defines, for example, the capability of the rock to exhibit strength-hardening and strain-hardening with increase in confining pressure, the type of limit curves, tendency to disintegrate under deformation etc. Elements having the same shear strength or fracture strength are assumed to be distributed evenly statistically throughout the material.

Formation of microshears in highly flexible structural elements is shown in scheme $0_{(1)}$. Microcracks are initiated at the tips of the shear elements. The minimum dimension of shear a_{min} and rupture b_{min} elements is determined by the dimension of the minimum structural elements (i.e., elements that cannot be crushed even under conditions of τ_{us}^{hor}). Larger elements of shear and rupture form through fusion of the minimum sized elements $\mathbf{a} = \mathbf{a}_{min} \cdot n$ and $\mathbf{b} = \mathbf{b}_{min} \cdot m$, where m and n are integers. Subsequently, for the sake of brevity, elements of shear will be labelled $\mathbf{a}$ and rupture elements $\mathbf{b}$. In experiments, formation of microcracks of rupture has been detected through acoustic emissions, at loads well below the elastic limit.

With increase in load, a number of micropairs of shear-rupture $\mathbf{a}$–$\mathbf{b}$ form and are stored in the solid; at a specific concentration, and given loading conditions, the micropairs merge to form the first macroscopic shear plane ω_1 oriented at angle α_1 to the σ_1-axis (scheme $0_{(2)}$). Under uniaxial loading, rupture cracks in rocks are greatly extended, $\mathbf{b} = \mathbf{b}_{min} \cdot m$. This is explained by the large difference between the shear strength and the tensile fracture strength. Growth of tensile cracks requires less energy than is needed to

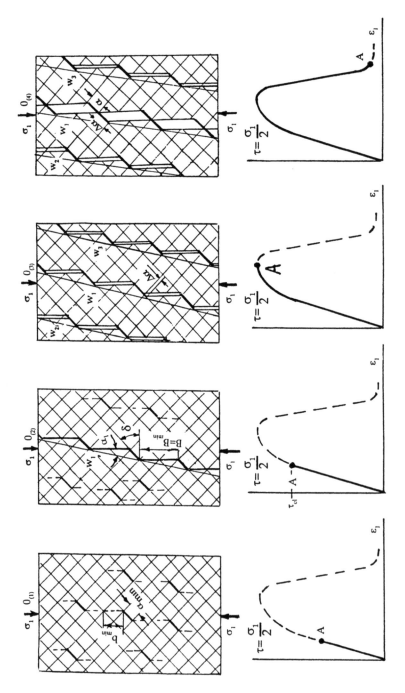

Fig. 1.52: Model showing the growth of deformation in a solid under uniaxial compression.

cause shear deformations. The macroscopic elastic limit τ_{el} is reached when the first plane ω_1 is formed.

Further deformation along the ω_1 plane occurs with strain hardening. This increases the stress level in the body and activates new and stronger micropairs **a–b** in the deformation process and new shear planes ω_2, ω_3 etc. are formed. The angle α of the newly formed planes is equal to the angle of the first plane formed at the elastic limit. The maximum number of shear planes is formed at the peak strength (scheme $0_{(3)}$). Traces of shear planes ω are seen on the surface of the specimen as slip lines (see Fig. 1.42). For simplicity, only one system of planes is shown in the scheme. In reality, the systems of planes intersect. Shear deformation, quantified by Δa, along the planes ω is accompanied by the formation of voids and dilatation. Stress hardening is accompanied by a reduction in contact area along the microareas **a**, i.e., the load-bearing area is reduced. These two processes quantify the peak strength.

When the post-failure strength is reached, deformation in the descending branch of the curve becomes localised in parts of the weak planes ω that are part of the planes formed in the pre-failure strength region. Planes which require higher stresses in order to develop cease to deform. If the planes ω have highly non-uniform properties, deformation in the post-failure zone may localise on a single plane. The number of planes participating in the deformation process in the post-failure strength zone defines rock brittleness, characterising the modulus of post-failure deformation or 'drop' modulus M. Localised deformation continues until complete slippage occurs in the microarea **a**, i.e., when the condition $\Delta a = $ **a** is reached (scheme $0_{(4)}$). The drop in load that occurs in the descending branch of the stress-strain curve is due to the reduction in the area **a** undergoing shear Δa.

The phenomenon described above is similar to the onset of *necking* in steel specimens loaded in tension, where the load in the post-failure strength zone falls due to reduction in the cross-sectional area of the specimen in the neck section. In both cases, the load-bearing capacity of the material is reduced although the stresses in the neck and in places of contact with slip elements **a** increase.

Once slip has developed over the entire area **a**, two parts of the specimen become completely disconnected and subsequent deformation occurs due to frictional sliding of the two parts of the specimen along rough surfaces. Dilatation ceases and the frictional resistance corresponds to the so-called *residual strength*.

The mechanical model proposed by the authors treats the process of deformation and failure as one of statistical selection of structural elements, governed by equilibrium of the forces acting on the structural element. A model of a single 'shear-rupture' micropair contained in the plane ω is shown in Fig. 1.53. The microarea of shear **a** is oriented at angle $\delta = 45°$; σ_1 and $\sigma_2 = \sigma_3$ are the principal external normal stresses; τ is the tangential

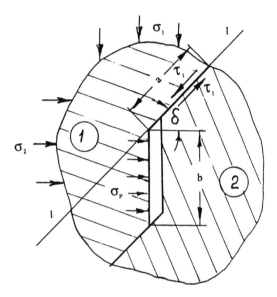

Fig. 1.53: Model of a single structural 'shear-rupture' pair and the system of stresses acting on the pair.

stress acting on the specimen and equal to half the difference in the principal normal stresses; τ_1 the ultimate shear strength along microarea **a**; and σ_p the fracture strength in microarea **b**.

Ultimate equilibrium of this pair is reached when the resultant of all the forces projected onto the I-I axis inclined at angle $\delta = 45°$ equals zero [69]:

$$(\tau a^2) - (\tau_1 \, a^2) - (\sigma_p \, ab \cos \delta) - (\sigma_2 \, ab \cos \delta) = 0. \qquad (1.17)$$

The first term in equation (1.17) is the total shear force trying to displace one portion of element 1 relative to element 2. The second term is the force corresponding to the peak shear strength in microarea **a**. The third term is the force corresponding to the fracture strength in microarea **b**. The fourth term is the force induced by the effect of stresses σ_2 in area **b**; a^2 and **ab** are the contact in microareas **a** and **b** respectively. The maximum shear resistance of the micropair is determined as:

$$(\tau a^2) = (\tau_1 \, a^2) + (\sigma_p \, ab \cos \delta) + (\sigma_2 \, ab \cos \delta). \qquad (1.18)$$

It is evident from equation (1.18) that the resistance of micropair **a–b** depends on the applied stress level σ_2. The higher the confining pressure σ_2 in the experiments, the higher the shear resistance of the micropair. Since micropairs **a–b** give rise to macroscopic shear planes ω, any increase in confining pressure σ_2 will increase the shear strength of the body as a whole. However, an increase in σ_2 induces specific changes in the deformation mechanics, which in turn produce changes in the parameters in equation (1.18). Thus the relationship between strength and confining pressure is multifactored, as described below.

1) Increase in confining pressure σ_2 in the rupture area **b** increases its fracture resistance. The stresses in the body build up and activate new and stronger microelements **a** into the deformation process, thus forming new micropairs **a–b**. With this increased concentration of micropairs in the body, the rupture cracks **b** become smaller.

2) Increase in σ_2, along with strength hardening of the body, augments its deformation capability. Increased deformation of the microelements **a** lead to considerable strength hardening. Of the total hardening of the body caused by increase of σ_2, strain hardening in microelements **a** may account for 70%.

The effects of both the first [69, 74] and second [97, 101] components of strength in materials have been well investigated. Discussion in this monograph is restricted to the qualitative effect of pressure σ_2 on the strength and deformation properties of rock.

Consider the development of the deformation process under triaxial compression. Figure 1.54 depicts typical deformation stages at points K, V, O, I, II, III, IV—all lying on the limit curve of the mechanical state diagram in Fig. 1.51. The stress states range from simple rupture (point K) to simple shear (point IV and points to the right of IV). Points K and V are in the tension-compression zone, points I, II, III and IV in the triaxial compression zone and point 0 in the uniaxial compression zone. The pattern of deformation at these points (at peak strength) under simple and complex paths of loading is essentially the same. The difference lies in the magnitude of linear and volume strains undergone by the material along different loading paths, as well as in the dynamics of variation in the mechanics of deformation as stresses in the body are increased. These questions will be discussed later. Diagrams in Fig. 1.54 illustrate the mechanics of deformation seen in uniaxial and triaxial compression for a series of typical $\tau - \varepsilon_1$ curves in rock for various values of C (simple loading) or for various confining pressures σ_2 (complex loading).

We shall now consider specific features of the development of deformation at each of the above-mentioned points. Taking all the diagrams together, we can distinguish the following features.

1) From left to right, the angle α of the planes of failure and deformation ω increase from 0 under simple rupture to 45° under simple shear.

2) From left to right, the density of deformation planes ω per unit volume of the solid increases and as a consequence, the degree of crushability of the material in the specimen also increases. At point K the specimen splits

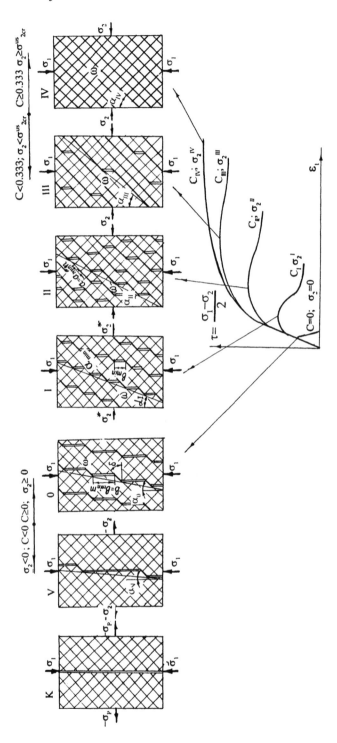

Fig. 1.54: Deformation of a heterogeneous solid subjected to various stress states, corresponding to points K, V, O, I, II, III, IV marked on the limit curve in Fig. 1.51.

into two parts, while at point IV deformation planes are activated on every structural element.

3) The degree of dilatation of the material increases from point K to point II, while from point II to point IV, dilatation decreases to zero. The degree of dilatation is determined by the density of microelements **b** and a possible increase in permanent deformation.

We shall consider the causes for development of the deformation process under conditions of triaxial compression. As mentioned earlier, according to equation (1.18) the strength of the material increases with increase in σ_2 while the dimension b of the rupture cracks reduces correspondingly. At a specific value of $\sigma_2 = \sigma_2^*$, the dimension of the rupture crack equals that of the structural element ($b = b_{min}$). This situation is shown in the model corresponding to point I. Under these conditions the angle α of the deformation planes ω is 30°. To understand the mechanics of development of the deformation process under high pressures, it is essential to consider the limit degree to which rock crushing is possible under high stresses (see Fig. 1.40) [87, 101]. The mechanics of deformation under conditions $\sigma_2 > \sigma_2^*$ has been considered in detail in several articles [97, 98, 101]. Its key features are as follows. The situation that rupture cracks **b** contained in the deformation plane ω might not be less than b_{min}, arises from the constraint imposed on the limit degree of crushability of the material. Finer cracks are not involved in the plane ω because convergence of the two adjacent shear planes **a** is restricted by the dimensions of the structural elements that have not failed. Finer rupture cracks are blind and closed and contribute little to increase voids in the material. Ignoring the principle of minimum dimensions of rupture microelements b_{min}, the model assumes continuous crushing of material and does not take into account cessation of hardening (and increased elastic limit) of the body with rise in σ_2, nor does it consider any conditions that preclude a possible increase in dilatation in the zone of permanent deformation A and an intense growth in plastic deformation.

Under conditions of confining pressures ($\sigma_2 > \sigma_2^*$), when the dimensions of all the initiated rupture cracks **b** take a constant value equal to b_{min} the equation of ultimate equilibrium for each individual microelement *shear-rupture* ($\mathbf{a}_{min} - \mathbf{b}_{min}$) is of the form:

$$\tau = \tau_1 + \sigma_p + \sigma_2. \tag{1.19}$$

In contrast to equation (1.18), the dimensions of the micropair are absent from equation (1.19) because in this case a_{min} and b_{min} belong to the same structural element; hence $b_{min} \cos \delta = a_{min}$. At $\delta = 45°$, the dimensions of areas contained in the equilibrium equation cancel out.

The possible initiation of a rupture crack in such an element is determined by the condition

$$\tau \geq \tau_1 + \sigma_p + \sigma_2. \tag{1.20}$$

In the situation under study, because of the higher confining pressure σ_2 the body hardens. This in turn leads to new and stronger microelements a_{min} becoming involved in the deformation process. Condition 1.20 indicates that this occurs without the formation of any microelements of rupture b_{min}. As a result, deformation planes ω form in which the shear elements a consist of several minimum (micro) elements ($a = a_{min} \cdot n$). This situation is depicted in model II. For such stress states, the material acquires enhanced plasticity (due to the large dimensions of the elements a) and increased ability to dilated (due to the large number of rupture elements b_{min} participating in the process, and the possible formation of internal voids with large Δa). In the experimental curve showing the relationship between the volume of dilated rock $\Delta\theta$ and the confining pressure σ_2 (Fig. 1.25), the mechanism under study corresponds to the condition $\Delta\theta_{max}$.

At still higher confining pressures σ_2, the number of elements a_{min} which satisfy condition (1.20) for the formation of rupture element b_{min} reduces even though the number of microelements a_{min} that participate in the process of permanent deformation is increased (due to the activation of stronger elements). Increase in the size of shear areas a leads to still higher plasticity compared to the preceding case. Reduction in number of microelements b_{min} reduces dilatation. This situation is illustrated in model III. The angle α of the deformation planes ω increases with rise in confining pressure σ_2 in all the aforementioned situations, and in the case under study is almost 45°.

Finally, when a certain confining pressure σ_2 is reached, the microelement rupture b_{min} cannot develop in a single microelement a_{min} participating in the process of permanent deformation (this situation is reflected in model IV). With the possibility of formation of microelements of rupture b eliminated, the dependence of strength and deformation characteristics of the material on the magnitude of the confining pressure σ_2 and the capability for dilatation with deformation close. The angle of the planes ω becomes equal to 45°, while the coefficient of permanent lateral deformation $\mu = 0.5$. In all the previous cases, during the process of deformation at different levels, a point was reached when the attained maximum strength began to drop (i.e., the onset of post-failure deformation). At low pressures σ_2 this is caused by slippage of planes ω along microelements a (see Fig. 1.52 (4)); at high values of σ_2 this situation corresponds to growth of the crack-pore space (see Fig. 1.54 (II and III)). In case (IV), the conditions necessary for initiation of post-failure deformation do not exist. The condition at which this situation occurs is determined by the following considerations. From equation (1.20), it follows that the weakest microelement a_{min} is the last can develop microelement rupture b_{min}. For heterogeneous solids which contain microelements a_{min} with a shear strength equal to zero ($\tau_1 = 0$), the critical condition at which the possible formation of microelements b_{min} no longer exists is described by the relationship (1.21).

$$\tau = \sigma_p + \sigma_2. \tag{1.21}$$

Taking into account the fact that the fracture strength of rock is negligibly low compared to the shear strength at high confining pressures σ_2, then the critical condition at which the dependence of strength on σ_2 no longer exists (i.e., the point at which the peak strength curve τ_{us} becomes horizontal) is as follows:

$$\tau = \sigma_2. \tag{1.22}$$

This condition applies also to the relationship between the elastic limits and σ_2 although we do not demonstrate proof of this in this monograph. In the mechanical state diagram shown in Fig. 1.51, which depicts typical diagrams for rocks, the coordinates of the transition of the peak strength and elastic limit curves to horizontal section are $\tau_{us}^{hor} = \sigma_{2cr}^{us}$ and $\tau_{us}^{el} = \sigma_{2cr}^{el}$ respectively. Possible explanations of how condition (1.22) can be violated are examined in [97, 101]. These are related, in part, to the structural characteristics of the material, established by the form of the distribution function of per cent composition of structural elements based on their shear strength.

The general condition for the onset of dilatation in heterogeneous solids, following from (1.22) is

$$\tau \geq \sigma_2. \tag{1.23}$$

This condition is satisfied in the shaded zone B in the mechanical state diagram.

Conditions (1.22) and (1.23) can also be established by examining the mechanical scheme shown in Fig. 1.55, in which part of a specimen subjected to non-uniform compression under principal (compressive) stresses σ_1 and σ_2 is shown. The structural element **a–b–c–d**, formed by shear planes ω_1, ω_2, ω_3 and ω_4, is subjected to the maximum tangential stresses τ across its faces. This situation is similar to that discussed earlier (Fig. 1.49) in which rupture cracks formed at the intersection of the slip planes in MgO and LiF crystals. The tangential forces T acting on faces **b–c** and **c–d** of a structural element where the area of the faces is F, are given by the expression

$$T = \tau F.$$

Projecting this force T onto the σ_2 stress axis, we obtain the resultant force T_1:

$$T_1 = 2\tau F \cos 45°.$$

The stresses induced by the force T_1 act along the diagonal segment **b–d**, dimension F_1, where $F_1 = 2F \cos 45°$.

The normal stress σ acting on the segment **b–d** is given by

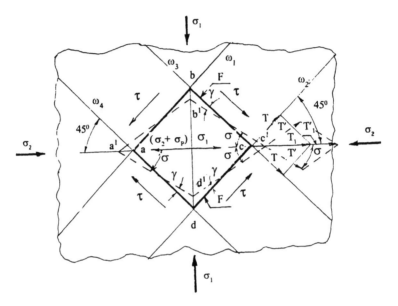

Fig. 1.55: Initiation of rupture cracks at the intersection of slip planes.

$$\sigma' = \frac{T_1}{F_1} = \frac{2\,\tau\,F\cos 45°}{2F\cos 45°} = \tau.$$

The maximum fracture resistance of segment **b–d**, which counteracts stress σ', is the sum $(\sigma_2 + \sigma_p)$, where σ_2 is the confining pressure and σ_p is the fracture strength.

Rupture and initiation of cracks along segment **b–d** occur when the condition $\sigma' = \tau = (\sigma_2 + \sigma_p)$ is satisfied.

Considering that the fracture strength σ_p for rock is usually negligible compared to the compressive strength at high confining pressures, we obtain the results

$$\tau = \sigma_2.$$

This is the required proof of conditions (1.22) and (1.23).

The validity of condition (1.23) has been established experimentally on several types of rocks. The results obtained from tests on marble specimens from the Urals are shown in Fig. 1.56 [102].

The onset of dilatation in specimens undergoing deformation was recorded for various test conditions using a U-shaped liquid manometer. One end of the manometer was hermetically attached at mid-height of the specimen being tested in a high-pressure cell, the other end was open to

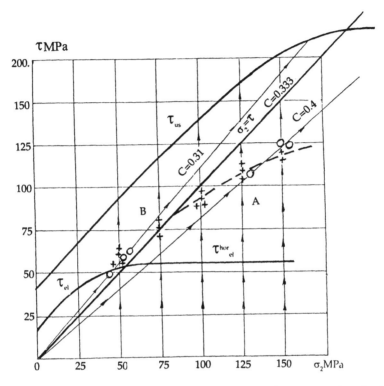

Fig. 1.56: Mechanical state diagram of marble from the Urals and experimental results showing the initiation of cracking processes under simple and complex loading paths.

the atmosphere. A hole 2 mm in diameter was drilled at mid-height of the specimen to establish a connection between the manometer and the mid-portion of the specimen. Investigations revealed that crack formation starts to develop first at mid-height of the specimen. The layout for this experiment is shown in Chapter 4, Fig. 4.4.

When the material starts to disintegrate, internal voids are generated. These voids are under a vacuum and hence suck air from the orifice and manometer tube, altering the liquid level in the latter. This feature allows recording initiation of cracking and determination of volume of internal voids formed. Complete curves showing the changes in pore volume in the material (including the volume reduction that occurs during the initial stages of loading) during testing are given together with other test results in Chapter 4, Figs. 4.8–4.11. The mechanical state diagram for marble from the Urals is shown in Fig. 1.56, in which the onset of volume increase is indicated by data points (crosses and open circles). Specimens were subjected to complex loading at various confining pressures σ_2 (i.e., following the paths indicated by the vertical lines with arrows) and simple loading at values of C equal to 0.31 and 0.4 (these paths are shown by the sloping lines (rays) from the origin, also marked with arrows). Stress values τ^*, at which cracking was initiated during the complex loading paths at different values of σ_2, are shown by crosses. Similar results obtained under simple loading are indicated by circles. For both paths of loading, cracking was initiated at almost the same values of tangential stress τ^*. This experiment provided direct proof of the fact that in zone A above the elastic limit (but below the critical line, determined theoretically by the condition $\tau = \sigma_2$ or $C = 0.333$), the permanent deformation process exhibits a truly plastic character, i.e., it occurs without dilatation, and the coefficient of permanent deformation μ, established experimentally, is equal to 0.5, while the angle of the deformation planes ω equals 45°. As can be seen from Fig. 1.56, the experimental critical boundary (dashed line) does not coincide with the theoretical critical line.

Analysis of the deformation scheme illustrated in Fig. 1.55 provides an explanation of the causes for deviation of the experimental critical boundary from the straight line boundary, given by the condition $\tau = \sigma_2$ or $C = 0.333$.

In the deformation process, especially at high confining pressures, when deformation grows rapidly, structural element a–b–c–d in Fig. 1.55 transforms from a square to a rhombus a′–b′–c′–d′. As a result, segment b–d is reduced to the value b′–d′ while segments b′–c′ and c′–d′ remain unchanged.

Reduction of segment b′–d′ is accompanied by a rise in the stress σ' acting on it, even though the value of the tangential stress τ remains constant. This can be demonstrated as follows.

The new resultant force T_1' is given by the equation:

$$T_1' = 2F\tau \cos (45° - \gamma).$$

Segment b′–d′, denoted as F' , has the value

$$F' = 2F \cos (45° + \gamma).$$

The stress σ_1' is then given as

$$\sigma_1' = \frac{2F \, \tau \cos(45°-\gamma)}{2F \cos(45°+\gamma)} = \tau\frac{\cos(45°-\gamma)}{\cos(45°+\gamma)}. \tag{1.24}$$

We shall denote $\dfrac{\cos(45°-\gamma)}{\cos(45°+\gamma)}$ as equal to K. K at $\gamma \neq 0$ is always more than unity, increasing with increase in deformation; γ is the relative shear.

From conditions (1.22) and (1.24), we find: $\sigma' = \sigma_2 = \tau$; $\sigma_1' = \tau K$. The stress required to form a rupture crack is given by: $\sigma'/K = \tau^*$.

According to (1.22), $\sigma' = \tau$; and hence

$$\tau/K = \tau^*, \tag{1.25}$$

where τ^* is the tangential stress at the start of crack initiation under high confining pressure σ_2 and at values of $C > 0.333$. Experimental values for τ^*, with mean values shown by the dashed line, are given in Fig. 1.56.

The value of the coefficient K varies from 1.06 to 1.2 as the shear angle γ varies from 2° to 5° and at pressures $\sigma_2 = 100$, 125 and 150 MPa. For these conditions, with such a variation of K, the computed values of τ^* are in good agreement with experimental values.

We shall now examine the differences exhibited in attaining the same final stress state along different loading paths [100, 101]. For simple loading with $C < 0.333$, all paths of loading, immediately upon passing the elastic limit, reach zone B (see the mehanical state diagram in Fig. 1.51). This indicates that the first shear plane ω, formed at the elastic limit, will contain microelements of rupture **b**. This implies that the volume of pore space would begin to increase immediately upon passing through the elastic limit, the value of the coefficient of permanent lateral deformation μ would be greater than 0.5 and the angle α of the planes of deformation and fracture would be less than 45°.

For $C > 0.333$, the rays representing the load paths pass through the elastic limit and enter zones A and D, where the development of permanent deformation is entirely plastic in nature—distintegration does not occur, $\mu = 0.5$ and $\alpha = 45°$. This is an idealised analysis, and does not take into account deviation of the critical line from the condition $\tau = \sigma_2$ shown in Fig. 1.56.

Under complex loading conditions, the pattern of deformation is similar to that of simple loading at values of $C < 0.333$ and holds good over a range of confining pressures $\sigma_p < \sigma_2 < \sigma_{2cr}^{el}$. A pattern of deformation similar to that considered for simple loading at $C > 0.333$ is observed over a range of confining pressures $\sigma_2 > \sigma_{2cr}^{us}$. The model for the growth of permanent deformation for pressures in the range $\sigma_{2cr}^{el} < \sigma_2 < \sigma_{2cr}^{us}$ is shown in Fig. 1.57. Here loading under constant confining pressure $\sigma_2 = \sigma_2^{III}$ (see diagram in Fig. 1.51) is conditionally considered. At the elastic limit (point L), the first macroscopic shear plane ω, which does not contain microelements of rupture **b**, appears. The loading path then intersects zone A where with increasing stress the number of such planes ω increases. In this zone plastic deformation only occurs, $\mu = 0.5$ and $\alpha = 45°$. With transition into zone B, microelements of rupture **b** begin to form in the deformation planes ω. The coefficient μ becomes greater than 0.5 and angle α less than 45°. Experimental curves revealing this change in the value of μ with transition from zone A to zone B, are shown in Fig. 1.58 [102]. These results were obtained for marble at three levels of confining pressures: $\sigma_2 = 75$, 125 and 150 MPa.

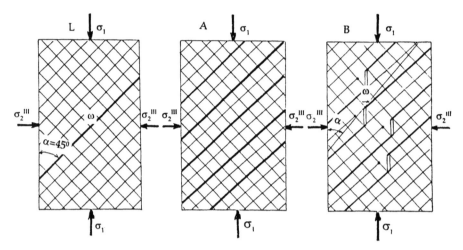

Fig. 1.57: Model for the development of deformation under complex loading when the loading path intersects either zone A and zone B in a mechanical state diagram.

Points corresponding to crossing of the boundaries between zones A and B are indicated by stars.

The discussion above leads to the important conclusion that it is possible to explain the deformation response of rock loaded along various paths in terms of its strongly varying structural states. Reaching a given point along a loading path that passes through stress zone B will cause intense disintegration of the rock and will augment its permeability to liquids and vapours. In contrast, in passing through zone A the rock retains to a large extent, its initial density and permeability. This topic will be dealt with in Chapter 4.

Experimental results confirming the large difference in the magnitude of permanent deformation $\Delta\varepsilon_1$ while arriving at the same final stress state through paths of either simple or complex loading, were shown earlier in Fig. 1.35. The reason for this difference can be explained from the experimental results given in Fig. 1.59 [66, 77]. The dependence of the stress values $\Delta\tau$ (continuous lines) and permanent deformation $\Delta\varepsilon_1$ (dashed lines) on the value of C for (a) marble from Kararsk, (b) marble from the Urals and (c) talc chlorite was obtained for the same final stress state through (1) simple and (2) complex loading. The meaning of $\Delta\tau$ is shown in the mechanical state diagram (Fig. 1.51) in which point I is reached by paths of simple ($C = C_1$) and complex (pressure σ_2') loading. Thus $\Delta\tau$ denotes the difference in stresses along the loading path between the elastic limit and the peak strength. A unique functional relationship exists between the stress $\Delta\tau$ and the deformation $\Delta\varepsilon_1$ for the two paths of loading in question [69, 67].

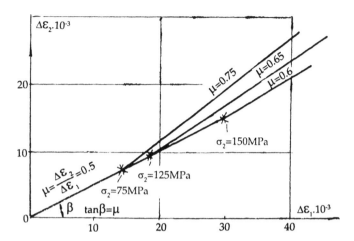

Fig. 1.58: Variation of the coefficient of permanent lateral deformation μ when the loading path crosses into either stress zone A or B.

This is demonstrated by the experimental curves shown in Fig. 1.60 for the same rocks. Points obtained under simple loading are indicated by circles and those obtained under complex loading by crosses. High values of deformation $\Delta\varepsilon_1$ correspond to high values of $\Delta\tau$, giving rise to differences in the magnitudes of deformation.

Typical complete relationships between the magnitudes of deformation $\Delta\varepsilon_{1 \text{ spl}}$, $\Delta\varepsilon_{1 \text{ clx}}$ and stresses $\Delta\tau_{\text{spl}}$ and $\Delta\tau_{\text{clx}}$ versus C (varying over a range from zero to 0.333), are shown in Fig. 1.61 (a) (spl = simple loading path; clx, complex loading path). These relationships were plotted by processing mechanical state diagrams and experimental relationships of $\Delta\varepsilon_1$ vs C. The relationship between the difference ($\Delta\tau_{\text{spl}} - \Delta\tau_{\text{clx}}$) and C is shown in Fig. 1.61 (b). It can be seen that the difference between $\Delta\varepsilon_1$ and $\Delta\tau$ for simple and complex loading is equal to zero at $C = 0$ and $C = 0.333$. The maximum difference is observed at $C \approx 0.23$ (see Fig. 1.61b).

Models similar to those shown in Figs. 1.52 and 1.54 allow mathematical relationships to be derived for the orientation of the macroscopic shear planes ω and the coefficient of permanent lateral deformation μ [69, 74, 101].

A model explaining the derivation of the relationship for the angle α, of the form given below, is shown in Fig. 1.62:

$$\tan \alpha = \frac{b}{2(bm + b/2)} = \frac{1}{2m + 1},\qquad (1.26)$$

where m is the number of structural elements **b** in rupture segment $b = b_{\text{min}} \cdot m$.

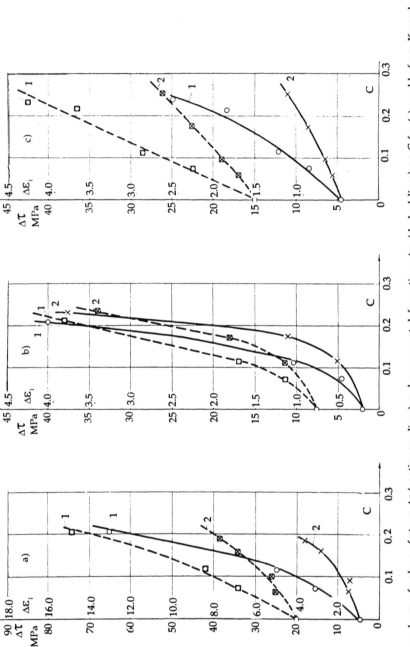

Fig. 1.59: Dependence of values of stress $\Delta\tau$ (continuous lines) and permanent deformations $\Delta\varepsilon_1$ (dashed lines) on C for (a) marble from Kararsk, (b) marble from the Urals and (c) talc chlorite, for the same stress state attained through (1) simple and (2) complex paths of loading.

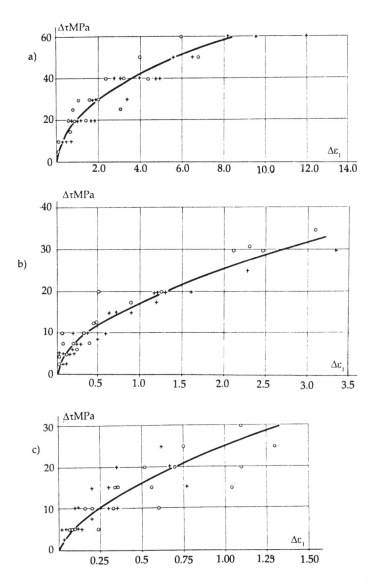

Fig. 1.60: General curves of permanent deformation for (a) marble from Kararsk, (b) marble from the Urals and (c) talc chlorite.

At $m = 1$, α is approximately equal to 18°, which coincides with the angle α obtained for many rocks tested under uniaxial compression. For cases in which the shear segment consists of n microelements of shear a_{min}, defined as $a = a_{min} \cdot n$, microelements a only are considered in the formula:

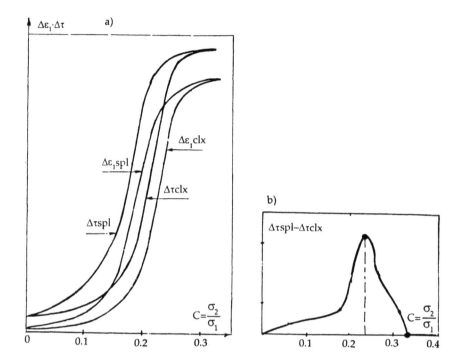

Fig. 1.61: (a) Typical complete relationships for permanent deformations ($\Delta\varepsilon_{1\,spl}$, $\Delta\varepsilon_{1\,clx}$) and stresses ($\Delta\tau_{spl}$, $\Delta\tau_{clx}$) under simple and complex paths of loading versus the parameter C. (b) Dependence of the stress difference ($\Delta\tau_{spl} - \Delta\tau_{clx}$) on the parameter C [101].

$$\alpha = 45° - \text{arc tan } (1/n + 1). \tag{1.27}$$

For $m = 1$ and $n = 1$, (1.26) and (1.27) yield the same result, i.e.: $\alpha \approx 18°$. If $m \to \infty$, then $\tan \alpha \to 0$, which corresponds to the condition of simple rupture. At $m \to 0$ and $n \to \infty$, angle $\alpha \to 45°$, which is the case of simple shear.

Derivation of the relationship to determine the coefficient of permanent lateral deformation μ is illustrated by the model in Fig. 1.63. Figure 1.63 (a) shows the case of shear on plane ω in a continuous uniform medium without microrupture formation. In this case, the coefficient μ_1 is defined by the equation:

$$\mu_1 = \frac{\Delta\varepsilon_2}{\Delta\varepsilon_1} = \frac{1}{2} \tan \alpha. \tag{1.28}$$

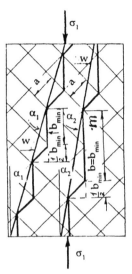

Fig. 1.62: Model for deriving a formula to determine the orientation of a shear plane relative to the axis of the specimen.

The multiplier 1/2 is introduced because the shear occurs in two perpendicular directions.

Derivation of the relationship to determine the coefficient μ for a heterogeneous medium, is explained in Fig. 1.63 (b) and (c). Here deformation occurs on stepped surfaces ω, consisting of segments of shear **a** and rupture **b**. It is first required to establish the number N_1 of planes ω per unit length in the direction of axial deformation $\Delta\varepsilon_1$ and the number N_2 of planes ω per unit length in the direction of later deformation $\Delta\varepsilon_2$:

$$N_1 = 1/b \text{ and } N_2 = 1/(b \tan \alpha).$$

Here b is treated as a segment of rupture as well as the distance between planes ω and ω_1 in the direction of axial deformation. Using the values of N_1 and N_2, the magnitudes of deformation $\Delta\varepsilon_1$ and $\Delta\varepsilon_2$ are found:

$$\Delta\varepsilon_1 = \Delta\varepsilon_1' \cdot N_1 \text{ and } \Delta\varepsilon_2 = \Delta\varepsilon_2' \cdot N_2.$$

Here $\Delta\varepsilon_1'$ is the displacement in the direction of the principal deformation $\Delta\varepsilon_1$, induced by shear Δa along a plane ω; $\Delta\varepsilon_2'$ is the corresponding displacement in the direction $\Delta\varepsilon_2$. These relationships are explained in Fig. 1.63 (c) in which a singular micropair 'shear-rupture' contained in the plane ω is shown. Since the angle of the microsegment a is taken to be 45°, we obtain the equality $\Delta\varepsilon_1' = \Delta\varepsilon_2'$ from which we have

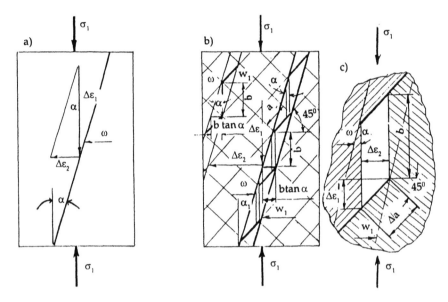

Fig. 1.63: Model for deriving a formula to determine the coefficient of permanent lateral deformation μ.

$$\mu = \frac{\Delta\varepsilon_2}{\Delta\varepsilon_1} = \frac{N_2}{N_1} = \frac{b}{b\,\tan\alpha} = \frac{1}{\tan\alpha}$$

Since deformation $\Delta\varepsilon_2$ occurs in two mutually perpendicular directions, the equation for μ should be multiplied by 1/2:

$$\mu = (\Delta\varepsilon_2 / \Delta\varepsilon_1)\ 1/2 = 1/(2\cdot\tan\alpha). \tag{1.29}$$

As can be seen from formulas (1.28) and (1.29), when microruptures are included in the model, a markedly different conclusion is reached from that obtained when it is assumed that the solid remains intact even when subjected to shear deformation. Here equation (1.29), which takes into account the violation of continuity, agrees with the test results. Equations (1.28) and (1.29) yield the same result, i.e., μ = 0.5 at α = 45° when simple shear deformation takes place. The coefficient μ depends on angle α only and is independent of the number of shear planes ω. The angle α depends on the parameter C and the confining pressure σ_2, as demonstrated by the structural statistical model.

Experimental relationships for the angle α and coefficient μ versus parameter C for marble from Kararsk and the Urals and talc chlorite were shown earlier in Figs. 1.45 and 1.37. These relationships are described well by the empirical equations:

$$\alpha = \alpha_0 \exp (DC) \tag{1.30}$$

$$\mu = \mu_0 \exp (YC), \tag{1.31}$$

where α_0 and μ_0 are the values of angle and coefficient respectively, obtained under uniaxial compression; D and Y are empirical constants.

The relationships shown in Figs. 1.45 and 1.37 are the same for both simple and complex paths of loading.

The variation of angles α and coefficients μ as a function of C for these three types of rocks is plotted in Fig. 1.64 on semi-logarithmic coordinates. The relationships between the coefficient m and C were plotted together, based on experimentally determined values (line 1), as approximated by equation (1.31) and equation (1.29) (line 2) and by equation (1.28) (line 3). It can be seen from the graphs that the model for equation (1.29) agrees best with the experiment. Equation (1.28) yields results that are qualitatively different from the experimental values.

1.4.4 Distribution Function of Number of Microelements of Shear Based on Shear Strength and Conditions for Selection of Structural Elements Involved in the Deformation Process

Rocks are complex structural formations containing various crystals, grains, cement, pores, cracks of different types and dimensions, inclusions and other defects. All these structural elements possess different mechanical properties. When loaded, these structural elements are subjected to shear and tensile (rupture) stresses. The characteristics of such heterogeneous solids are best described by statistical distributions for the percentage composition of different elements based on their shear strength and fracture strength. Rocks typically have a much higher shear strength than tensile strength. This is particularly evident under triaxial compression. The strength characteristics of various structural elements based on shear strength show wide scatter compared to that based on fracture indices. Thus, it is the distribution of percentage composition of shear elements based on shear strength that exerts the greatest effect on variation in strength and deformation properties of rocks under various conditions of loading.

The shape of this distribution function specifically determines the ability of the material to increase its strength (hardness) and deformation properties with increase in confining pressure, the shape of limit curves, the ability to disintegrate under deformation etc. Hence a study of the relationships between the key structural features (in the given case, this is the distribution function) and the properties of the material over a wide range of loading

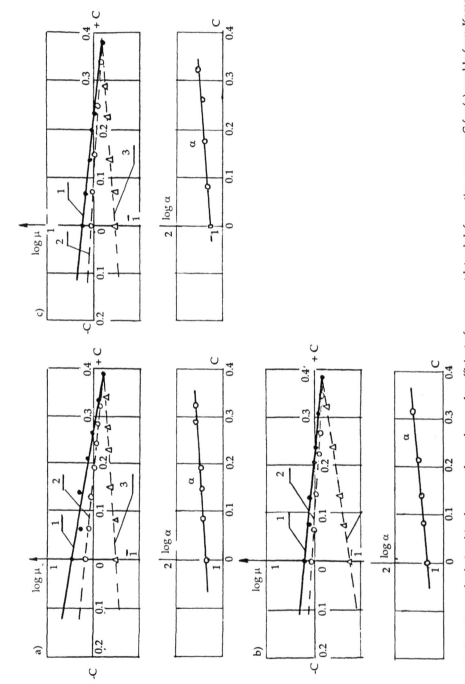

Fig. 1.64: Experimental relationships between the angle α and coefficient of permanent lateral deformation μ versus C for (a) marble from Kararsk, (b) marble from the Urals and (c) talc chlorite, plotted on semi-logarithmic coordinates.

conditions is a vital aspect in studies on the nature of strength and deformation in heterogeneous materials.

A statistical model for a heterogeous solid is used to determine the statistical distribution function [69, 99, 101].

Figure 1.65 shows a model in which the macroscopic shear planes ω, oriented at angles α, vary from 45° under simple shear to 0° under simple rupture, depending on the stress state. According to this model, the angles α of the planes ω and their density in the solid are determined by the concentration of microelements of shear a_{min} involved in the process of inelastic deformation. The higher the concentration, the larger the angle α and the higher the density. At 100% concentration of participating microelements of shear, the deformation planes w are oriented at 45° and attain a limit density. These planes ω are separated from each other by a distance indicated by the structural element a_{min}. The concentration of structural elements a_{min} involved in the deformation process when the planes ω are positioned at different angles α, can be established from the following considerations.

Any one of the planes ω which traverse the model from one lateral face to the other at any arbitrary angle α (except when α = 0), contains the same number of microelements of shear a_{min}. This means that the same number of shear microelements a_{min} will occur in any volume of the body described by this model and restricted in length by the dimension l_{α}, in which the plane ω extends at angle α. Assuming the dimension of the upper faces AB of the model to be unity, the concentration of elements K_a when the plane ω is at an angle α, compared to 100% concentration at α = 45°, can be established from the ratio:

$$K_a = \frac{l_{45}}{l_{\alpha}} \cdot 100\% = \frac{1}{l_{\alpha}} \cdot 100\% = \tan \alpha \cdot 100\%. \tag{1.32}$$

Equation 1.32 is shown graphically in Fig. 1.65 (b). Using this graph and arranging the experimental data pertaining to the relationships of angle α and magnitudes of limit stresses τ_{us} in real materials sequentially, an integral function of the distribution (ΣK vs τ_{us}) is plotted, which reflects the percentage dependence of the number of microelements of shear **a** involved in the deformation on the magnitude of the externally applied shear stress, which depends in turn on the stress state. Fig. 1.66 shows graphs of such relationships for a series of rocks [99, 101].

Differential statistical functions for the distribution (K vs τ) of the number (per cent) of microelements of shear **a** based on shear strength can be determined by differentiating the integral functions. As an example, the differential statistical function for marble from the Urals is shown in Fig. 1.67 (a). Marble is a monomineral rock. If it is assumed that due to scale effect, the strength characteristics of the structural elements are inversely

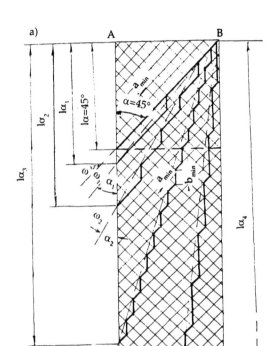

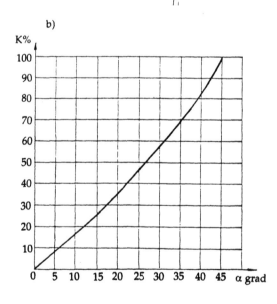

Fig. 1.65: Model for finding the statistical distribution function of percentage composition of structural elements based on shear strength.

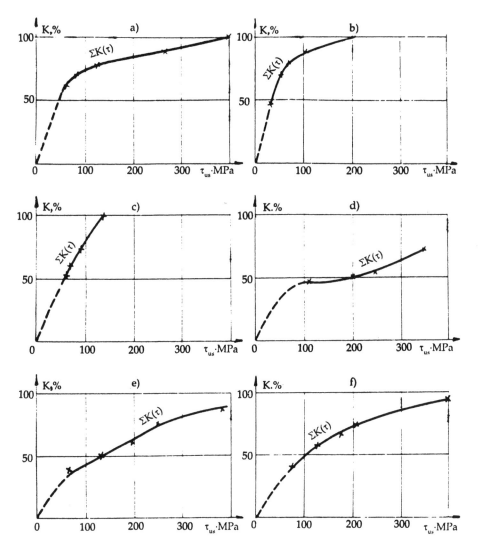

Fig. 1.66: Integral functions for the distribution of microelements of shear based on shear strength for (a) marble from Kararsk, (b) marble from the Urals, (c) talc chlorite, (d) diabase, (e) BP sandstone and (f) NBP sandstone from Donbass.

related to their dimensions, then the petrographic function of the distribution (per cent) of the composition of structural elements based on their linear dimension ($n\%$ vs d) may to some extent be a reflection of the distribution function (K vs τ). The petrographic distribution function ($n\%$ vs d) for marble from the Urals is shown in Fig. 1.67 (b). A mirror image of

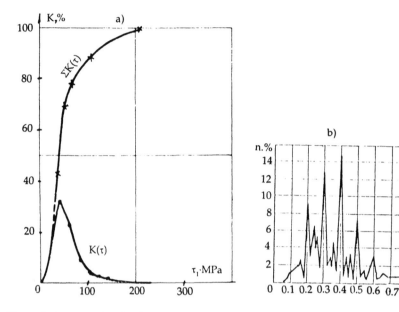

Fig. 1.67: (a) Integral and differential statistical distribution functions of structural microelements of shear based on shear strength for marble from the Urals. (b) Petrographic function of the per cent distribution composition of grains, depending on their dimensions, for the same type of marble.

this function, viewed externally, compares reasonably well with the distribution function (K vs τ) obtained.

From analysis of the distribution function (K vs τ) it appears that a great number of structural elements are involved in the process of permanent deformation under a pressure σ_2 approaching 100 MPa ($C \approx 0.27$). This corresponds to the maximum permanent volumetric strain attained by this marble (see Fig. 1.25).

The features mentioned above help in understanding the factors that influence the maximum permanent volumetric strain. The maximum is obtained when the stresses are sufficient to activate the maximum number of structural elements of shear in the deformation process. These shear elements initiate microruptures and produce dilatation. The distribution function also determines other special features of the behaviour of heterogeneous solids, as described in the literature [99, 101].

Induction of new, stronger structural elements into the deformation process with increase in σ_2 or in the value of C, occurs as a result of the statistical selection of elements. The condition of violation of ultimate equilibrium of forces acting on a structural element (1.18) is the basis for this selection. The general principle of such a selection is considered below [69, 74].

Dividing all terms in equation (1.17) by a^2 and putting $b/a = \chi$ and after a series of simple transformations and considering that $\cos 45° \approx 0.7$, we arrive at the equation:

$$\tau_1 = \tau - [0.7 \, \chi \, (\sigma_2 + \sigma_p)]. \tag{1.33}$$

The expression within brackets is the magnitude of the resistance offered to relative displacement of elements in the 'shear-rupture' pair induced by fracture strength σ_p and confining pressure σ_2 acting on segment **b**. Let us put

$$P = 0.7 \, \chi \, (\sigma_2 + \sigma_p), \tag{1.34}$$

where P equals zero in two cases, i.e., when $\chi = 0$ and when $-\sigma_2 = \sigma_p$. The first case conforms to the condition of deformability through simple shear without initiation of microruptures, since rupture segments **b** disappear. The second case occurs when stresses σ_2 become tensile, wherein only rupture segments of the **b** type are formed and failure is of the simple rupture type due to the effect of tensile stress σ_p. This picture is illustrated in Fig. 1.54.

Differentiating equation (1.33) with respect to C, while considering (1.34), we obtain

$$\frac{\partial \tau_1}{\partial C} = \frac{\partial \tau}{\partial C} - \frac{\partial P}{\partial C} \quad \text{and} \quad \frac{\partial \tau}{\partial C} = \frac{\partial \tau_1}{\partial C} + \frac{\partial P}{\partial C}. \tag{1.35}$$

The condition for involvement of new microsegments **a** with higher shear strength than those previously involved in the deformation process, is of the form:

$$\frac{\partial \tau_1}{\partial C} \leq \frac{\partial P}{\partial C} \tag{1.36}$$

i.e., new, higher shear strength microsegments become involved in the deformation process only when the increment in shear strength of these microsegments with respect to C is less than or equal to the corresponding increment in P. Substituting (1.36) in (1.35), we obtain

$$\frac{\partial P}{\partial C} \geq \frac{1}{2} \frac{\partial \tau}{\partial C}. \tag{1.37}$$

Assuming the value of τ in expression (1.37) to be the elastic limit, and taking it from equation (1.6), we obtain

$$\frac{\partial P}{\partial C} \geq \frac{1}{2} \frac{\partial \tau_{el}}{2\partial C}. \tag{1.38}$$

This inequality is the condition for the selection of structural elements that form the first macroscopic shear plane ω at the elastic limit under different stress states and values of the parameter C. After formation of the first plane ω at the elastic limit, subsequent planes ω are formed at a specific value of C as a result of strain hardening. This leads to an increase in the average level of stresses in the body, facilitating the involvement of shear microelements of higher shear strength in the deformation process. Here the angle of orientation of planes ω at a given value of C remains unchanged in all sections of the diagram, from the elastic limit to the peak residual strength. This has been confirmed by experimental results establishing the invariability of the coefficient of lateral residual deformation μ for constant values of the parameter C (see Fig. 1.36).

When the conditions of limit elastic states (1.6) become horizontal at $C \approx 0.333$, the value of P in equation (1.38) becomes zero, since the rupture elements **b** disappear at this point and the derivative of the elastic limit with respect to C in the horizontal section is also equal to zero.

Figure 1.68 shows the relationships between the limit elastic states, values of P for two types of marble and their derivatives with respect to C. It can be seen from these graphs that the absolute values of the elastic limits and P differ considerably numerically; however, the values of their derivatives with respect to C are quite similar, which indicates the authenticity of the condition for statistical selection (1.38).

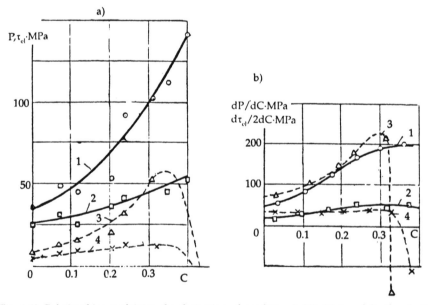

Fig. 1.68: Relationships explaining the derivation of condition (1.38). (a) 1 and 3—the elastic limits τ_{el} and values of P for marble from Kararsk; 2 and 4—the same for marble from the Urals versus parameter C. (b) Dependence of derivatives P and τ_{el} on C for two types of marble.

2

Time-dependent Properties of Rocks

2.1 INTRODUCTION

The duration of loads acting on materials, including rocks, is a key factor affecting their mechanical response. This is particularly so in mining and geological activities for which time may vary over very wide ranges, from geological time to fractions of a second. It is convenient to study the effect of time on rock properties by changing a parameter such as the rate of deformation (or strain rates) in laboratory testing of rock. It is necessary to cover a wide range of deformation rates during experiments in order to obtain an adequate amount of authentic information and to establish time-dependent relationships with reasonable accuracy. The authors of this monograph have developed a set of special instruments that allow investigations to be conducted over a range of strain rates from 10^{+2} s^{-1} to 10^{-10} s^{-1}, under conditions of uniaxial compression and triaxial compression at hydrostatic pressures up to 300 MPa. This equipment is mechanically very stiff and allows complete 'stress-strain' diagrams, including the descending (post-failure) zone to be obtained for highly brittle rocks. Such a wide range of conditions for experimental studies has allowed the authors to identify a series of new phenomena that accompany irreversible deformations of rocks, both in the pre-failure strength and post-failure strength regions, especially on the descending branch of the 'stress-strain' diagram.

The apparatus developed by the authors is described below in detail together with the results of experiments conducted with this apparatus on a wide series of rock types.

2.2 METHOD FOR STUDYING THE EFFECT OF DEFORMATION RATE ON ROCKS

2.2.1 Stiff Machine for Dynamic Testing under Triaxial Compression

A schematic design and photograph of the machine assembly for dynamic testing under triaxial compression (variant a) are shown in Fig. 2.1 [76, 103, 113, 101].

The machine consists of a high-pressure cell containing a cylinder (1), piston rod (2) and a cover (3). After placing a test specimen (12) in the cell, the cell is positioned in the frame (4). The piston (5) of the hydraulic jack, a node for regulating deformation rate (6), a fast-acting valve (7) and receiver of variable volume (8) are also positioned in the frame. The press has two pressure sources: (13) for subjecting the specimen to triaxial stress and pressure transmission to the cell (15) for unloading and retracting the piston rod of the jack; (14) for subjecting the specimen to longitudinal strain.

The experiment is conducted as follows. The required level of hydrostatic pressure is developed in the cell (1) by the working fluid from the source (13). The pressurised working fluid flows into the compensating cavity (9) through grooves (18) in the piston rod (2) forcing open the cell, compressing the cover (3) and bottom of the cylinder (1) against the frame. At this juncture, the piston rod (2) is in a suspended state, not transmitting the pressure load acting on it to the loading piston (5). The specimen is subjected to a uniform state of stress $\sigma_1 = \sigma_2 = \sigma_3$. The Elastic energy of the compressed fluid is used to subject the specimen to dynamic axial loading. This elastic energy is accumulated, with the help of source (14), in receiver (8). When the fast-acting valve (7) is opened, high pressure fluid flows rapidly into the cavity above the piston rod (5) and by moving it together with piston rod (2), the specimen is dynamically loaded. Movement of the piston rod (2) in cylinder (1) does not increase the pressure in it, since excess fluid flows into the compensating cavity (9) through grooves (18) in piston rod (2).

To obtain a constant rate of deformation in a specimen in the post-failure region, the initial pressure in the receiver and its volume are computed such that the fluid flowing from the receiver into the hydraulic jack will develop known rate of pressure change in the hydraulic jack, depending on the load-bearing ability of the specimen, in the post-failure region. In this case, the characteristic force in the loading system ABD (see Fig. 2.1a) and post-failure strength part of curve OBD coincide or are observed to be parallel.

The Initial pressure P in the receiver and its volume V, taking the stiffness characteristic of the press into account, are selected in the manner outlined in Chapter 1 when describing variant (b) of the stiff press for static

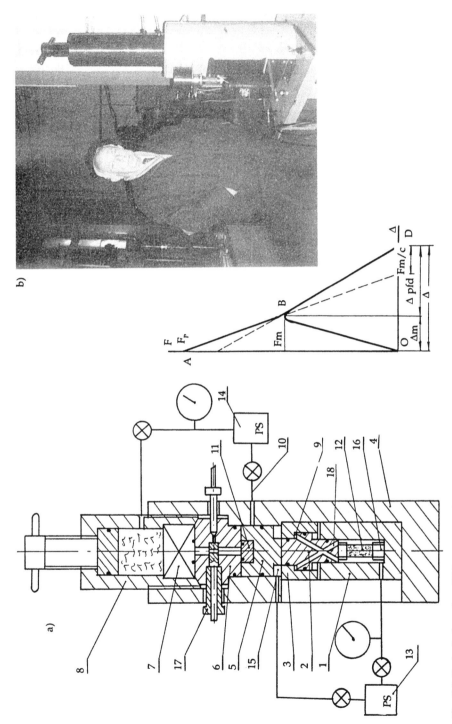

Fig. 2.1: Basic design of a stiff dynamic machine for triaxial compression testing (variant a). The photograph shows an external view of the machine, with Professor Andrey Stavrogin.

testing. The difference in this case, lies only in that there is no transmission of displacement and force through wedges and hence in formulas (a) and (b) given below, the corresponding factor which taken these into account is absent:

$$V = S_d (\Delta - F_m/C)/\beta;$$ (a)

$$P = F_m (\Delta - F_m/C)/S_d (\Delta_{pfd} - F_m/C),$$ (b)

where C is the stiffness of the frame structure; S_d the piston area in the hydraulic jack; β the compressibility of working fluid; Δ the total absolute specimen deformation; Δ_{pfd} the post-failure specimen deformation; and F_m the maximum load on the specimen.

Preliminary parameters of the complete 'load-strain' diagram of the material being tested are obtained by testing a similar specimen statically in the same press. If necessary, these initial values of P and V can be corrected after the first dynamic experiment. The rate of dynamic deformation is regulated by a throttle (17) and by varying the magnitude of pressure P.

Static testing is carried out by feeding pressurised fluid from the source (14) through the tube (10) into the hydraulic jack. In this case, a valve (11) automatically closes the groove connecting the jack and receiver.

This press allows specimens to be tested in various modes of loading. Using a static drive, it is possible to subject a specimen to different stages of pre-failure and post-failure deformation, and later to implement impulse loading by means of a dynamic drive. Extensometers affixed to the specimen record longitudinal and transverse deformation during the test. Force is measured by a dynamometer (16), which is in contact with the lower end of the specimen.

Error in measurement of the above indicated parameters in the dynamic mode ranges from 2 to 5%.

The extensometer for measuring strain in dynamic tests consists of a tensioned within a thin high-resistance wire a frame, which is affixed to the lower thrust-bearing-dynamometer (16). A movable contact, which glides along the wire when the specimen is deformed, is fixed to the upper thrust bearing on the side of the specimen. Such a gauge has low inertia and allows the dynamic deformation to be measured at rates up to $\dot{\varepsilon}_1 = 10^2$ s^{-1}. In several experiments, strain resistance gauges were glued directly to the surface of the specimen. Such a method is the only effective way of studying wave processes in a specimen subjected to dynamic loading. Here the gauge simultaneously records stress and strain states in a specimen at the point of loading. In where experiments, the specimen is not tested to failure, dynamic pulses on the specimen may be delivered at a frequency of 10 pulses per hour, with the possibility for altering the initial stress state between tests.

This press allows testing at deformation rates varying from $\dot{\varepsilon}_1 = 10^{-6}\ s^{-1}$ to $\dot{\varepsilon}_1 = 10^{+2}\ s^{-1}$.

The pressure in the cell may be varied from 0 to 300 MPa.

Force developed by the press up to 1500 kN.

Dimensions of the press: 1500 × 500 × 250 mm.

Specimen size: diameter 30 mm, length 60 mm.

A special fast-acting pressure control valve, developed by the authors, is a part of these dynamic machines. It is very reliable and gives a fast response. The principles mentioned in the inventions [111, 115, 101] were used in designing this valve. The fast-acting (pressure-control) valve is schematically illustrated in Fig. 2.2.

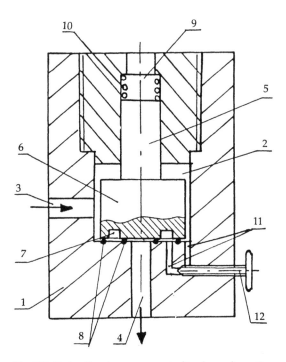

Fig. 2.2: Fast-acting (pressure-control) valve, schematic.

The valve consists of a housing (1), working cell (2) with inlet (3) and outlet (4), and a plunger (5) with a head (6). The lower face of the head (6) is equipped with a circular groove (7) that is hermetically isolated by packing rings (8). The free end of the plunger (5) goes into the cell (9) which is externally exposed. The end of the plunger is spring-loaded. The drain mechanism consists of conduits (11) linking the working chamber with the cavity below the circular groove and the cock (12). The valve works as follows. Initially, the spring (10) presses the head (6) to the packing ring (8). The cock (12) is closed. The cavity of the receiver (not shown in Figure 2.2) is connected to the working cell (2) through the inlet (3). When the receiver is pressurised, the same pressure is generated also in the working cell of the valve. This pressure presses the head (6) against the packing rings (8) and hermetically covers the outlet (4). Valve control is effected by the cock (12). When the cock is opened, pressure from the cavity (2) is transmitted to the circular cavity below the head (6) and is discharged, thereby opening the outlet (4).

The valve functions reliably both under low and high pressures. The force clamping the head to the packing rings is regulated automatically: the higher the pressure, the more intense the clamping force, providing a dependable hermetic sealing.

Another type of machine for dynamic testing under triaxial compression [109, 101] is shown in Fig. 2.3. Compared to the machine described above, the unique feature of this machine is the addition of a high-pressure cell and positioning of a dynamic drive. Cell and drive are fabricated as an independent dynamic attachment to any static press or even to a simple loading frame.

The cell and loading hydraulic jack together make a compact assembly, which is vital if the system is to be very stiff. The loading jack is incorporated in piston (4) of the high-pressure cell (2). The piston of the jack (8) and loading piston rod (9) are integrated into a single part. A compensating cell (6), connected to the working cell (5) through grooves (7), is situated in the piston rod (4). The compensating cell is intended to maintain contact pressure in the working cell during dynamic loading, accompanied by entry of the piston (9) into the cell (5). Working fluid, drained from the system at this juncture, flows through conduits (7) into the empty space (6). The receiver (10), variable in capacity, and fast-acting valve (14) are taken out of the frame of the press, thereby reducing the linear dimensions of parts of the system that undergo elastic deformation during loading the specimen. These parts lower the stiffness of the loading system.

In the machine developed by the authors, such a dynamic attachment was intended for a stiff press with frame (1) and an independent static loading drive (11). The stiffness of this type of machine was higher than that in the case of variant (a). Provision of two independent drives makes it possible to test the specimen (3) under static, dynamic, of a variable mode of deformation. The third mode allows the static deformation process to reach any stage of deformation, including the post-failure region, and later induction of dynamic rates of loading. Gauges (12) and (13) attached to the specimen record forces and strains in it.

The low cross-sectional area of the hydraulic jack makes it necessary to use high pressures in the receiver in order to develop heavy loads on the

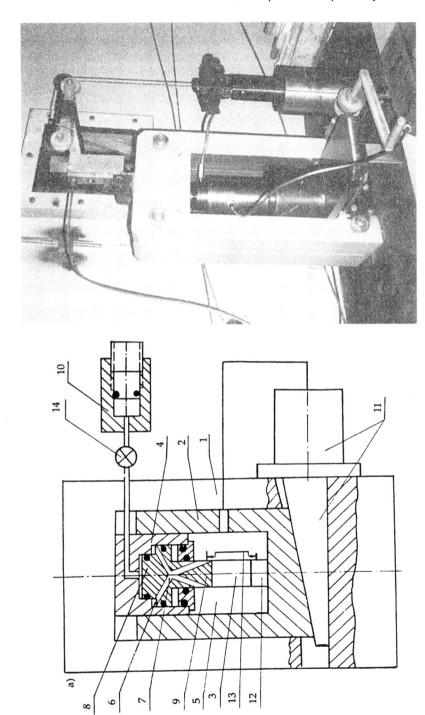

Fig. 2.3 (a) Schematic diagram and (b) external view of a stiff dynamic machine for testing under triaxial compression (variant b).

specimen—a drawback of this machine. High reliability and convenient operation were exhibited when low-strength rocks were tested. The parameters of this machine, used by the authors for conducting experimental studies, were as follows:

— stiffness of the machine: 10^{10} N/m;
— strain rate $\dot{\varepsilon}_1$: from 10^{-6} s^{-1} to 10^{+2} s^{-1};
— maximum pressure in the cell: 50 MPa;
— maximum force developed by the hydraulic jack: 750 kN.

2.2.2 Stiff Machine for Dynamic Testing under Uniaxial Compression

Figure 2.4 shows a machine developed for studying the properties of rocks in the pre-failure and post-failure region under uniaxial compression over a range of strain rates from $\dot{\varepsilon}_1 = 10^{-6}$ s^{-1} to $\dot{\varepsilon}_1 = 10^{+2}$ s^{-1}, [80, 101, 103, 110].

As in the case of the two preceding variants of stiff dynamic machines, this machine is equipped with two independent loading drives: static and dynamic. The first drive is a stiff W-drive (1), the principles of operation of which were discussed in detail while describing static stiff machines in Chapter 1. This drive allows strain rates ranging from $\dot{\varepsilon}_1 = 10^{-6}$ s^{-1} to $\dot{\varepsilon}_1 = 10^{-3}$ s^{-1} to be generated. The drive has a high stiffness (2×10^{10} N/m), which permits highly brittle rocks to be tested in the post-peak strength zone. This drive also prepares the specimen for dynamic testing.

The dynamic loading drive is situated in the stiff power frame (5) of the press. It includes a variable capacity receiver (6), regulated by a screw (12), fast-acting valve (9), fluid flow control panel (10), and dynamic hydraulic jack with loading piston head (4) positioned in the power frame (5). The receiver is intended to store energy of the elastically compressed fluid, which is pumped into it by a high pressure pump (7). Pressure is controlled by a manometer (8).

For testing, the specimen (2) is seated on a wedge (3) that also serves as a dynamometer, and with the help of the hydraulic drive (1), the specimen is compressed against the piston rod of the dynamic hydraulic jack (4). The dynamic drive works on the principle of expansion of a compressed fluid. When the fast-acting valve (9) is opened, the confined fluid in the receiver flows rapidly into the cavity of the hydraulic jack (4), thereby subjecting the specimen to a dynamic load. The applied to the specimen load is measured by means of a dynamometer (3), while strain in the specimen is recorded by the extensometers glued to it. The rate of dynamic deformation of the specimen is controlled by a panel (10) and also by the initial pressure created in the receiver at the beginning of the test. The maximum strain

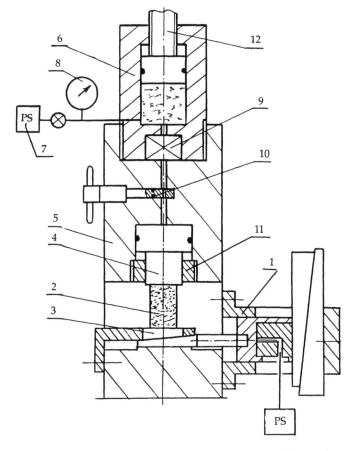

Fig. 2.4: Basic design of a stiff machine for dynamic testing under conditions of uniaxial compression.

rate was observed to be $\dot{\varepsilon}_1 = 10^{+2}\ \text{s}^{-1}$. To provide a constant rate of deformation in a specimen in the post-failure strength zone, the volume of the receiver and the initial pressure in it are selected based on the principle discussed in the preceding section.

This machine can also test specimens in a variable loading rate regime. Using the static drive, a specimen can be subjected to strain at low speeds to attain any stage of pre-failure or post-failure deformation; impulsive loading is then developed using the dynamic drive. It is possible to provide a reverse sequence of variation in speed regime: first, the specimen is subjected to dynamic action followed by a sharp decrease in rate of deformation until the static rate is reached. In this case, the magnitude of dynamic deformation is controlled by a special arrester (11). This method of conducting tests at variable loading rates is used in particular for studying the effect of deformation rate on the mechanical properties of material in the post-failure zone.

The parameters of the machine used by the authors in experimental studies are as follows:
— stiffness of the machine: 2×10^{10} N/m;
— strain rate $\dot{\varepsilon}_1$: 10^{-6} s^{-1} to 10^{+2} s^{-1};
— maximum force: 750 kN;
— dimensions of the press: $1200 \times 500 \times 250$ mm;
— specimen dimensions: diameter 30 mm, length 60 mm.

2.2.3 Spring-type Press for Studying Creep under Conditions of a Constantly Assigned Load

Creep tests were conducted on press-type machines, intended for long-duration testing (LDT), capable of developing a maximum force of a 1000 kN [67]. Constancy of the applied load is assured by the stored elastic energy of a stack of compressed plate-type springs. The press is schematically illustrated in Fig. 2.5. The specimen (2) (or high-pressure cell with specimen) is seated on a bench situated on a mobile traverse (3). Force is generated from the hydraulic jack (9) and transmitted through a screw (6) to the stack of plate-type springs (5) and further through a cross-piece (3) to the specimen (2). The magnitude of applied load is controlled by means of a permanently installed indicator device (not shown in the diagram) which measures the deflection on the stacked of springs, both during the loading process and the creep process. Each press has its own standard calibration curve showing the relationship between the span of sag and the load. After the pre-assigned load is attained, the nut (7) on the screw (6) tightened against the cross-piece (8). The jack (9) is then disconnected and the load "transferred" to the columns (4) between the cross-pieces (1) and (8). The maximum span of sag in the stack of springs is observed to be 100 mm. During the creep process, the load on the press will tend to decrease. In order to avoid this, a correction in load is peridically made by the hydraulic jack (9) and setting the nut (7) at a new position.

A general view of the testing room with a set of LDT machines is shown in Fig. 2.6.

A general view of a prismatic specimen for creep testing, equipped with dial indicators for measuring strains, is given in Fig. 2.7. The specimen is $150 \times 150 \times 300$ mm in size. The compressive load P is transmitted through a spherical hinge. The system of indicators allows measurement of both longitudinal and transverse deformations. Indicators (3) and (4) (the latter not shown in Fig. 2.7) record longitudinal deformations up to 100 mm over

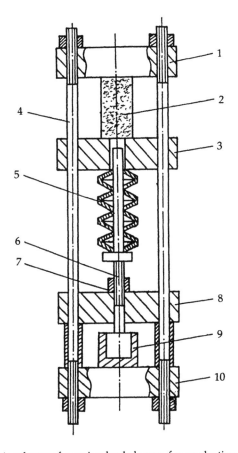

Fig. 2.5: Basic scheme of a spring-loaded press for conducting creep tests.

the mid-length of the specimen, while indicators (1) and (2) measure longitudinal deformation in the entire specimen. Indicators (5) and (6) record transverse strains in the middle part of the specimen, while indicators (7) and (8) record transverse strains in the lower part.

In uniaxial creep tests prismatic specimens of rock salt, sylvinite, concrete and Cambrian clays were used, i.e., materials of relatively low strength.

Stronger rocks were tested by using specimens of cylindrical shape cored from monolithic blocks of rocks. The design of a cylindrical specimen for long-term testing under uniaxial compression is shown in Fig. 2.8. The specimen (5) is equipped with two mirror extensometers (10) for measuring longitudinal strains and a dial indicator (8) with a precision of 0.001 mm for recording transverse deformation. The indicator is attached to the frame

Fig. 2.6: General view of the room equipped with machines for testing creep and long-term strength (LDT).

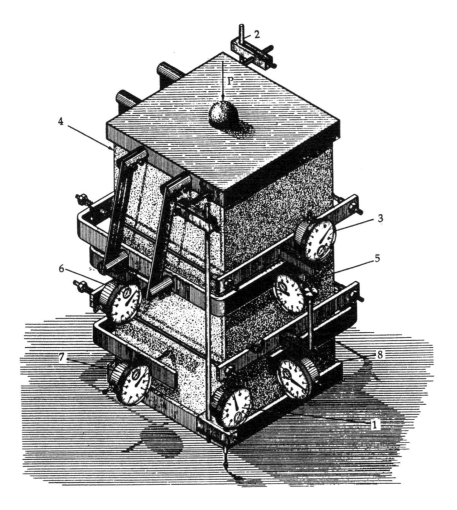

Fig. 2.7: Design of a prismatic specimen for long-term testing.

(7) with elastic hinges (11), which provide a reliable interface between the indicator base, backstop (3) and specimen surface. Elastic hinges (11), with the help of part (12) are stiffly connected to the lower thrust bearing (13) of the specimen.

Knives of the mirror extensometers are attached to the middle of the specimen from two opposite sides and connected to it by an elastic strap (not shown in the Fig.) When deformation occurs a movable knife (4) turns the mirror (10) by means of levers (9) rigidly linked to the movable knife. Depending on the specimen dimensions, the measurement base varies from 30 to 50 mm. During deformation, the rotation of the mirror is recorded

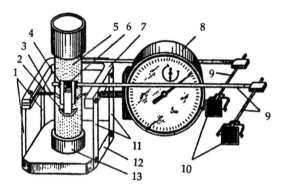

Fig. 2.8: Design of a cylindrical specimen for long-term testing.

by a theodolite using Morten's method. Using the indicators and mirror extensometers, deformations were measured for a range of relative rates varying from $\dot{\varepsilon}_1 = 10^{-10}$ s^{-1} to 10^{-7} s^{-1}.

Mechanical systems of recording forces and strains are much preferred in long-term tests because they offer highly reliable measurements compared to electric and electronic systems. They are constantly operable (in very prolonged tests—over a period of more than five years) and do not require periodic switch-off from the power circuit (as in electrical systems). Mechanical systems are also devoid of errors associated with *zero drift* typically found in electric and electronic instrumentation.

2.3 STRENGTH AND PLASTICITY OF ROCKS UNDER WIDELY VARYING DEFORMATION RATES AND STRESS STATES

As already mentioned, rocks are a classical example of an inhomogeneous solid. The degree of inhomogeneity can be quantified by the magnitude of dispersion of a distribution function of the number N of structural elements constituting the given rock versus the value of τ, which represents the strength properties of these elements. The higher the dispersion, the more inhomogeneous the rock.

Rocks can be classified into three categories based on the magnitude of dispersion (Fig. 2.9): low dispersion, when the distribution function has a sharp maximum (curve 1); moderate dispersion, when the distribution function has a diffuse (not sharp) maximum (curve 2); and high dispersion, when the distribution function has no distinct maximum (curve 3).

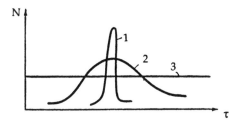

Fig. 2.9: Statistical curves showing the distribution of number of elements N versus their resistance (strength) τ for various values of dispersion.

The magnitude of dispersion affects several aspects of rock behaviour under different experimental conditions, e.g., the increase in strength and plasticity with rise in confining pressure; the coefficient of permanent lateral strain; the number and orientation of slip planes (Chernov-Luders lines) generated as a function of stress; the dependence of the ultimate volumetric strain on confining pressure for conditions of deviatoric triaxial compression and the presence of a maximum for dilatant volumetric strain at a specific ratio between the principal components of stress; loss in strength under conditions of dynamic (impact) deformation etc. The influence of inhomogeneity is particularly marked in experiments in which deformation rates and stress state are varied over a wide range.

Figure 2.10 shows the results of determination of (a) peak strength and (b) elastic limits for marble from the Urals, as a function of the logarithm of strain rate, for various values of confining pressure σ_2 (indicated in a table in the diagram). Each point on the curve is plotted by averaging 5–12 independent test results obtained on similar specimens [76].

After statistical analysis the results were plotted in the appropriate coordinates in the form of a series of rays, emanating from a common origin [75, 76] at the coordinate for marble of $\log \dot{\varepsilon}_1 = \overline{\overline{33}}$ s^{-1}. Each ray represents a specific value of confining pressure. Uniaxial compression is indicated as $\sigma_2 = 0$. The strain rate $\dot{\varepsilon}$ was varied over 12 orders of magnitude (from $\dot{\varepsilon}_1 = 10^{-10}$ s^{-1} to $\dot{\varepsilon}_1 = 10^{+2}$ s^{-1}). The confining pressure was varied from 0 to 150 MPa.

These results, can be described by a kinetic equation, proposed by S. N. Zhurkov for computing strain rates.

$$\dot{\varepsilon}_1 = \dot{\varepsilon}_0 \exp\left(- (U_0 - \gamma\tau)/KT\right). \tag{2.1}$$

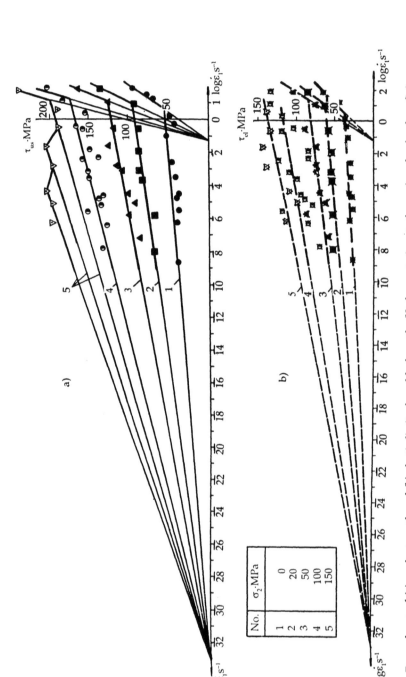

Fig. 2.10: Dependence of (a) peak strength and (b) elastic limit of marble from the Urals on strain rate for various levels of confining pressure σ_2.

An equation similar to the Zhurkov equation for time decay is of the form

$$t = t_0 \exp((U_0 - \gamma\tau)/KT), \tag{2.2}$$

where $\dot{\varepsilon}_0$, t_0, U_0 are constants independent of σ_2; K is Boltzman's constant; T the temperature in the Kelvin scale; γ a structurally sensitive coefficient dependent on σ_2 that determines the angle of inclination of rays (as σ_2 increases, γ decreases); τ the tangential stress, as computed by the formula

$$\tau = (\sigma_1 - \sigma_2)/2.$$

The coordinate of the origin is determined by setting the value of stress τ in equation (2.1) to zero.

Conditions for the peak strength τ_{us} and elastic limits τ_{el}, and the stress state characterised by $C = \sigma_2/\sigma_1$, are described by equations (1.6) and (1.7) derived earlier:

$$\tau_{us} = \tau_{us}^0 \exp(AC); \tag{1.7}$$

$$\tau_{el} = \tau_{el}^0 \exp(BC). \tag{1.6}$$

In equations (1.6) and (1.7), τ_{us}^0 and τ_{el}^0, as well as coefficients A and B, depend on the deformation rate. Thus when using equations (1.6) and (1.7), it is necessary to specify the rate of deformation at which the indicated parameters were obtained. This question will be discussed in detail later. A combined solution to equations (2.1) and (1.6) and equations (2.1) and (1.7) was obtained for the relationship between γ and C at the ultimate strength γ_{us} (σ_2) and elastic limit γ_{el} (σ_2):

$$\gamma_{us}(\sigma_2) = \gamma_{us}^0 \exp(-AC) \tag{2.3}$$

$$\gamma_{el}(\sigma_2) = \gamma_{el}^0 \exp(-BC) \tag{2.4}$$

where γ_{us}^0 and γ_{el}^0 are structurally sensitive coefficients for uniaxial compression.

It follows from equations (2.3) and (2.4) that the structurally sensitive coefficients do not depend on the deformation rate. The experimental points obtained in tests conducted with a confining pressure $\sigma_2 = 150$ MPa did not lie on the ray emanating from a common origin. At this pressure, marble exhibited no increase in strength; instead the strength fell by roughly 7% for strain rates varying from 10^{-7} s^{-1} to 10 s^{-1}. Thus the approximation using a single kinetic equation is limited by the confining pressure σ_2.

Data from tests carried out at high strain rates (from 10 to 10^{+2} s^{-1}) did not lie on the ray from the origin at coordinate $\log \dot{\varepsilon}_1 = \overline{33}$ s^{-1} but lay on a ray having a high slope angle with the coordinate of the origin located at log $\dot{\varepsilon}_1$ = $\overline{1.25}$ s^{-1}. A similar result was obtained in studies cited in [37, 45] conducted using the Hopkinson bar method. The reason for such a sharp rise in strength lies in the mechanics of energy transmission to the tip of the rupture crack; at high rates, not all of the energy reaches the crack tip.

Dependence of permanent $\Delta\varepsilon_1^{us}$ and volumetric $\Delta\varepsilon_1^{us}$ strains of marble at ultimate strength on the strain rate are shown in Fig. 2.11 (a). The confining pressures σ_2 MPa for which these results were obtained, are indicated by numbers in the tables shown in the diagram. The largest strains were observed at high strain rates $\dot{\varepsilon}_1$ = 10 s^{-1} with pressures σ_2 = 20 and 50 MPa. At confining pressures of σ_2 = 100 and 150 MPa, maximum deformations were obtained at low strain rates of $\dot{\varepsilon}_1$ = 10^{-6} s^{-1}. Volumetric dilatational strains $\Delta\theta^{us}$ for marble are shown in Fig. 2.11 (b). Here the curves also show the changing nature of the relationship between $\Delta\theta^{us}$ and strain rate at different levels of confining pressure.

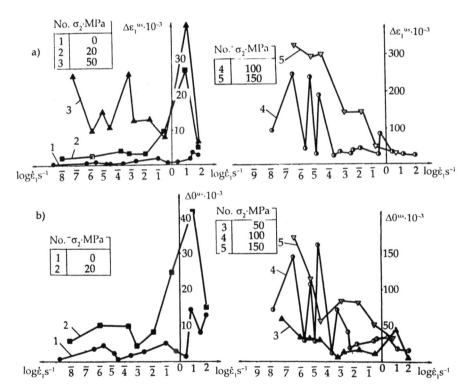

Fig. 2.11: Relationship between permanent (a) axial and (b) volumetric strains and strain rates in marble specimens at Peak strength, for different values of confining pressure σ_2.

The limits of peak strength and elasticity for diabase for varying strain rates and confining pressures σ_2 are shown in Fig. 2.12 (a) and (b). Experimental points obtained as a result of averaging 6–12 readings on similar specimens are plotted along the rays emanating from an origin at coordinate $\log \dot{\varepsilon}_1 = \overline{33}$ s^{-1}. It can be seen that the results are adequately described by the kinetic equation (2.1). A stronger relationship between strength and strain rate, similar to the one established for marble, was obtained in this case for uniaxial compression tests ($\sigma_2 = 0$) at high strain rates ($\dot{\varepsilon}_1 = 1 - 10$ s^{-1}).

Figures 2.13 and 2.14 show the dependence of axial and volumetric permanent strains on strain rate and confining pressure for diabase. It can be seen that the relationship between the magnitudes of the axial and volumetric strains and strain rate for various levels of confining pressure is complicated. As in the experiments with marble, maxima and minima appear at certain strain rates. The absolute values of strain for diabase are very low.

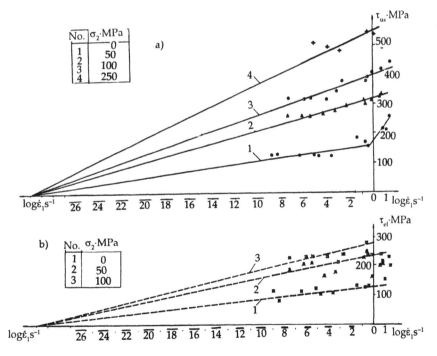

Fig. 2.12: Relationship between the limits of (a) strength and (b) elasticity and strain rate for various levels of confining pressure σ_2 for diabase specimens.

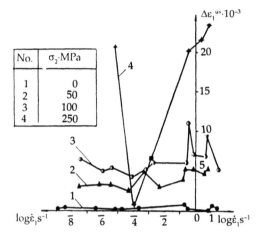

No.	$\sigma_2 \cdot$MPa
1	0
2	50
3	100
4	250

Fig. 2.13: Dependence of permanent axial strains at Peak strength on the strain rate at various confining pressures σ_2 for diabase specimens.

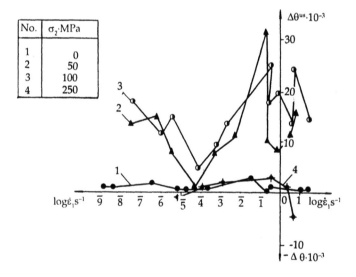

No.	$\sigma_2 \cdot$MPa
1	0
2	50
3	100
4	250

Fig. 2.14: Relationship between dilatant volumetric strains at peak strength and strain rate at different confining pressures σ_2 for diabase specimens.

For marble, axial strains at $\sigma_2 = 50$ MPa amount to 4%, while at $\sigma_2 = 150$ MPa, the axial strains reach 30%. For diabase, the maximum linear strains for strain rates of $\dot{\varepsilon}_1 = 10^{-1}$ to 10 s^{-1} and $\sigma_2 = 50$ and 100 MPa were 0.5% and 1.5% respectively. The maximum dilatation strains over the same range of strain rates, and pressures of $\sigma_2 = 50$ and 100 MPa, are –2.5% and

3% respectively. At σ_2 = 50 and 100 MPa and in the range of creep strain rates $\dot{\varepsilon}_1 = 10^{-7}$ s^{-1} to 10^{-6} s^{-1} the second maximum of the dilatant volumetric strains was observed to be at 1.5% ~ 2% respectively. Under uniaxial compression (σ_2 = 0), axial deformation did not exceed 0.1% over the entire range of strain rates, while volumetric strains showed weak maxima compared to the preceding tests over the same range of strain rates. At high strain rates, volumetric strains amounted to 0.5% and 0.3% at creep rates.

Figure 2.15 shows the relationships between the peak strength and elastic limits versus strain rate and confining pressure σ_2 obtained from tests on highly porous (20–30%) specimens of quartzitic sandstone. As in the preceding cases, each point on the graph was obtained as the result of averaging 6–12 independent results by testing similar specimens. It can be seen that the experimental points are located quite well on the ray emanating from an origin at coordinate $\log \dot{\varepsilon}_1 = \overline{19.5}$ s^{-1}.

As distinct from marble and diabase, in which the initial porosity did not exceed 0.1–0.15%, new microcracks and voids were initiated simultaneously with the closure of existing pores during the process of permanent deformation in porous sandstone, such that the resultant volumetric strain was the algebraic sum of two counteracting processes. The observed axial strains $\Delta\varepsilon_1^{us}$ at peak strength are shown in Fig. 2.16 (a) and (b). The maximum magnitudes of strain (1%) at σ_2 = 50 MPa were obtained in the zone of high strain rates. At σ_2 = 100 MPa, a maximum plasticity of 4.5% was obtained for creep rates $\dot{\varepsilon}_1 = 10^{-7}$ to 10^{-6} s^{-1}.

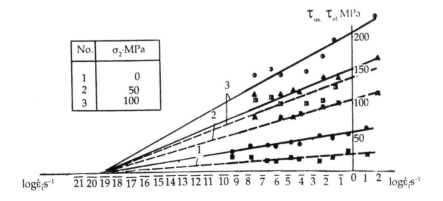

No.	σ_2·MPa
1	0
2	50
3	100

Fig. 2.15: Dependence of the peak strength (continuous lines) and elastic limits (dashed lines) of highly porous quartzitic sandstone on strain rates for different levels of confining pressure σ_2.

Volumetric strains in sandstone are shown in Fig. 2.17. The maximum dilatation of 1% was obtained at $\sigma_2 = 0$ at low strain rates ranging from $\dot{\varepsilon}_1 = 10^{-8}$ to 10^{-7} s^{-1}. At $\sigma_2 = 50$ MPa, dilatation was roughly 0.5%. Dilatant volumetric strain was observed at a pressure of $\sigma_2 = 100$ MPa at high strain rates and compression at low strain rates. Very large permanent linear strain deformation was observed at low strain rates (see Fig. 2.16b). Under high strain rates and all levels of σ_2, dilatation was roughly the same, about 0.5%.

Limestone from 'Leningradslanets' was also tested. The effects of strain rate and confining pressure on the peak strength and elastic limits of this limestone are illustrated in Fig. 2.18 (a) and (b). All experimental points lie

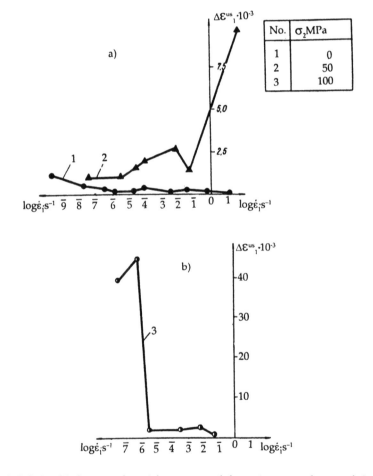

Fig. 2.16: Relationship between the axial permanent deformations at peak strength in highly porous quartzitic sandstone and strain rates for various confining pressures σ_2.

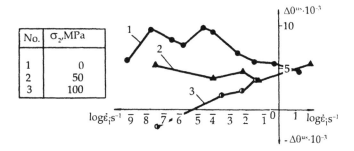

No.	σ_2,MPa
1	0
2	50
3	100

Fig. 2.17: Relationship between volumetric permanent strains at peak strength in highly porous quartzitic sandstone and strain rates for various confining pressures σ_2.

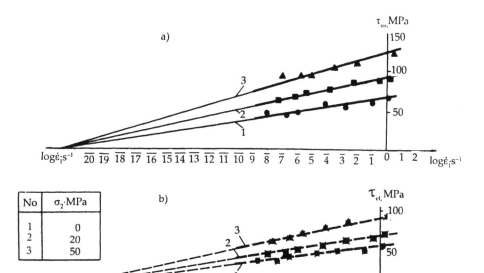

No	σ_2·MPa
1	0
2	20
3	50

Fig. 2.18: Dependence of (a) the peak strength and (b) elastic limit of limestone on the strain rate for tests at various confining pressures σ_2.

on rays emanating from a common origin at the strain rate coordinate $\log \dot{\varepsilon}_1 = \overline{22}$ s^{-1}. Each point on the graph was plotted by averaging data from 6–10 tests on similar specimens. The linear and volumetric strains of the limestone observed at various strain rates and pressures are plotted in Fig. 2.19 (a) and (b) respectively. The relationship for the axial deformation of limestone is seen to be smoother and more monotonic with a pronouned

a)

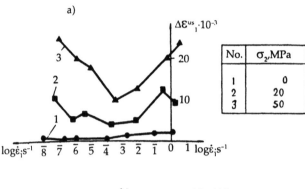

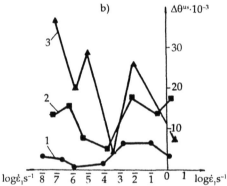

b)

Fig. 2.19: Dependence of (a) axial and (b) volumetric permanent strains at peak strength in limestone on the strain rate for various confining pressures σ_2.

minimum at strain rates around $\log \dot{\varepsilon}_1 = \bar{5}$ to $\bar{3}$ s^{-1}. Volumetric strains at dilatation agree qualitatively with the relationships obtained for the linear deformations, but exhibit a more erratic variation.

Figure 2.20 shows the results of tests conducted to determine the peak strength of talc chlorite under uniaxial compression and under triaxial compression at a confining pressure $\sigma_2 = 250$ MPa. Experimental points are seen to lie quite well on straight-line rays with a common origin at the coordinate $\log \dot{\varepsilon}_1 = \overline{17}$ s^{-1}. Each point on the graph was obtained by averaging data from 6–10 tests on similar specimens.

Specimens of NBP and BP sandstones from the Donbass region gave significantly different results from those discussed above. Figure 2.21 shows the relationships observed between the peak strength of NBP sandstone

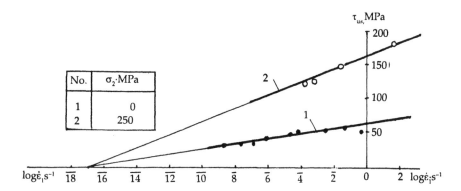

Fig. 2.20: Relationship between peak strength of talc chlorite and strain rate for tests at confining pressure $\sigma_2 = 250$ MPa and for uniaxial compression tests.

and strain rates for various confining pressures σ_2. It is evident from these graphs that the experimental results cannot be approximated by straight-line rays emanating from a common origin and hence cannot be described by the kinetic equation (2.1). Results obtained under uniaxial compression can be explained by 'stretching' the interpretation of the kinetic equation (2.1). In all remaining cases, increase in strain rate results in a non-dependence of strength on the strain rate (at pressures $\sigma_2 = 100$ and 150 MPa) or a reduction in strength (at pressures $\sigma_2 = 100$ and 150 MPa).

Figure 2.22 (a) and (b) shows the linear and volumetric strains observed in NBP sandstone at peak strength at various strain rates and confining pressures. Two maxima are observed in the linear deformations: one in the low strain rate range (log $\dot{\varepsilon}_1 = \bar{7}$ to $\bar{5}$ s^{-1}), the other in the high range (log $\dot{\varepsilon}_1$ = 1 s^{-1}). Dilatant volumetric strains show maxima at log $\dot{\varepsilon}_1 = \bar{6}$ s^{-1}. In the range of high strain rates at log $\dot{\varepsilon}_1 = \bar{1}$ to 1 s^{-1} and under confining pressures $\sigma_2 = 100$ and 150 MPa, a 1% reduction in volume (contraction) of material was observed. Contraction occurred due to the closure of pores in the sandstone. Effective porosity in NBP sandstone was initially about 6–7%; high linear deformation produced pore closure. Volumetric strain contraction in this interval exceeded the dilatant volumetric strain, resulting in an overall volumetric reduction.

The peak strength of BP Donbass sandstone, as a function of strain rate and confining pressure, are shown in Fig. 2.23. The strengths obtained by testing BP sandstone agree almost completely with those obtained for NBP sandstone, both qualitatively and quantitatively. This fact was highlighted

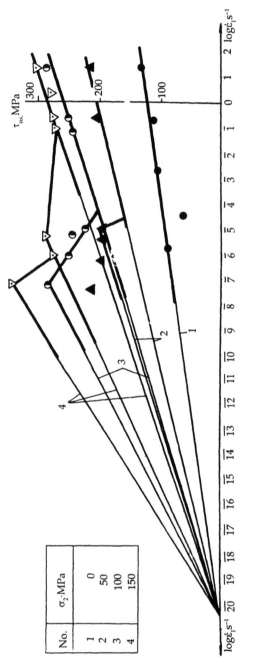

Fig. 2.21: Dependence of the peak strength of NBP Donbass sandstone on strain rates for various confining pressures σ_2.

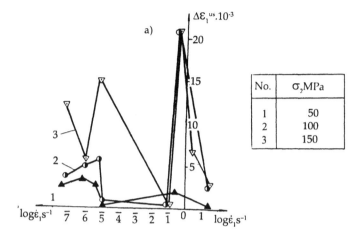

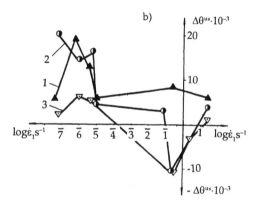

Fig. 2.22: Relationships between (a) axial and (b) volumetric strains at peak strength in NBP Donbass sandstone and strain rate for various confining pressures σ_2.

in Chapter 1 when analysing test results for these two types of sandstones, which are markedly different petrographically (see Appendix II). The (a) axial and (b) volumetric strains observed at peak strength in BP sandstone for various strain rates are illustrated in Fig. 2.24 (a) and (b). As in the case of relationships obtained for NBP sandstone, linear deformations for this rock also show two maxima: one at low strain rates (log $\dot{\varepsilon}_1$ $\bar{7}$ to $\bar{5}$ s^{-1}), the other at high rates (log $\dot{\varepsilon}_1$ = $\bar{1}$ to 1 s^{-1}).

At low strain rates (log $\dot{\varepsilon}_1$ = $\bar{7}$ to $\bar{6}$ s^{-1}), maxima in dilatant volumetric strains were observed. At high rates (log $\dot{\varepsilon}_1$ = $\bar{1}$ to 1 s^{-1}), however, as in the case of NBP sandstone, volume reduction, i.e., contraction was

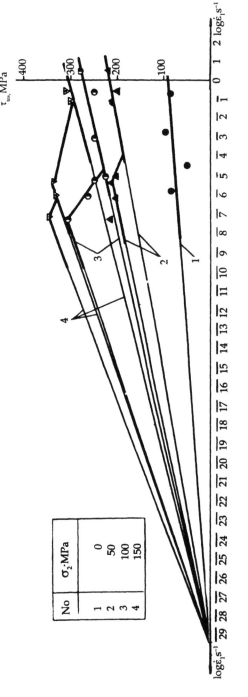

Fig. 2.23: Dependence of peak strength of BP Donbass sandstone on strain rate for various confining pressures σ_2.

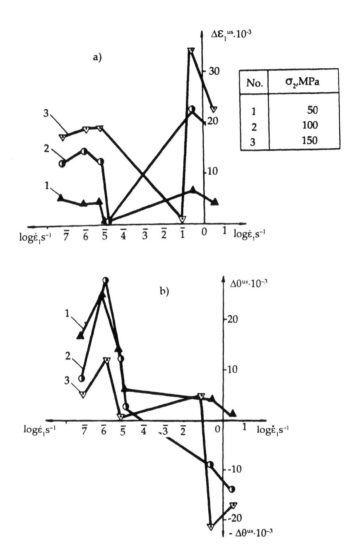

Fig. 2.24: (a) Axial and (b) volumetric permanent strains at peak strength in BP Donbass sandstone as a function of strain rate at various confining pressures σ2.

observed: at σ_2 = 100 MPa the volume was reduced by 1–1.5% and, at σ_2 = 150 MPa, by about 2%. The reason for volumetric contraction in BP sandstone was the same as for NBP sandstone. The total porosity in the BP sandstone was 10–12% and the effective porosity 6–7%. BP sandstone compacted more (twice as much), therefore, due to a higher total porosity compared to NBP sandstone. Thus it may be concluded that the mechanical characteristics of NBP and BP sandstones cannot serve as dependable criteria in defining the 'burst-proneness' of these rocks. Petrographic indices and permeability characteristics are the dependable criteria here. The permeability of these two types of sandstone differs by more than an order of magnitude. Of the two, the BP sandstone exhibits higher permeability, as confirmed by several studies conducted in different regions of Donbass. This topic is discussed in detail in Chapter 3.

2.3.1 Bituminous Coals

Since 1952, scientists from the Institute of Mine Geomechanics and Mine Survey (VNIMI), Leningrad have sought to solve the problem of intermittent coalbursts in the coal basins of the former Soviet Union by studying the mechanical properties of this bituminous coal, for the purpose of predicting and preventing such occurrences.

Specific studies were conducted by the authors into the mechanical properties of coals at varying deformation rates and confining pressures. The results are given here.

Figure 2.25 shows the experimental relationships observed between peak strength, strain rate and magnitude of confining pressure for lignite from the Shurabsk coal basin in Tadzhikistan. Considering the huge scatter in experimental results that is typical of tests on coals in general, the results are considered to be satisfactorily approximated by rays from a common origin at coordinate $\log \dot{\epsilon}_1 = \overline{15} \ \text{s}^{-1}$.

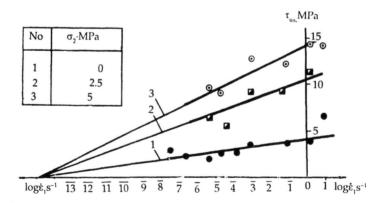

Fig. 2.25: Relationships between the peak strength and strain rate at various confining pressures σ_2 for lignite from the Shurabsk coal basin.

The (a) linear and (b) volumetric strains at peak strength in lignite from Shurabsk, as a function of strain rate and magnitude of confining pressure, are illustrated in Fig. 2.26 (a) and (b) respectively. The linear deformations show a distinct maximum in the range of strain rates $\log \dot{\varepsilon}_1 = \bar{1}$ to $1\ s^{-1}$. Maximum dilatant volumetric strains occur in the interval of rates $\log \dot{\varepsilon}_1 = \bar{4}$ to $\bar{2}\ s^{-1}$. In the range of strain rates $\log \dot{\varepsilon}_1 = \bar{1}\ 10\ s^{-1}$, volumetric strains are minimal, despite the very large linear strains in this range. This observation is explained by closure of pores in lignite whose initial porosity was about 30%. Thus in this range of strain rates and pressures $\sigma_2 = 2.5$ and 5.0 MPa, contraction predominates over dilatation, which also takes place.

Specimens of hard coal (anthracite) from the Donetsk coal basin were also tested. Peak strength of anthracite as a function of strain rate and magnitude of confining pressure, are plotted in Fig. 2.27. Anthracite is stronger than lignite by almost one order of magnitude. The scatter in strength values of anthracite is much greater than in lignite. In all cases, with increase in the strain rate and confining pressure, strength hardening was observed; however, approximation of experimental points by rays proceeding from a common origin was found to be highly unreliable and conditional.

Figure 2.28 (a) shows the relationships of linear deformations at peak strength in anthracite for various strain rates and confining pressures. In the range of strain rates $\log \dot{\varepsilon}_1 = \bar{4}$ to $\bar{3}\ s^{-1}$, a distinct maximum was noticed in the linear deformations. The absolute strain values in anthracite were

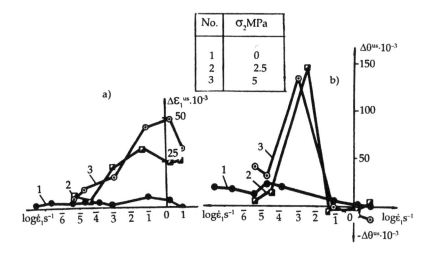

Fig. 2.26: (a) Linear and (b) volumetric permanent strains at peak strength for lignite from the Shurabsk coal basin, as a function of strain rate at various confining pressures σ_2.

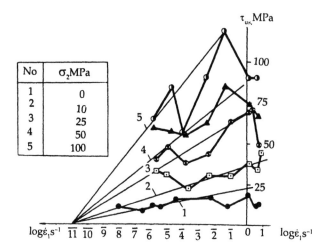

Fig. 2.27: Relationships between peak strength of Donbass anthracite and strain rate at various confining pressures σ_2.

observed to be 5–10 times less than strains in lignite from Shurabsk. Anthracite is distinguished by high brittleness and strength.

Figure 2.28 (b) shows the corresponding relationships for volumetric strains. It can be seen that these strains were predominantly dilatant. The porosity of the anthracite was about 1.5%. High values of volumetric strains were obtained under uniaxial compression and in compression under a confining pressure of 100 MPa at low strain rates (log $\dot{\varepsilon}_1 = \bar{6}$ to $\bar{5}$ s^{-1}). At high rates (log $\dot{\varepsilon}_1 = 10$ s^{-1}), intense dilatation was observed under uniaxial compression and for compression at a confining pressure of 10 MPa.

Figure 2.29 shows the results of tests concerning uniaxial compressive strength at various strain rates, conducted on specimens of bituminous coals from the Kizelovsk coal basin, situated in the Urals of the former Soviet Union, where coalbursts of high intensity occurred for the first time.

Bituminous coal from Kizelovsk is quite strong and very brittle. In all cases, with rise in strain rate, a reduction in strength was observed. A maximum value of strength (15 MPa) was obtained under low strain rates (log $\dot{\varepsilon}_1 = \bar{8}$ s^{-1}); the minimum strength (3 MPa) was recorded at high deformation rates (log $\dot{\varepsilon}_1 = 0.5$ s^{-1}). The strength results obtained are thus contrary to the kinetic concept of failure in solids. The experimental curve cannot be described directly by a kinetic equation of the type (2.1). Rays drawn from a common origin are purely speculative.

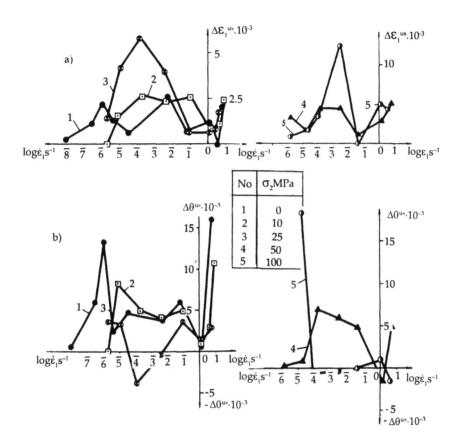

Fig. 2.28: Relationships between (a) axial and (b) volumetric strains at peak strength in anthracite and strain rate for various confining pressures σ_2.

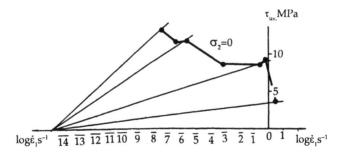

Fig. 2.29: Uniaxial compressive strength of bituminous coal from the Kizelovsk basin at varying strain rates.

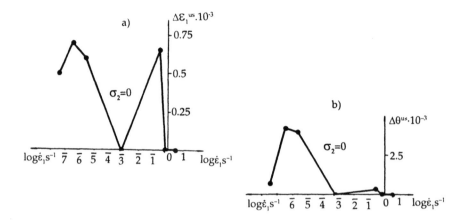

Fig. 2.30: Relationship between (a) axial and (b) volumetric permanent strains at peak strength in specimens of bituminous coal from Kizelovsk basin and strain rate for uniaxial compression.

Figure 2.30 shows the relationships between (a) linear and (b) volumetric strains at peak strength in specimens of bituminous coal from Kizelovsk basin versus strain rate.

Linear deformations exhibit two maxima: one under low deformation rates, the other under high rates. Volumetric strains show only one maximum at low strain rates.

2.3.2 Salt

Experiments were conducted on specimens of rock salt (halite) [75] from the Starobinsk deposit in Byelorussia and other deposits in Tadzhikistan as well as specimens of sylvinite from the Starobinsk deposit. Tests were conducted to investigate the causes of rockbursts and outbursts occurring in these deposits.

Figure 2.31 shows the results of tests conducted on specimens of rock salt from the Starobinsk deposit under uniaxial compression at various strain rates. Prismatic specimens of $150 \times 150 \times 300$ mm were tested. Each experimental point was obtained by averaging data from 3–4 independent tests on similar specimens. Values of ultimate strength are indicated by black circles and values of elastic limit by crossed circles.

Strength hardening was observed with increase in strain rate over a range of low strain rates (log $\dot{\varepsilon}_1 = \bar{8} - \bar{6}$ s^{-1}); these results accord with the kinetic concept. In the high velocity range (log $\dot{\varepsilon}_1 = \bar{6}$ s^{-1} to log $\dot{\varepsilon}_1 = 2$ s^{-1})

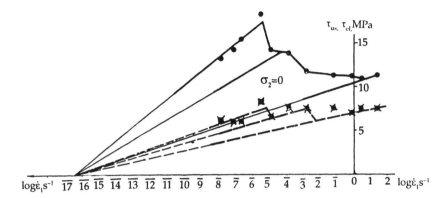

Fig. 2.31: Dependence of peak strength (continuous line) and elastic limits (dashed line) on strain rate under uniaxial compression for specimens of rock salt from the Starobinsk deposit.

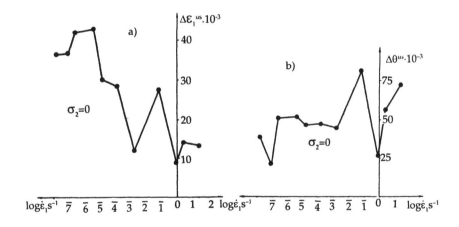

Fig. 2.32: Relationships between (a) axial and (b) volumetric permanent strains and strain rate under uniaxial compression for rock salt specimens from the Starobinsk deposit.

the strength of the rock salt decreased and the elastic limit dropped quite significantly.

Linear and volumetric strains of rock salt at peak strength are shown in Fig. 2.32 (a) and (b) respectively. Linear strains decreased with increase in strain rate. Maximum values of strain were recorded over a range of low strain rates (log $\dot{\varepsilon}_1 = \bar{8} - \bar{6}$ s^{-1}). On the contrary, volumetric strains were

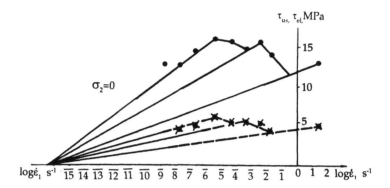

Fig. 2.33: Relationships between peak strength (continuous lines) and elastic limits (dashed line) and strain rate under uniaxial compression for specimens of rock salt from Tadzhikistan.

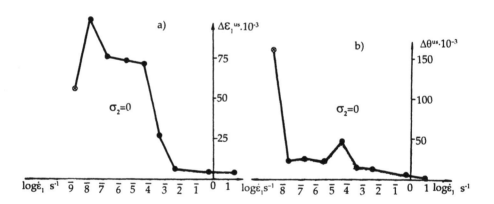

Fig. 2.34: Dependence of (a) axial and (b) volumetric permanent strains at peak strength on strain rate under uniaxial compression.

highly dilatant at high strain rates. As can be seen from the graphs, the relationships obtained for strains are unstable.

Uniaxial compression tests, similar to those just described, were conducted on rock salt specimens from Tadzhikistan. Core specimens 90 mm in diameter and 200 mm in height were tested.

Test results of strength and elasticity of this rock salt under various strain rates are shown in Fig. 2.33. Complete agreement, both qualitative and quantitative, was observed with the results obtained on rock salt from the Starobinsk deposit. Strain relationships as a function of strain rate (for these specimens) differed to some extent, as depicted in Fig. 2.34 (a) and (b). Plasticity and dilatation maxima are clearly seen at low strain rates.

The next set of uniaxial compression tests was conducted on specimens of sylvinite from the Starobinsk deposit. Prismatic specimens 150 × 150 × 300 mm in size and cores of 90 mm diameter and 200 mm height were tested. Tests were conducted at strain rates extending over a range of 10^{12} (from $\dot{\varepsilon}_1 = 10^{-10}$ s^{-1} to $\dot{\varepsilon}_1 = 10^{+2}$ s^{-1}). The duration of experiments at the very low strain rates, which coincide with creep rates, extended to five years and more. The sylvinite consisted of 68–70% sylvine and 30–32% rock salt.

Figure 2.35 shows the results of tests conducted to determine the peak strength and elastic limits of sylvinite under uniaxial compression. Each point on the graph was obtained by averaging data from 4–5 tests conducted on similar specimens.

Experimental points corresponding to peak strength values are indicated by black circles; those corresponding to elastic limits are shown by crossed circles. These data were obtained by testing core specimens at a particular strain rates. Points obtained by testing prismatic specimens of sylvinite under conditions of creep, are indicated by black triangles. Points obtained under conditions of creep by testing specimens of sylvinite from the Verkhnekamsk potash deposit in the Urals, are shown by open triangles. In the creep tests, specimens were tested to failure; hence points on the graph correspond to peak strength under creep conditions. Sylvinites from

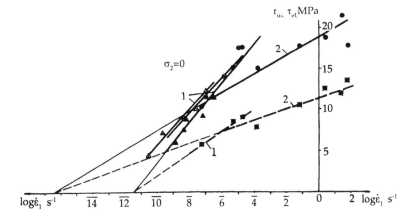

Fig. 2.35: Limits of peak strength and elasticity versus strain rate for specimens of sylvinite from Starobinsk and Verkhnekamsk (the Urals) deposits.

Starobinsk and Verkhnekamsk are very close in composition and texture; hence their strength indices are very similar.

The experimental points in Fig. 2.35 were divided into two groups: the first group was obtained under conditions of a specific strain rates approximated by rays (2) emanating from a common origin at coordinate $\log \dot{\varepsilon}_1 = \overline{17}$ s^{-1}; the second group was obtained under conditions of creep, approximated by rays (1) emanating from a common origin at coordinate $\log \dot{\varepsilon}_1 = \overline{12}$ s^{-1}. Experimental points for sylvinite from the Verkhnekamsk deposit deviate to some extent from those obtained for sylvinite from the Starobinsk deposit.

Results in the range of strain rates in which the assigned rates were equal to creep rates could not be differentiated because the constant load on the spring-loaded presses was regularly corrected in the creep zone. This experiment is essentially equivalent to one with a pre-assigned very low strain rate, varied by small steps.

The presence of two systems of rays emanating from different origins provides a basis for assuming that the process of deformation in sylvinite occurs mainly in silvine grains, in the range of low rates since very low stresses are required for this. At rates starting from $\dot{\varepsilon}_1 = 10^{-6}$ s^{-1} to $\dot{\varepsilon}_1 = 10^{+2}$ s^{-1}, rock salt grains were encompassed by the deformation process, since in this range still lower stresses were required for the occurrence of the deformation process; in this case, the rays of sylvine pass above those for rock salt. A detailed analysis of this aspect is presented later.

Figure 2.36 shows the linear and volumetric strains in sylvinite at peak strength, depending on strain rate. Distinctly expressed maxima were obtained at low strain rates. At high strain rates, sylvinite exhibited increased brittleness. When rockbursts occur in sylvinite deposits, rocks fail under a dynamic regime and the increased brittleness of sylvinite facilitates a more

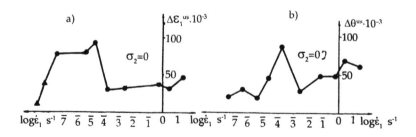

Fig. 2.36: Dependence of (a) axial and (b) volumetric permanent strains at the peak strength level on strain rate under uniaxial compression in specimens of sylvinite from the Starobinsk deposit.

violent and fierce process of failure. Increase in brittleness and reduction in strength with rise in deformation rate are typical of rock salt, bituminous coal and anthracite. All these rocks are prone to manifestation of rockbursts of high intensity. Reduction of strength in the band of high strain rates can be explained to some extent by the specific texture of the rocks.

2.3.3 Effect of Structural Factor on Strength Reduction in Rock Salt

A special series of experiments was designed to understand the role of the structural factor in rock salt. As already mentioned, pure rock salt, with very few impurities, exhibits considerable reduction in strength at high deformation rates.

Rock salt from the Starobinsk deposit was tested. Petrographic investigations were conducted first and the percentage composition of structural elements (crystals) versus their linear dimension determined, as shown in Fig. 2.37. An extraordinarily wide range of variation in the dimensions of constituent grains in rock salt, from a fraction of a millimetre to 15 mm or more, is a special structural feature of natural rock salt.

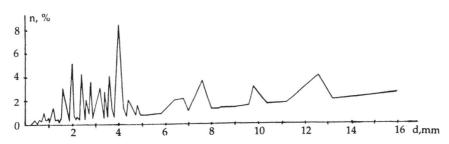

Fig. 2.37: Distribution function of number of grains (in per cent) in natural rock salt from the Starobinsk deposit versus their linear dimension.

Brief Petrographic Description of Natural Rock Salt from the Starobinsk Deposit

This salt consists of crystalline grains, coarse grains and very coarse grains. Rock salt grains are cuboid, rarely parallelepiped or irregular in shape (from 0.24 to 16 mm with fragments of 1 to 12 mm predominant, about 84%).

In the interstices between the rock salt grains and sometimes at accumulations along the grains and grain edges, a finely dispersed clayey material of macrogranular structure and microaggregate polarisation is encountered. One can distinguish micrograined crystals (0.01–0.05 mm) of carbonate and anhydride (0.1–0.15 to 0.4 mm) of prismatic and entwined fibrous structure in the clayey material. Bulk weight is 2.08–2.09 g/cm^3. Exposed (open) total porosity is 0.78–1.06%. Similar petrographic studies were conducted on all the rocks tested. The results of these studies are given in Appendix II.

The essential features of these special experiments were as follows: natural rock salt was mechanically pounded and later sorted through a set of sieves; artificial specimens were prepared by compacting the pounded fractions under high pressure. These specimens were later subjected to tests under widely varying rates of deformation.

Three fractions of rock salt were obtained: particle size $d \leq 0.24$ mm, $d = 0.34$–0.7 mm and $d = 1.1$–2.1 mm. Specimens of cylindrical shape were prepared from these fractions in a special mould under 200–300 MPa pressure. Later, the process of compacting was continued in a hydrostatic cell under a confined pressure of 700 MPa. Before placement in the hydrostatic cell, the specimens were encased in a protective sheath to preclude fluid entry. Such a method allowed preparation of specimens of 30 mm diameter and 80 mm length, with uniform mechanical properties. Under such high pressures, intermolecular cohesion occurred between salt particles, while porosity practically vanished; specimens attained a high strength value. The specimens were annealed in a thermostat to remove residual stresses induced by compaction.

Figure 2.38 (a) shows the dependence of the uniaxial compressive strength of artificial rock salt specimens on the strain rate [81, 83]. Rays 1, 2 and 3 were obtained from specimens prepared with fractions $d \leq 0.24$ mm, $d = 0.34$–0.7 mm and $d = 1.1$–2.1 mm respectively. An increase in strength with increase in strain rate was observed in all specimens throughout the entire range of rates investigated. Experimental points were observed to lie fairly well on rays emanating from a common origin at coordinate $\log \dot{\varepsilon}_1 = \overline{16}$ s^{-1}, which almost coincides with the coordinate of the origin obtained for specimens of natural rock salt. The strength of specimens prepared from the fine fractions was observed to be higher than that of specimens prepared from coarse fractions.

A fourth line was plotted through these experimental points obtained by testing specimens prepared from a mix of 33% by weight of each of the three fractions. These results did not conform to the features obtained from testing specimens prepared from individual fractions. At a strain rate of $\dot{\varepsilon}_1 = 10^{-6}$ s^{-1}, the strength of mixed specimens was observed to be nearer that of specimens prepared from very fine rractions; at a rate $\dot{\varepsilon}_1 = 10^{-4}$ s^{-1}, the strength of mixed specimens was comparable to that of specimens prepared from the fraction $d = 0.34$–1.7 mm. At the strain rate $\dot{\varepsilon}_1 = 0.5$ s^{-1}, the strength of mixed specimens was nearer that of specimens made from the very coarse fraction. Thus the deformation rate appeared a factor in the

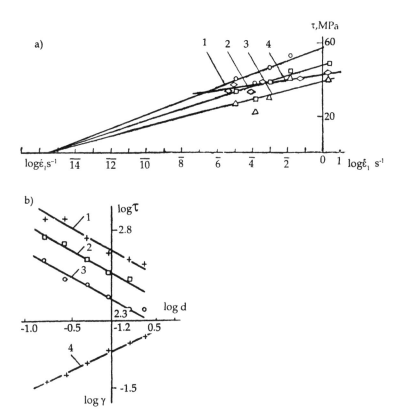

Fig. 2.38: Relationship between strength of artificial specimens of rock salt and (a) strain rate and (b) grain size.

selection of structural elements based on their strength. The difference in strength between specimens prepared from fractions with different particle dimensions is explained by the scale effect. The scale effect was first studied experimentally by Russian scientists A. P. Alexandrov and S. N. Zhurkov, whose studies were published in 1933. A higher strength, on average, of thin glass threads compared to that of thicker threads, was established. The dispersion of certain strength values during failure of thin threads was observed to be significantly higher than the dispersion of strength for failure of thicker threads. These results served as the basis for invoking statistical methods to discuss the nature of the strength of solids.

Relationships of strength versus linear dimension of fragments for artificial specimens of rock salt, plotted in logarithmic coordinates, are shown in Fig. 2.38 (b). Line 1 was obtained at strain rate $\dot{\varepsilon}_1 = 0.5$ s^{-1}, line 2 at $\dot{\varepsilon}_1 = 10^{-3}$ s^{-1} and line 3 at $\dot{\varepsilon}_1 = 10^{-5}$ s^{-1}. Line 4 represents the relationship of structurally sensitive coefficient γ in equation (2.1) with dimension of fragments. The relationships thus obtained are described quite well by equations of the type:

$$\tau_{us} = Kd^b; \qquad (2.5)$$

$$\gamma_{us} = K_1 d^{-b}, \qquad (2.6)$$

where K, K_1, b, $-b$ are constants depending on rock properties. Constants b and $-b$ are equal in magnitude but have opposite signs, $b = -b$.

In this manner, the proposition about the participation of very coarse grains with low shear strength in the deformation process at high strain rates is proven. The physical meaning of this feature is considered below by using the statistical mechanical model.

Equation (2.5) coincides with the equation proposed by Weibull in his statistical theory of the scale effect, the fundamental aspects of which were published in 1939:

$$\tau_{us} = LV^{1/m}, \qquad (2.7)$$

where L is a constant dependent on the property of material and type of stress state; m is a constant taking into account defects distributed in the material; V the specimen volume.

In order to consider the dimensional effect of structural fragments, it is necessary to replace specimen volume V in formula (2.7) by the linear dimension of fragments, which gives us equation (2.5). As can be seen, formula (2.7) is identical to formula (2.5).

Thus, loss of strength with increase in the deformation rate, is induced solely by structural features of natural rock salt, consisting of crystals whose dimensions vary from fractions of a millimetre to 15 mm and more, as is evident from Fig. 2.37. During tests on artificially prepared specimens, no drop in strength was observed because preparation of specimens by the above-described method of compaction gave a more uniform structure and excluded the use of fractions larger than 2 mm, since coarser fragments disintegrated under pressure.

The highly transparent result mentioned above, concerning deviation from the general pattern of strength of marble from the Urals at a pressure $\sigma_2 = 150$ MPa (Fig. 2.10), where a small drop in strength with increase in deformation rate was observed, is explained by structural features. As with rock salt, marble is a monomineral rock with no alien impurities. This feature justifies the assumption that even in the case of marble, the structural factor, similar to that established for rock salt, is valid.

Deviations from the laws described by the kinetic equation were also observed in BP and NBP sandstones from Donbass, bituminous coal from Kizelovsk basin and in other rocks. As the results of the latter are not all that large, they are not discussed here. These deviations can probably also be explained by structural features of the materials, which by and large are heterogeneous, consisting of structural elements with different thermodynamic parameters. A study of such structures would require a vast programme of experimental investigations under a wide variation of thermodynamic conditions; hence it has not been undertaken to date.

Among the tests conducted by us on heterogeneous rocks, the results obtained for sylvinite from the Starobinsk deposit were more comprehensive. As already mentioned, the tested sylvinite consisted of 30–32% rock salt (NaCl) and 68–70% sylvine (KCl), i.e., two substances with different thermodynamic parameters, such as activation energy U_0—indicating energy of intermolecular bonds in a crystalline grid of rock salt and sylvine. Investigations into the diffusion processes in these two substances [63] made it possible to obtain values of activation energies Q of diffusion processes in atoms of Na and K in crystalline grids of rock salt and sylvine.

2.4 KINETIC THEORY OF DEFORMATION AND STRENGTH AND JUSTIFICATION FOR ITS POSSIBLE APPLICATION IN ROCK MECHANICS

The kinetic theory of gases developed by L. Boltzman and subsequently extended to fluids encompasses fundamental postulates in classical physics. The theory describes and predicts physical and physicochemical processes occurring in these media with a high degree of reliability.

The kinetic concept, covering interaction, separation and mutual displacement of elementary particles in gases and liquids, differs markedly when applied to solids, mainly because of uncertainty in the number of elementary particles involved in the deformation process and the percentage of each in relation to the total particles contained in the body under consideration. During permanent deformation of a solid, most of the elementary particles play no active role in the process of deformation and failure. Relative displacements of particles are known to differ, ranging from complete rupture of bonds to simple substitution and induction movements within the crystalline grid, double rate diffusion etc. Each movement is accompanied by removal of a potential barrier typical for the given case, representing the activation energy of the given process. These barriers can notably differ quantitatively, which leads to an additional indeterminate state concerning the concept of gram-moles and quantities of U_0, γ, $\dot{\varepsilon}_0$, t_0 in the kinetic equations (2.1) and (2.2).

A kinetic equation can describe only those cases in $\tau - \log \dot{\varepsilon}_1$ coordinates with solely formal signs in which the magnitude of stress increases with increase in strain rate. It is apparent from the experimental material presented above that such cases are encountered quite often; however, it is not possible to predict their occurrence without preliminary experimental studies. The experimental results presented above have also revealed cases wherein, due to the need for corrections and additions, a kinetic equation is not suitable. Such cases are also widely encountered and their prediction is also not possible without preliminary experimental investigations.

The statistical theory and model of a heterogeneous medium discussed in Chapter 1, allow us to distinguish those microvolumes within a solid whereby its distortion and rupture occur relatively more evenly, and thus the number of elementary particles participating in the process can be determined. This strengthens the application of a kinetic approach to the problem of deformation and rupture of a solid such as rock. We shall consider in some detail the terms contained in the kinetic equations (2.1) and (2.2), only summarily defined earlier.

U_0 represents the activation energy, characterising the bond energy between elementary particles in a solid. It is not a universal constant however, since the mechanics involved in separation of particles in a solid differ, requiring a different quantity of U_0 to activate them. Thus each solid has a particular value of energy U_0, which is a thermodynamic characteristic unique to the given body.

Parameter γ is a structurally sensitive coefficient, interpreted in kinetic theory as the product of an activated volume and a coefficient of localised overstress. Its magnitude depends on body structure, structural composition, dimensions of structural elements, thermal treatment the body has undergone, annealing (for example, of metals) etc. With increase in grain size, the coefficient γ increases, as demonstrated in experiments with artificial specimens of rock salt. Although the coefficient γ plays a very important role in the kinetic equation, it still remains a highly enigmatic characteristic of a solid body.

By *activated volume*, we mean that part of the volume of a crystalline grid 10^{-23} cm^3 in size in which reconstruction and failure take place. This process is induced by applied external stress τ, augmented by the coefficient of localised overstress.

The quantity $(U_0 - \gamma\tau)$ is termed the *effective activation energy* and is denoted by U_{eff}. The applied stress τ decreases U_{eff} and thereby reduces the potential barrier preventing an elementary particle from leaving its state of equilibrium. The lower the value of U_{eff}, the lower the waiting time required by an elementary particle or atom, at a given temperature, to exit from the *potential well*.

In equation (2.2) a quantity t_0, equal to the period of natural oscillations of an atom around the position of equilibrium, namely 10^{-12} to 10^{-13}, is termed the frequency multiplier. In equation (2.1) for strain rates, the quantity $\dot{\varepsilon}_0$ s^{-1} is represented as the ultimate possible strain rate strain in a solid and also serves as a frequency multiplier. This is discussed in detail later. For now, let us return to the historical developments in this field.

Prior to the research activities taken up by S. N. Zhurkov and his school, a group of scientists under the guidance of N. N. Davidenkov [64] had started studying plastic deformations in metals at different deformation rates, including the aspect of propagation of plasticity waves [54]. The kinetic concept of plastic deformation in metals was developed during these studies. The equation relating relaxation time (t) to external stress τ and temperature took the form:

$$t = t_0 \, \exp \frac{U_0 - U_1 \, \tan h \, (\gamma \, \sigma)}{KT},\tag{2.8}$$

where U_0 is activation energy, as defined in equations (2.1) and (2.2); U_1 is a constant indicating the collective action of diffusion; t_0 is the period of natural oscillations of an atom around the equilibrium position; γ is a structurally sensitive coefficient, conveying the same meaning as in equations (2.1) and (2.2); K is Boltzman's constant; T is the absolute temperature; $KT = 2463.7$ J/mole at 20°C.

As can be seen, equation (2.8) differs from the equation derived for longevity (2.2) by inclusion of the term $\tan h$ (hyperbolic tangent) of the product ($\gamma \, \sigma$). In a physical sense, the two equations are very close. In both cases, U_0, the activation energy, is present. The difference between U_0 and work done by external stresses, termed *effective activation energy*, U_{eff}, is overcome by an atom at the expense of energy of thermal fluctuations, which in one case leads to failure of the body, and in the other to formation of residual deformation in the body. Tan h is introduced to reflect more precisely the shape of the potential curve describing the bond between two atoms. Let us consider certain experimental studies conducted at the level of dislocations, to explain the physical meaning of the frequency multiplier $\dot{\varepsilon}_0$.

Studies by Joystone and Gilmon [19] on the mobility of dislocations through direct methods of experimentation with crystals of lithium fluoride (LiF) yielded interesting results. They established a relationship between the speed of movement of dislocations and magnitude of applied stress, and obtained a complex curve consisting of two sections: section of weak influence in the zone of low speeds and section of strong influence at high speeds.

The graph depicted in [19] is qualitatively similar to the relationships between strength and strain rate for marble (Fig. 2.10) and for diabase

(Fig. 2.12) discussed earlier in this Chapter. A weak relationship between strength and strain rate in the range of low strain rates and a strong relationship in the range of high rates were observed in the case of these rocks.

Investigating the relationship between speed of dislocations V_s and stress σ, Joystone and Gilmon (op. cit.) obtained the following empirical relation:

$$V_s = \sigma^w \exp(-U_0/KT), \tag{2.9}$$

where w and U_0 are constants of σ.

Equation (2.9) can be replaced by another equation, similar in meaning and used in our studies:

$$V = V_0 \exp(-U_{eff}/KT) \tag{2.10}$$

Here V_0 is the highest speed of propagation of dislocations.

Bessonov [4] gave an expression for the rate of relative deformation $\dot{\varepsilon}$

$$\dot{\varepsilon} = bmV_s, \tag{2.11}$$

where b is the Burger vector; m is the density of dislocations; V_s is the speed of movement of dislocations.

Substituting (2.10) in (2.11), we obtain

$$\dot{\varepsilon} = bmV_0 \exp(-U_{eff}/KT). \tag{2.12}$$

At $U_{eff} = 0$, we have

$$\dot{\varepsilon} = bmV_0. \tag{2.13}$$

The value of $\dot{\varepsilon}_0$ (from equation (2.1)) can be computed by equation (2.13). Physically, $\dot{\varepsilon}_0$ implies the maximum possible rate at which the process of deformation and rupture takes place. The maximum possible speed of dislocations V_0 is nearer to the propagation speed of a transverse wave and is of the order of 10^5 cm/s. Since $\dot{\varepsilon}_0$ reflects the process of rupture in a solid, the density of dislocations should attain a critical value. The value of $m = 10^9$ can be taken as such a critical value, as was done in [19]. Substituting this value in (2.13) and putting the value of Burger's vector $b = 10^{-8}$, we have

$$\dot{\varepsilon}_0 = 10^{-8} \cdot 10^9 \cdot 10^5 = 10^6 \text{ s}^{-1}. \tag{2.14}$$

The value of $\dot{\varepsilon}_0$ can be assessed by other means if the empirical relationship suggested in [4] is used:

$$t\,\dot{\varepsilon}_0^{\,k} = \varepsilon_0 \tag{2.15}$$

The quantity of ε'_0 taken as a constant signifies strain and is of the order 10^{-3}; t is the period of natural oscillations in an atom amounting to 10^{-12} s^{-1}; exponent k is nearer to unity (see Table 2.1).

Substituting the above values in (2.15), we obtain

$$\dot{\varepsilon}_0 = 10^{-3} \cdot 10^{12} = 10^9 \text{ s}^{-1}$$

i.e., we obtain a quantity higher than that in the first evaluation. Such a difference in evaluation is explained by the non-availability of precise data about the ultimate (critical) density of dislocations m in the first evaluation. Depending on various factors [19], m can range from 10^9 to 10^{12}. Substituting the last value (10^{12}) would have made the two results closer. On the other hand, the value of $t = 10^{-12}$ s^{-1} used in the second evaluation is also quite random. To study this aspect, more thorough fundamental investigations are required in future. It is advisable to make calculations for the entire range of possible values $\dot{\varepsilon}_0$ from 10^6 to 10^9 s^{-1} in case such data are not available. This will enable assessment of the corresponding variation in the quantity U_0.

Values of U_0 were calculated according to equation (2.1) for the condition $\tau = 0$. Here it did not matter which of the experimental rays related to various values of pressure σ_2 was taken in calculations, since equation (2.1) for this computation was of the form:

$$\dot{\varepsilon}_1 = \dot{\varepsilon}_0 \exp (-U_0/KT), \tag{2.16}$$

where $\dot{\varepsilon}_1$ corresponds to the coordinate of the origin from which the rays emanate in the graphs.

Calculated values of U_0 obtained using equation (2.16) for a series of rocks are given in Table 2.1. It can be seen from the Table that U_0 did not vary by more than 10% for a change of three orders of magnitude in the value of $\dot{\varepsilon}_0$. This is due to the structure of the kinetic equation. Values of the activation energy for diffusion processes Q of molecules and atoms in crystalline grids of marble, rock salt and sylvine are also listed in the Table. The activation energy for diffusion of $CaCO_3$ molecules in the calcite grid of marble obtained by the method of tagged atoms was taken from [31]. The activation energy for diffusion of Na atoms in the NaCl grid for rock salt and diffusion of K atoms in the KCl grid for sylvine were taken from [63].

As can be seen from Table 2.1, the values of activation energy U_0 obtained through mechanical testing and the values of activation energy for diffusion Q obtained by physical methods, are quite comparable. The closeness of the values of U_0 and Q is encouraging and suggests that the processes of deformation and rupture of the rocks tested are controlled by kinetic phenomena at the microlevel. Shear deformations and failure due to shear

Table 2.1

(1)	(2)	(3)	(4)	(5)	(6)	(7)	(8)	(9)	(10)	(11)
$\dot{\varepsilon}_0\ s^{-1}$	k	$U_0 \cdot 10^3$, J/mole	Constant		γ_{us}, $\frac{J/cm^2}{mole \cdot kg}$	$\dot{\varepsilon}_{A=0}\ s^{-1}$	$U_{0p} \cdot 10^3$, J/mole	γ_p, $\frac{J/cm^2}{mole \cdot kg}$	$Q \cdot 10^3$ J/mole	γ_{el}, $\frac{J/cm^2}{mole \cdot kg}$
			ρ_{us}	ρ_{el}						
					MARBLE					
10^6	1.1	227.1	19	22.7	377.1	3.2×10^{-42}	389.7	11,522	250	—
10^7	1.1	227.1	19	22.7	377.1	3.2×10^{-42}	397.2	11,522	250	528
10^8	1.1	232.5	19	22.7	377.1	3.2×10^{-42}	402.5	11,522	250	—
10^9	1.1	238.8	19	22.7	377.1	3.2×10^{-42}	408.5	11,522	250	—
					DIABASE					
10^6	1	204	8.4	12.9	105.6	4×10^{-24}	252.6	3163	—	130
10^7	1	209.5	8.4	12.9	105.6	4×10^{-24}	257.7	3163	—	130
10^8	1	215.8	8.4	12.9	105.6	4×10^{-24}	263.1	3163	—	130
10^9	1	221.2	8.4	12.9	105.6	4×10^{-24}	269.4	3163	—	130
					ROCK SALT I					
10^6	1.2	127.8	—	—	387.6	—	—	—	150.8	1022
10^7	1.2	133.2	—	—	387.6	—	—	—	150.8	1022
10^8	1.2	139.1	—	—	387.6	—	—	—	150.8	1022
10^9	1.2	145	—	—	387.6	—	—	—	150.8	1022
					ROCK SALT II					
10^6	1	127.8	—	—	358.2	—	—	—	—	846
10^7	1	133.2	—	—	358.2	—	—	—	—	846
10^8	1	139.1	—	—	358.2	—	—	—	—	846
10^9	1	145	—	—	358.2	—	—	—	—	846
					SYLVINITE					
10^6	1.1	99.7	—	—	240.9	—	—	—	125.7	427
10^7	1.1	105.1	—	—	240.9	—	—	—	125.7	427
10^8	1.1	111.0	—	—	240.9	—	—	—	125.7	427
10^9	1.1	116.5	—	—	240.9	—	—	—	125.7	427
					QUARTZITIC SANDSTONE					
10^6	1.1	144.5	6	9.1	192.7	6.2×10^{-24}	310	10,014	—	248
10^7	1.1	149.6	6	9.1	192.7	6.3×10^{-24}	318.4	10,014	—	248
10^8	1.1	155	6	9.1	192.7	6.3×10^{-24}	324.7	10,014	—	248
10^9	1.1	161	6	9.1	192.7	6.3×10^{-24}	329.3	10,014	—	248
					SANDSTONE (BP)					
10^6	0.8	105.6	—	—	—	—	—	—	—	—
10^7	0.8	155	—	—	—	—	—	—	—	—
10^8	0.8	158	—	—	—	—	—	—	—	—
10^9	0.8	166.8	—	—	—	—	—	—	—	—

are controlled by phenomena closer to the diffusion processes in crystals of rocks.

Consider some results obtained by the authors concerning the effect of deformation rate on tensile strength of rocks from the concept of the dual nature of the strength of solids. Four rocks were tested: marble, diabase, quartzitic sandstone (highly porous) and limestone. These rocks are described in detail in Appendix II. They were also subjected to confining pressure at various rates of strain. The results of these tests are given in Table 2.1.

Cylindrical specimens, prepared as described earlier (Fig. 1.17a), were subjected to uniaxial tension. The rupture strengths (under uniaxial tension) obtained at different strain rates for the four rock types are shown in Fig. 2.39. Each point on the graph was obtained by averaging data from 3–4 tests on similar specimens. The results were distributed within a narrow range which allowed them to be represented reliably by straight-line relationships extending to the strain rate axis. The values of activation energy at rupture U_{0p} and the corresponding coefficients γ_p were computed on the basis of these results. These values are also given in Table 2.1.

Comparison of the values of activation energies U_0 and U_{0p} revealed that the value at rupture U_{0p} was twice as high as the activation energy U_0. The reason for this difference may be found in the difference between the processes of deformation and failure due to shear and due to tensile rupture. Tensile rupture processes are associated with complete snapping of bonds between elementary particles. Diffusion phenomena play only a

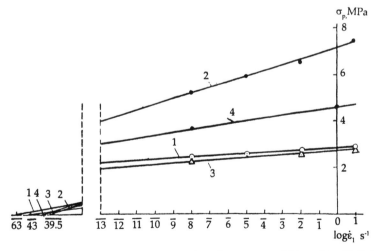

Fig. 2.39: Relationship between rupture strength and strain rate of (1) marble, (2) diabase, (3) highly porous quartzitic sandstone and (4) limestone.

secondary role in this rupture process. The nature of failure of elementary bonds due to tensile rupture requires a very high energy barrier to be overcome. The barrier is higher than that encountered during diffusion and under plastic deformation. With tensile rupture, the value of U_{0p} should approach the value of the sublimation energy or the energy for complete destruction of a crystalline grid. This is always twice the activation energy for diffusion processes.

Torsion tests were conducted on metals and polymers prior to failure [56]. It was found that the activation energy under torsion is approximately half that for rupture, which agrees with the results of tests on rocks.

2.5 SYNTHESIS OF STATISTICAL AND KINETIC THEORIES OF ROCK STRENGTH

The statistical model of a heterogeneous medium, discussed in Chapter 1, is based on the fact that the properties of individual constituent elements of the medium differ only in the elastic limit and peak strength. A medium of this type was designated by the authors as a *unicomponent medium*, typical of uniform rocks.

It follows from this model that the magnitude of permanent deformation at peak strength is directly proportional to the number n_1 of macroscopic shear planes ω involved in this process and the magnitude of shear along

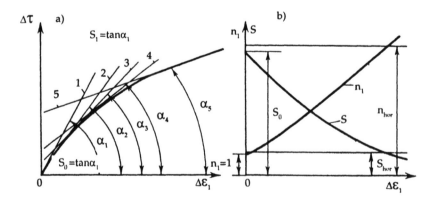

Fig. 2.40: (a) General curve of residual strain and (b) relationship between plasticity modulus and number of planes ω with principal linear permanent deformation.

each of the planes. The first ω plane is formed at the elastic limit level. Strain hardening along this plane results in the selection of new planes ω, the total number of which reach a maximum at the peak strength for a given confining pressure σ_2.

This process of selection of planes ω occurs at any deformation rate but with a change in the latter, the strain hardening index also varies. To be specific, given the kinetic nature of permanent deformation, an increase in deformation rate will cause an increase in the plasticity modulus, i.e. strain hardening intensifies. This, in turn, leads to a further increase in the stress applied to the specimen. The higher stress involves stronger elements in the deformation process and new planes ω can now form in addition to those existing at the same stress state but at a lower deformation rate. This process explains the phenomenon of increased plasticity with increased deformation rate in the experiments described above.

To explain the various features mentioned above, we make use of a general curve of residual deformation for rocks under any stress state, schematically shown in Fig. 2.40 (a) in the coordinate system ($\Delta\tau - \Delta\varepsilon_1$). This schematic curve was plotted from experimental curves shown in Fig. 1.59, 1.60 (Chapter 1).

Here $\Delta\tau = (\tau - \tau_{el})$, where $\tau_{el} = \tau_{el}^0 \exp (BC)$, the elastic limit, and τ the external stress applied in excess of the elastic limit. As a limiting case, τ can attain the value of peak strength $\tau_{us} = \tau_{us}^0 \exp (AC)$, since the limiting value of stresses also lies on the general curve.

The second coordinate, $\Delta\varepsilon_1$, represents the principal residual deformation, including that at peak strength. The principal component of linear residual deformation is used as the second coordinate instead of the shear magnitude since, due to the phenomenon of opening of microcracks along the microareas of rupture **b** in the body, the relationship between $\Delta\tau$ and shear magnitude does not follow the same (general) curve for different values of C. Here, a different curve is obtained for each value of C. The general curve of residual strain is convex upwards and at high deformations becomes almost a straight line. The maximum curvature is observed near the origin. Angles $\alpha_1, \alpha_2, \alpha_3, \alpha_4, \alpha_5$ (see Fig. 2.40a) are formed between the tangents to the general curve (1, 2, 3, 4, 5) and the deformation axis. The tangents of these angles may be termed *moduli of plasticity* (*S*) at their points of intersection with the general curve. Since the plasticity modulus is a variable quantity, as is seen in Fig. 2.40 (b), it varies from the maximum value $\tan \alpha_1 = S_0$ to the minimum value $\tan \alpha_5 = S_{hor}$. The modulus S_{hor} corresponds to the zone of transition of the limit curves to a horizontal position.

Accepting the foregoing assumption regarding the sequential involvement of planes ω in total deformation, then the modulus S_0 becomes the proportionality coefficient between $\Delta\tau$ and $\Delta\varepsilon_1$ for the case in which a single plane ω is formed in the body. It should be noted that, in this model, S_0 does not depend on the number of shear elements **a** giving rise to the given plane ω, i.e., S_0 does not depend on the value of $\chi = $ **b/a**, nor on the parameter C, which indicates the stress state.

After the first plane ω has formed, similar planes develop, leading to a reduction in modulus S. If it were assumed that, for any single plane ω, the modulus S_0 remains the same and is also independent of the magnitude of deformation $\Delta\varepsilon_1$ in this plane, then the reduction in the current modulus S would be inversely proportional to the number n_1 of planes ω. The relationship between $\Delta\tau$ and $\Delta\varepsilon_1$ can be written as:

$$\Delta\tau = (S_0/n_1) \cdot \Delta\varepsilon_1, \tag{2.17}$$

where $S_0/n_1 = S$, the current modulus of plasticity.

As long as n_1 continues to change during the process of deformation in the body, the modulus S will also vary; when n_1 becomes constant, S also becomes constant and, to be precise, becomes S_{hor}, which corresponds to the zone where the limit curves are horizontal, when all the remaining deformation planes ω become involved in the deformation process. Results of deformation in marble at a pressure $\sigma_2 = 250$ MPa (shown in Fig. 1.15) demonstrate the constancy of the plasticity modulus of marble under a pressure corresponding to the passage of the limit curves to the horizontal. Studies conducted by Bridgman [10] regarding the deformability of different metals under high pressures in the pre-strain stage, reaching several thousands of per cent, also showed invariability of the plasticity modulus under the entire range of large strains. Thus, the constancy of the plasticity modulus S_{hor} is experimentally confirmed. The straight-line section of the plasticity curve can be described by:

$$\Delta\tau = S_{hor} \Delta\varepsilon_1. \tag{2.18}$$

The relationship S versus $\Delta\varepsilon_1$, shown in Fig. 2.40 (b), was obtained by differentiating the general curve of residual strain. At a specific large value of strain $\Delta\varepsilon_1$, wherein the curves of S and n intersect the horizontal lines S_{hor} and n_{hor}, the values of n_1 and S become constant simultaneously and take the values n_{hor} and S_{hor}. At the origin of the coordinate system, we have $\Delta\varepsilon_1 = 0$, $S = S_0$ and $n_1 = 1$.

The plasticity modulus can be expressed as a function of the deformation by the equation:

$$S = S_0 \exp{(-L)} \Delta\varepsilon_1. \tag{2.19}$$

Here, $\exp(-L)\,\Delta\varepsilon_1 = n_1$, where L is an experimental constant.

Relationship (2.19) is almost linear since the exponential index is very small compared to unity. This equation can be rewritten as:

$$S = S_0 \exp(-L)\,\Delta\varepsilon_1 \approx S_0 - K\Delta\varepsilon_1, \qquad (2.20)$$

where K is a new experimental constant.

The expression for n_1 can be written similarly in the form of a linear relationship:

$$n_1 \approx 1 + K_1\,\Delta\varepsilon_1. \qquad (2.21)$$

Figure 2.41 shows the relationships between the plasticity modulus S and number n_1 of shear planes ω and the magnitude of deformation $\Delta\varepsilon_1$, computed while processing the experimental relationships obtained on two types of marble and two types of sandstone. It can be seen from Fig. 2.41 that the relationships are very close to linear, which is a special case, typical for the particular rocks tested. General curves of permanent strains in rock

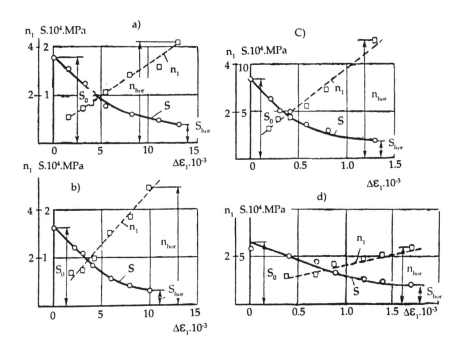

Fig. 2.41: Dependence of modulus of plasticity and number of shear planes on the magnitude of residual strain for (a) marble from Kararsk, (b) marble from the Urals, (c) BP sandstone and (d) NBP sandstone.

can be approximated quite well over a large section by a second-degree parabolic equation

$$\Delta\varepsilon_1 = K_2\Delta\tau^2, \tag{2.22}$$

where K_2 is a new constant.

General curves of residual deformation for two types of marble, talc chlorite and BP sandstone from Donbass are shown in Fig. 2.42. Linear sections are quite distinct in the curves for marbles where the maximum values of permanent deformation were obtained.

The equation for the ultimate equilibrium (1.17) for a single micropair, *shear-rupture*, several of which constitute the macroscopic shear plane ω, considers both shear resistance (strength) τ_1 and fracture resistance (strength) σ_p. For microareas **b**, just after occurrence of rupture, the term involving σ_p can be dropped; equation (1.17) is thereby simplified and takes the form:

$$\tau a^2 - \tau_1\, a^2 - \sigma_2\, ab\, \cos\, \delta = 0. \tag{2.23}$$

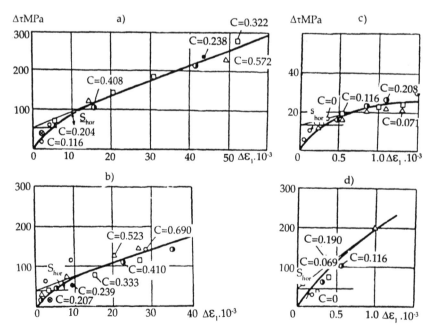

Fig. 2.42: General curves of residual deformation for (a) marble from Kararsk, (b) marble from the Urals, (c) talc chlorite and (d) BP sandstone.

After rupture in microarea **b**, the microarea of shear is reduced by Δa. The dimension of the microarea is determined by:

$$a \, (a - \Delta a). \tag{2.24}$$

Strain hardening occurs along these areas, as a result of which the shear resistance τ_1 (see Fig. 1.53) increases. This increase is taken into account by the expression

$$\tau_1 = \tau'_{el} + S_0 \, (\Delta a/a), \tag{2.25}$$

where τ'_{el} is the limit of elasticity along microscopic plane ω.

The first and the third terms in equation (2.23) also change with increase in shear Δa, which considered by the expressions

$$\tau_{el} \, a^2 + S_0 a^2 \, (\Delta a/a) \tag{2.26}$$

$$\left(\sqrt{\frac{2}{2}}\right) \sigma_2 \, a^2 \, \chi + \left(\sqrt{\frac{2}{2}}\right) \sigma_2 \, a \, \Delta \, a \tag{2.27}$$

where τ_{el} is the macroscopic limit of elasticity along plane w; $\chi = b/a$.

Considering the above-mentioned aspects, the equation for ultimate equilibrium can be written in a new expanded form

$$[\tau_{el} \, a^2 + S_0 \, (\Delta a/a) \, a^2] - [\tau'_{el} + S_0 \, (\Delta a/a)] \, a \, (a - \Delta a) -$$
$$- \left(\sqrt{\frac{2}{2}}\right) \sigma_2 \, a^2 \, (\chi + \Delta a \, /a) = 0. \tag{2.28}$$

Since $\Delta a/a$ is smaller compared to χ, it can be neglected. We then have:

$$[\tau_{el} \, a^2 + S_0 \, (\Delta a/a) \, a^2] - [\tau'_{el} + S_0 \, (\Delta a/a)] \, a(a - \Delta a) -$$
$$- \left(\sqrt{\frac{2}{2}}\right) \sigma_2 \, a^2 \, \chi = 0. \tag{2.29}$$

By differentiating the equation with respect to Δa and equating the derivative to zero, we obtain the extreme value of shear in the microelement **a**. By a series of transformations, we obtain

$$\Delta a/a = \tau'_{el}/2S_0. \tag{2.30}$$

The value of extreme shear in microelement **a** thus obtained can be considered a deformation criterion for attaining the peak strength limit. The value of extreme shear is of the order 10^{-3}, as can be demonstrated from the following approximate computations.

As shown in Fig. 2.41, the value of S_0 is of the order 10^4 MPa and τ'_{el} of 10 MPa. Hence we get $\Delta a/a = 10^{-3}$, which coincides with the value of maximum strain $\dot{\varepsilon}_0$, suggested in [4] and included in condition (2.15).

With increase in deformation rate, the values of the plasticity modulus S_0 and the elastic limit τ'_{el} also increase. In this case, the value of maximum strain in equation (2.30) may either increase or decrease. Since the authors do not have reliable results concerning the variation of extreme strain, we shall defer further discussion of this topic to the future.

We next consider several models of limit states in ($\log \dot\varepsilon - \tau$) coordinates. The simplest model, reflecting the series of experimental relationships considered above, is shown in Fig. 2.43. Here ray 1 corresponds to the strength at $\sigma_2 = P_1 = 0$ and ray 2 to $\sigma_2 = P_2$, ray 3 to $\sigma_2 = P_3$ and ray 4 to $\sigma_2 = P_4$. Further, $P_2 < P_3 < P_4$. Rays plotted with dashed lines indicate the elastic limits at the corresponding pressures P. All these rays intersect at a common point having the coordinate $\dot\varepsilon_1$ at $\tau = 0$. Such a fan of rays, as mentioned earlier, is described by the kinetic equation (2.1) in which the structural coefficient γ, determines the slope of the rays, is a function of the parameter C and pressure σ_2.

At any particular strain rate $\dot\varepsilon_1$, the product ($\gamma \cdot \tau$) is constant for all rays in the fan. For example, the products ($\gamma_{us} \cdot \tau_{us}$) and ($\gamma_{el} \cdot \tau_{el}$) at all points a, b, c, d, e, f, g and h have the same value for all rays of peak strength and elastic limit at any pressure value σ_2.

With the product ($\gamma \cdot \tau$) being constant at a given strain rate, as γ is reduced, the values of peak strength and elastic limit will increase proportionately. As the strain rate increases, $\dot\varepsilon_1^1 < \dot\varepsilon_2^2 < \dot\varepsilon_1^3 < \dot\varepsilon_1^4$ etc., the value of ($\gamma \cdot \tau$) will also increase since τ_{us} and τ_{el} increase.

The values of peak strength and elastic limit, at a particular strain rate and various confining pressures σ_2, are described by the equations of limit states (1.4) and (1.5). Variation of the strain rates in these equations leads to changes in the quantities τ_{us}^0 and τ_{el}^0 as well as A, B, C, which are functions of the strain rate. At a constant rate of deformation, these quantities are considered as constants that characterise the properties of a given solid.

Quantities $\tau_{us}^0(\dot\varepsilon_1)$ and $\tau_{el}^0(\dot\varepsilon_1)$, functions of the strain rate, are obtained from the kinetic equations:

$$\tau_{us}^0\ (\dot\varepsilon_1) = [\ln(\dot\varepsilon_1/\dot\varepsilon_0)\ KT + U_0]/(1/\gamma_{us}^0); \qquad (2.31)$$

$$\tau_{el}^0\ (\dot\varepsilon_1) = [\ln(\dot\varepsilon_1/\dot\varepsilon_0)\ KT + U_0]/(1/\gamma_{el}^0), \qquad (2.32)$$

where γ_{us}^0 and γ_{el}^0 are coefficients related to the rays (of peak strength and elastic limit respectively) at $\sigma_2 = 0$.

Graphs of stresses at peak strength and elastic limits in ($\log \tau - C$) coordinates were plotted for various strain rates, $\dot\varepsilon_1^4 < \dot\varepsilon_2^4 < \dot\varepsilon_1^2 < \dot\varepsilon_1^1$, as vertical sections of the fan of rays given in Fig. 2.43, at the values of strain rates indicated. These graphs are shown in Fig. 2.44. It was presumed that ray 4 in Fig. 2.43 corresponds to the point of transition of the limit curves to the horizontal position.

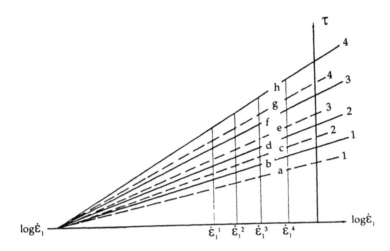

Fig. 2.43: Model showing the relationship between the strength and logarithm value of strain rate for a medium with moderate dispersion of the properties of its structural elements.

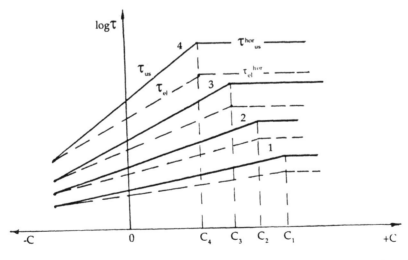

Fig. 2.44: Limits of peak strength and elasticity for various strain rates. Continuous lines indicate peak strength and dashed lines the elastic limits.

The stress state parameter C varies with change in strain rate, as is evident from the following relation:

$$C = (\sigma_2/\sigma_1) = \sigma_2/[\sigma_2 + \Delta\sigma_1 \, (\dot{\varepsilon}_1)] = \sigma_2/[\sigma_2 + 2\tau \, (\dot{\varepsilon}_1)]. \qquad (2.33)$$

In this expression $\Delta\sigma_1$ and τ are functions of the strain rate. At constant pressure σ_2, C decreases as the strain rate increases.

Stress values at which the limit curves become horizontal can be determined as functions of the strain rates from equations similar to (2.31) and (2.32):

$$\tau_{us}^{hor} \, (\dot{\varepsilon}_1) = [\ln (\dot{\varepsilon}_1/\dot{\varepsilon}_0) \, KT + U_0]/(1/\gamma_{us}^{hor}); \qquad (2.34)$$

$$\tau_{el}^{hor} \, (\dot{\varepsilon}_1) = [\ln (\dot{\varepsilon}_1/\dot{\varepsilon}_0) \, KT + U_0]/(1/\gamma_{el}^{hor}). \qquad (2.35)$$

Here γ_{us}^{hor} and γ_{el}^{hor} are structurally sensitive coefficients of the ray at which the limit curves become horizontal, and the confining pressures σ_2 attain values $\sigma_2 = \sigma_{2cr}^{us}$ for the peak strength and $\sigma_2 = \sigma_{2cr}^{el}$ for the elastic limits.

The slope of the limit lines varies with change in strain rate (see Fig. 2.44), indicating that coefficients A and B in equations (2.3) and (2.4) are functions of the strain rate. The following relationship between the aforesaid coefficients and the strain rate was obtained as a result of processing the experimental relations that agree qualitatively with the scheme shown in Fig. 2.43:

$$A(\dot{\varepsilon}_1) = \ln [\dot{\varepsilon}_1/\dot{\varepsilon}_{1 \text{ at } A = 0}] \cdot (1/\rho_{us}); \qquad (2.36)$$

$$B(\dot{\varepsilon}_1) = \ln [\dot{\varepsilon}_1/\dot{\varepsilon}_{1 \text{ at } A = 0}] \cdot (1/\rho_{us}), \qquad (2.37)$$

where ρ_{us}, ρ_{el} are additional experimental constants, which define the slope of the rays in semi-logarithmic coordinates ($\ln \dot{\varepsilon}_1 - A$) and ($\ln \dot{\varepsilon}_1 - B$); and $\dot{\varepsilon}_{1 \text{ at } A = 0}$ is the coordinate on the axis of strain rate at which A and B simultaneously become equal to zero.

Thus, the parameters C, A and B depend individually on the deformation rate but their products (AC) and (BC) do not, as follows from equations (2.3) and (2.4) in which only the structurally sensitive coefficient γ is involved, which is independent of the strain rate.

To derive complete equations for limit states that take into account the stress state and strain rate, it is necessary to solve simultaneously the equations mentioned above: (1.4), (1.5), (2.3), (2.4), (2.31), (2.32), (2.36), (2.37). When this is done the following equations are obtained:

$$\tau_{us} (\dot{\varepsilon}_1; \, C) = [\ln (\dot{\varepsilon}_1/\dot{\varepsilon}_0) \, KT + U_0] \, (1/\gamma_{us}^0) \times$$

$$\times \, \exp[(C/\rho_{us}) \ln (\dot{\varepsilon}_1/\dot{\varepsilon}_{1 \, \text{at} \, A \, = \, 0})] \qquad (2.38)$$

$$\tau_{el} (\dot{\varepsilon}_1; \, C) = [\ln (\dot{\varepsilon}_1/\dot{\varepsilon}_0) \, KT + U_0] \, (1/\gamma_{el}^0) \times$$

$$\times \, \exp[(C/\rho_{el}) \ln (\dot{\varepsilon}_1/\dot{\varepsilon}_{1 \, \text{at} \, A \, = \, 0})]. \qquad (2.39)$$

These equations allow graphs of strength and elasticity to be obtained numerically for any rate of strain. The coordinate of any point defining the transition of the strength-elasticity graphs to the horizontal for a given strain rate can be found from the relations:

$$C_{us}^{hor} (\dot{\varepsilon}_1) = \frac{\sigma_{2cr}^{us}}{\sigma_1^{us} (\dot{\varepsilon}_1)} = \frac{\sigma_{2cr}^{us}}{\sigma_{2cr}^{us} + 2\tau_{us}^{hor} (\dot{\varepsilon}_1)}; \qquad (2.40)$$

$$C_{el}^{hor} (\dot{\varepsilon}_1) = \frac{\sigma_{2cr}^{el}}{\sigma_1^{el} (\dot{\varepsilon}_1)} = \frac{\sigma_{2cr}^{el}}{\sigma_{2cr}^{el} + 2\tau_{el}^{hor} (\dot{\varepsilon}_1)}. \qquad (2.41)$$

The values of $C_{us}^{hor} (\dot{\varepsilon}_1)$ and $C_{el}^{hor} (\dot{\varepsilon}_1)$ are equal, i.e., $C_{us}^{hor} (\dot{\varepsilon}_1) = C_{el}^{hor} (\dot{\varepsilon}_1)$. By substituting these values of C into equations (2.38) and (2.39), τ_{us}^{hor} and τ_{el}^{hor} can be computed.

At the same time, the confining pressure σ_2 at which the graphs of strength and elasticity transit to a horizontal position, differ considerably from each other:

$$\sigma_{2cr}^{us} > \sigma_{2cr}^{el}.$$

This topic was covered in detail in Chapter 1 while discussing the role of loading path on the strain in rocks but without taking into account the strain rate.

After substituting expressions (2.34) and (2.35) into equations (2.40) and (2.41), we obtain

$$C_{us}^{hor} (\dot{\varepsilon}_1) = \frac{\sigma_{2cr}^{us}}{\sigma_{2cr}^{us} + 2[\ln (\dot{\varepsilon}_1/\dot{\varepsilon}_0) \, KT + U_0] \cdot \dfrac{1}{\gamma_{us}^{hor}}}; \qquad (2.42)$$

$$C_{el}^{hor} (\dot{\varepsilon}_1) = \frac{\sigma_{2cr}^{el}}{\sigma_{2cr}^{el} + 2[\ln (\dot{\varepsilon}_1/\dot{\varepsilon}_0) \, KT + U_0] \cdot \dfrac{1}{\gamma_{el}^{hor}}}. \qquad (2.43)$$

It is evident from these equations that when the strain rate is increased, the values of the parameters $C_{us}^{hor} = C_{el}^{hor}$ decrease. The relationship between the strength or elasticity limits and the parameter C are shown in Fig. 2.45 for the three rock types tested. The strain rates at which these curves were obtained are indicated alongside the respective curves.

The magnitude of the confining pressure σ_2 can be substituted for the parameter C in equations (2.3) and (2.4). After treatment of experimental results, the following relationships for the coefficient γ as a function of pressure σ_2 were obtained:

$$\gamma_{us} = \gamma_{us}^0 \, \sigma_2^{-K_{us}};$$ (2.44)

$$\gamma_{el} = \gamma_{el}^0 \, \sigma_2^{-K_{el}}.$$ (2.45)

In these equations γ_{us}, γ_{el}, γ_{us}^0 and γ_{el}^0 assume the same values as in equations (2.3) and (2.4); K_{us} and K_{el} are new constants based on observed

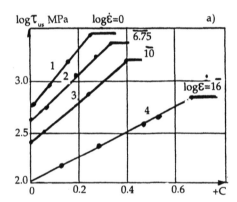

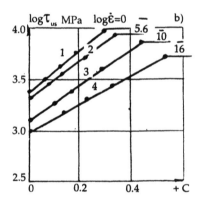

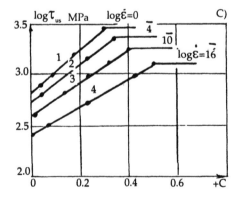

Fig. 2.45: Relationships of peak strength versus parameter C for different strain rates for (a) marble from the Urals, (b) diabase and (c) quartzitic sand.

rock properties. The combined solution of equations (2.44) and (2.45) along with equations (2.3) and (2.4), taking into account the invariability of the product $(\gamma \cdot \tau)$ for a specific strain rate, yields the following:

$$AC = K_{us} \ln \sigma_2; \qquad (2.46)$$

$$BC = K_{el} \ln \sigma_2. \qquad (2.47)$$

It follows from (2.46) and (2.47) together with (2.3) and (2.4) that the products (AC) and (BC) are independent of strain rate.

Substituting (2.46) and (2.47) into equations (2.38) and (2.39), we obtain

$$\tau_{us} (\dot{\varepsilon}_1 \sigma_2) = [\ln (\dot{\varepsilon}_1/\dot{\varepsilon}_0) KT + U_0] (1/\gamma_{us}^0) \times \exp (K_{us} \ln \sigma_2); \qquad (2.48)$$

$$\tau_{el} (\dot{\varepsilon}_1 \sigma_2) = [\ln (\dot{\varepsilon}_1/\dot{\varepsilon}_0) KT + U_0] (1/\gamma_{el}^0) \times \exp (K_{el} \ln \sigma_2). \qquad (2.49)$$

These equations are similar to equations (2.38) and (2.39), the difference being that (2.38) and (2.39) involve the constants ρ_{us} and ρ_{el}, while (2.48) and (2.49) involve the constants K_{us} and K_{el}. The choice is decided by which of the two—C or σ_2—it is convenient to use in a given situation. Equations (2.48) and (2.49) have the disadvantage that they cannot be used for negative values of σ_2, since σ_2 is part of a logarithmic expression. Since equations (2.38) and (2.39) do not have this disadvantage, they can be used to determine strength under the effect of tensile stresses σ_2.

2.5.1 Model for the Mechanics of Loss in Strength with Increase in Deformation Rate

Loss of strength with increase in deformation rate, observed in the experiments discussed earlier, is typical of rocks that exhibit considerable dispersion in the properties of their constituent structural elements.

The results of petrographic studies regarding structural features, presented in the form of percentage size distribution of the grains (structural elements), are given in Appendix II together with a brief petrographic description of the rocks tested.

We shall consider the results obtained for natural rock salt as well as for artificially prepared specimens.

A model to explain the loss in strength with increase in strain rate is shown in Fig. 2.46. The process of deformation and fracture with a growth in the strain rate is accompanied by a selection of structural elements with lower values of shear resistance, as observed in experiments with rock salt. To be specific, it was assumed that the test material had an peak long-term strength τ_{el}^1 corresponding to the level of stresses applied at point **a**. The spectrum of structural elements yielded a fan of rays 1, 2, 3, 4, 5 emanating from a common origin at $\dot{\varepsilon}_{1\tau = 0}$ In section **a–b**, following ray 1,

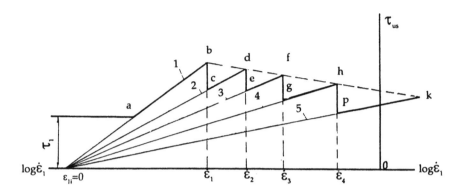

Fig. 2.46: Model to explain observed loss in strength with increase in strain rate.

the process occurs in accordance with the kinetic equation. At point **b** the process switches over to ray 2 where a jump **b–c** occurs, after which the process continues along section **c–d** on ray 2. The same process with jumps is repeated in sections **d–e–f, f–g–h** and **h–p–k,** and the strength declines progressively. The product ($\gamma \cdot \tau$) remains constant at a given strain rate for all the rays emanating from a common origin. The process switches to ray 2 at point **b** because deformation following ray 1 beyond point **b** involves further increase in stress τ, whereas the deformation of other structural elements can continue the process at lower stress. At point **c** the stress τ required to deform the specimen drops but the product ($\gamma \cdot \tau$) at points **b** and **c** is the same, which does not alter the effectiveness of the activation energy $U_{eff} = (U_0 - \gamma \tau)$ in the kinetic equation. The same is true for all the other rays and sections exhibiting jumps. Experimental results from tests on rock salt confirmed that rays 1 to 5 are related to structural elements of different properties, which result in a drop in strength.

Further experiments that take material structure into account are required to establish the validity of this model.

It follows from the statistical model of a heterogeneous solid proposed by the authors that the number of possible shear planes ω depends on the number of structural elements N per unit volume of the body; the higher this quantity, the higher the number of possible planes ω.

The number n of planes ω determines the total modulus of plasticity in the general strain curve, which reduces from the maximum value S_0 on a single plane ω to a value $S = S_0/n$ when n planes are involved.

We shall now determine for specimens artificially prepared from a mixture of three different-size fractions of particles (see 2.3.3), the possible number of planes ω formed from particles of each fraction, i.e., from grains of the same size. Assuming the average size of particles in each fraction d_1 = 0.3 mm, d_2 = 0.7 mm, d_3 = 2 mm and using the formula N = $1/d^3$, we obtain the following number of particles in a volume of mm^3: N_1 ≈ 40 particles; N_2 ≈ 3 particles; N_3 ≈ 0.1 particles. The quantity thus obtained differs by more than two orders of magnitude. The number n of possible planes ω formed from particles of each fraction, would differ roughly by the same amount. Thus, in a specimen made from a mixture of three fractions, the number of planes ω formed from the very fine fraction would be roughly 10 times more than the number of planes formed from the intermediate fraction and 400 times more than the number of planes formed from the very coarse fraction. Such a difference in quantity of possible planes can significantly influence which planes become activated under accelerated strain hardening.

General strain curves for the various fractions, over different ranges of strain rates, are shown schematically in Fig. 2.47. The three fractions are indicated by the numbers 1, 2, 3. The ranges of strain rates **a, b** and **c** are taken from the experimental graphs given in Fig. 2.38, at the point where relationship 4 intersects the relationships 1, 2, 3 obtained by testing specimens prepared from three uniformly sized fractions. Elasticity limits τ_{el}^1, τ_{el}^2, τ_{el}^3 of planes ω, consisting of particles from fractions 1, 2, 3 respectively, are plotted on the vertical axis. The magnitudes of the maximum strains $\Delta\varepsilon_{1us}$ on the plane ω, at peak strength, are plotted on the horizontal axis. Strain rates are shown under the graphs.

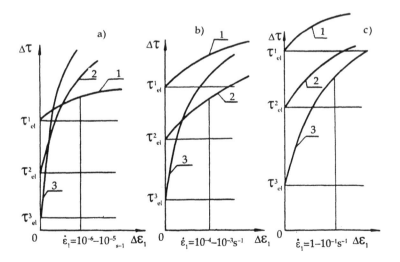

Fig. 2.47: Simulated general strain curves concerning to the planes ω formed from fractions 1, 2, 3 at different ranges of strain rates.

Consider an experiment conducted at low strain rates, (in the interval 10^{-6} to 10^{-5} s^{-1}). As the stress increases, fraction 3 participates first in the process of permanent deformation since it has the lowest elastic limit. Deformation moves along the general curve 3 until it intersects deformation curve 2, formed by particles of fraction 2. Fraction 2 participates in the general deformation once the stresses at the elastic limit τ^2_{el} are reached. Until the intersection of curves 3 and 2 is reached, both fractions participate in the deformation process. The plasticity moduli are maximum on those planes consisting of the highly coarse fraction 3, since the number of planes ω are minium for this fraction. The moduli of plasticity for fractions 2 and 1 are consequently reduced by factors of 20 and 400. When the stress level in the specimen reaches τ^1_{el} the planes ω formed by fraction 1 become involved in the deformation process. The contribution of deformation in fraction 1 to the total permanent deformation is distinctly different, and after the deformation curve 1 intersects with curves 3 and 2; the entire deformation occurs along the general curve 1 until the peak strength is attained and the permanent deformation $\Delta\varepsilon^{us}_1$ attains its maximum value. Here it should be noted that the localised rate of relative strain attained in a unit plane ω depends on the number of these planes involved. In the case wherein a single plane ω is formed at the elastic limit, the rate of localised strain may exceed the average strain rate by several times. The latter leads to an increase in the plasticity modulus.

We shall now consider a test conducted at strain rates in the interval 10^{-4} to 10^{-3} s^{-1} (see Fig. 2.47b). The elasticity limits increased in all three fractions. The highest increase was observed in τ^1_{el}, followed by τ^2_{el}, and the smallest increase in τ^3_{el}.

When the specimen is loaded, the process of permanent (irreversible) deformation starts in fraction 3 and continues along the general curve 3. When a stress level equal to τ^2_{el} is attained, fraction 2, which is subjected to deformation along the general curve 2, becomes involved. When curves 3 and 2 intersect, fraction 3 is excluded from the process, and deformation continues along curve 2. Fraction 1 does not participate at all in the process, if the values of peak strength at peak deformation $\Delta\varepsilon^{us}_1$ in fraction 2 are attained at stresses below the elastic limit of the first fraction τ^1_{el}.

Finally, we shall consider an experiment conducted at a deformation rate between 1 and 10 s^{-1} (Fig. 2.47c). In this case, it is seen that the elastic limits increased for all fractions, reaching the levels shown in the diagram.

When the specimen is loaded, fraction 3 participates first in the process of permanent deformation since it has the lowest elastic limit τ^3_{el}. Deformation proceeds along the general curve 3. When the stress level reaches the elastic limit τ^2_{el} of fraction 2, planes ω consisting of particles from fraction 2 become involved in the deformation process. However, peak strength and peak permanent deformation are attained in the ω planes involving particles from fraction 3, which has a lower strength than either of the other two fractions.

Loss in strength, with increase in strain rate, is accompanied by a significant reduction in the permanent deformation at peak strength level, as shown in Figs. 2.32 and 2.34 which relate to specimens of natural rock salt. The final permanent deformation $\Delta\varepsilon_1^{us}$ decreases with increase in strain rate, due to a reduction in the number of macroscopic shear planes ω that participate in the deformation process. The number of these planes is determined by the dimensions of crystals (grains); as the crystal grain size increases, the number of planes ω drops sharply. This analysis makes no claims of high accuracy and is entirely qualitative. However, it does help in understanding the physical nature and mechanics of the phenomenon associated with loss in strength and plasticity of rocks as the deformation rate increases. This analysis relates directly to rock salt.

A small loss in strength of marble specimens observed at a confining pressure $\sigma_2 = 150$ MPa may, perhaps, be explained by a similar scheme since marble, like rock salt, is monominerallic. If the analytical method described above is to be applied to polyminerallic rocks, it must be developed further and 'fleshed out' with additional features.

Among the entire set of experimental results concerning the effect of deformation rate on properties of polyminerallic rocks discussed above, tests on specimens of sylvinite, consisting of 68–70% pure sylvine and 30–32% pure rock salt, are easier to understand. Sylvinite and rock salt differ in thermodynamic parameters, in particular in activation energies U_0. This difference prompted the authors to formulate a model from two rays with different origins on the strain rate axis (see Fig. 2.35). Tests conducted on specimens of BP and NBP sandstones from Donbass (see Figs. 2.21 and 2.23) and with specimens of rock salt from Kizelovsk basin (see Fig. 2.29), gave results that could not be explained using the hypotheses outlined above. These rocks are polyminerallic with a complex petrographic composition (see Appendix II). For these cases, an extensive series of investigations, under widely varying thermodynamic conditions, is essential.

2.6 EFFECT OF MOISTURE CONTENT ON THE PHYSICOMECHANICAL INDICES OF ROCKS FOR VARIOUS PRESSURES AND DEFORMATION RATES

Water and water-vapours are surface-active substances (according to A. A. Rehbinder) whose action on rock results in loss of strength, elasticity (Young's modulus) and other characteristics.

A series of investigations was designed to obtain exhaustive information about variations in physicomechanical indices for a wide range of stresses, loading (deformation) rates and degree of water saturation of several rocks [88].

Tests were conducted in high-pressure cells, following the methodology described above, under triaxial compression of the type $\sigma_1 > \sigma_2 = \sigma_3$. The strain rate varied over the range 10^{-8} s^{-1} to 10^{+2} s^{-1}, i.e., from creep rates to impact rates.

Specimens were saturated with water in a vacuum. The specimens were initially dried until the weight became constant and then placed directly in a bath situated within a vacuum cell. Each specimen was vacuum-treated in advance to remove air from the pores. Water was poured into the bath containing the specimen and the latter kept in the cell for a specific time until the moisture saturation attained a given level [70]. Maximum moisture absorption is indicated by a constant steady-state weight. In experiments conducted on specimens of quartzitic sandstone and limestone by the authors, complete saturation was observed at 3% and 8% moisture content respectively. Here the moisture content was determined as a weight percentage of the water fraction compared to the weight of the specimen.

The results of studies relating to peak strength τ_{us} and elastic limits τ_{el} of quartzitic sandstone specimens, plotted in ($\tau - \ln \dot{\varepsilon}_1$) coordinates, are illustrated in Fig. 2.48 (a) and (b). The values of confining pressure σ_2 and moisture content $W\%$ and the legend are given in Tables next to the diagrams. The rays obtained for dry specimens ($W = 0$) are indicated by continuous lines; the dashed lines are approximate values for wet specimens.

Figure 2.49 (a) shows similar results, obtained on specimens of limestone (Estanoslanets deposit). The rays obtained with dry specimens (continuous lines) are described by equations (2.38) and (2.39). Moisture content alters the slope of the rays. The strength and elasticity are reduced compared to those for dry specimens. Moreover, the extent of this reduction increases with increase in moisture content. The slope of the rays in the given coordinate system determines the coefficient γ used in equations (2.3) and (2.4). Variation in the value of γ with change in moisture content is taken into account by introducing new equations of the type:

$$\gamma_{us}(W) = \gamma_{us}^0 \exp - (AC + ZW); \qquad (2.50)$$

$$\gamma_{el}(W) = \gamma_{el}^0 \exp - (BC + Z_1 W), \qquad (2.51)$$

where Z and Z_1 are new constant coefficients obtained experimentally; W is the moisture content.

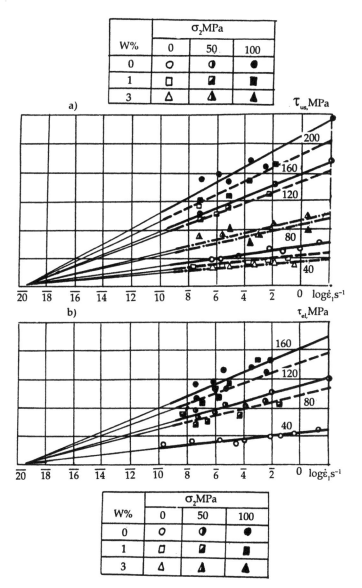

Fig. 2.48: Dependence of (a) peak strength and (b) elasticity limits of highly porous quartzitic sandstone on strain rate at different confining pressures σ_2 and moisture contents W.

Equations (2.38) and (2.39), revised to include the effect of moisture, now appear as follows:

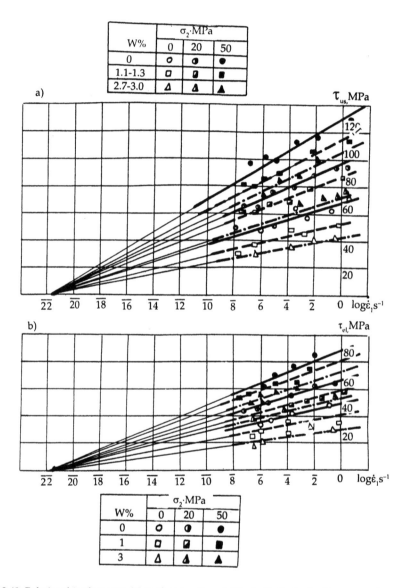

W%	σ₂·MPa		
	0	20	50
0	○	◑	●
1.1-1.3	□	◪	■
2.7-3.0	△	◮	▲

a)

$\tau_{us,}$MPa

b)

$\tau_{el,}$MPa

W%	σ₂·MPa		
	0	20	50
0	○	◑	●
1	□	◪	■
3	△	◮	▲

Fig. 2.49: Relationships between (a) peak strength and (b) elastic limits for limestone (Estanos-lanets deposit) and strain rate at various confining pressures σ₂ and moisture contents W.

$$\tau_{us} (\dot{\varepsilon}_1 \cdot W) = [\ln (\dot{\varepsilon}_1/\dot{\varepsilon}_0) \, KT + U_0] \, (1/\gamma_{us}^0) \times$$

$$\times \exp \left[\frac{C}{\rho_{us}} \ln (\dot{\varepsilon}_1/\dot{\varepsilon}_{A=0}) - ZW \right]; \qquad (2.52)$$

$$\tau_{el} (\dot{\varepsilon}_1 \cdot W) = [\ln (\dot{\varepsilon}_1/\dot{\varepsilon}_0) KT + U_0] (1/\gamma_{el}^0) \times$$

$$\times \exp \left[\frac{C}{\rho_{el}} \ln (\dot{\varepsilon}_1/\dot{\varepsilon}_{B=0}) - ZW \right]. \tag{2.53}$$

Conditions for the peak strength and elastic limits at any specific strain rate (i.e., any section of the fan of rays in Figs. 2.48 and 2.49 at a given value of $\dot{\varepsilon}_1$) are described by equations of the type (1.4) and (1.5), with correction for moisture incorporated in them:

$$\tau_{us} (W) = \tau_{us}^0 \exp (AC - ZW); \tag{2.54}$$

$$\tau_{el} (W) = \tau_{el}^0 \exp (AC - Z_1 W), \tag{2.55}$$

where Z and Z_1 have the same values as in equations (2.50) and (2.51).

Equations (2.54) and (2.55) are represented as straight lines in the $(\ln \tau - C)$ coordinate system, the slope of which is established by the coefficients A and B, which are independent of moisture content. With rise in moisture content, the position of the rays with respect to the $\ln \tau$ axis decreases, staying parallel to the initial ray obtained for dry specimens. The magnitude of the descent of a ray $\delta\tau$ due to change in moisture content can be expressed as follows:

$$\delta\tau_{us} = \tau_{us}^0 \exp (-ZW); \tag{2.56}$$

$$\delta\tau_{el} = \tau_{el}^0 \exp (-Z_1 W). \tag{2.57}$$

With change in the strain rate, rays for a given moisture content also change in inclination, i.e., the coefficients A and B depend on strain rate, as described by equations (2.36) and (2.37).

The above discussion of the conditions for limit states describes the transition to such states due to the action of tangential stresses, inducing shear in the material. In rocks this shear is realised through the effect of compressive stresses.

Rocks subjected to tensile stresses fail in a brittle manner through fracture, as mentioned in Chapter 1. Fracture strength, just like shear strength, depends on the rate of deformation and moisture content.

Fracture strength σ_p can be determined for these cases from equations (2.52), (2.53), (2.54) and (2.55). The methodology and rationale for this approach have been detailed by the authors in the literature [74, 65, 77]. The essence of the given definition lies in equating the peak strength and elastic limits in the aforesaid equations

$$\tau_{us} (\dot{\varepsilon}_1 W) = \tau_{el} (\dot{\varepsilon}_1 W) \tag{2.58}$$

which implies equating the macroscopic plastic deformation to zero. Condition (2.58) enables the value of fracture strength σ_p under different levels of moisture and strain rates to be determined. Recognising that in the fracture zone the maximum normal tensile stress theory of strength is valid, the magnitude of tangential stress $\tau_p(\dot{\varepsilon}_1 W)$, which itself is the ultimate fracture strength at different moisture levels and deformation rates, can be expressed as follows:

$$\tau_p(\dot{\varepsilon}_1 W) = [\sigma_p(\dot{\varepsilon}_1 W)]/2. \tag{2.59}$$

Values for the fracture strengths of quartzitic sandstone and limestone are given in Fig. 2.50. Comparing Figs. 2.48, 2.49 and 2.50, one can recognise the complete qualitative similarity between them, except for the large differences in location of the origins of each ray. The physical meaning of this difference in coordinates of the origins was outlined in the discussion earlier of results from determination of the fracture strength at various strain rates (see Fig. 2.39).

The strength results thus obtained can be represented in the coordinate system by envelopes of principal Mohr circles. The key characteristics of Mohr circles are: coefficient of cohesion F and angle of internal friction ρ. In equation (2.54), τ_{us}^0 is close in meaning to F.

The dependence of the cohesion coefficient on the strain rate at different moisture contents is shown in Fig. 2.51. It is evident that the experimental curves agree well qualitatively with the relationships illustrated in Figs.

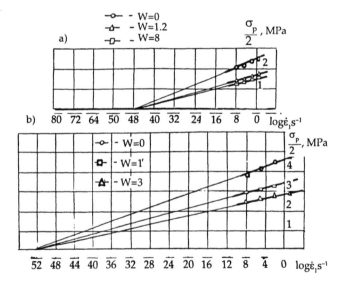

Fig. 2.50: Dependence of fracture strength limits of (a) quartzitic sandstone and (b) limestone on strain rate at different levels of moisture content W.

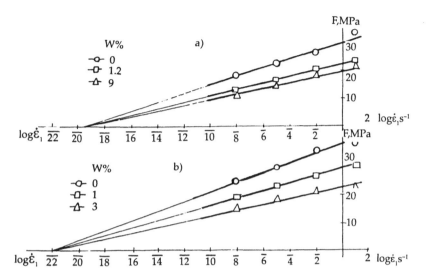

Fig. 2.51: Coefficients of cohesion for (a) quartzitic sandstone and (b) limestone versus strain rates at various levels of moisture content W.

2.48 and 2.49. The relationships plotted in Fig. 2.51 can be expressed analytically by an equation of the type:

$$F\left(\dot{\varepsilon}_1\ W\right) = \left(KT/\gamma_0\right)\ln\left(\dot{\varepsilon}_1/\dot{\varepsilon}_{us\ F = 0}\right)\exp\left(-ZW\right),\qquad(2.60)$$

where γ_0 is the structurally sensitive coefficient for rock in its dry state.

The relationship between the coefficient of cohesion and moisture content at a given strain rate is as follows:

$$F\left(W\right) = F_0\exp\left(-ZW\right),\qquad(2.61)$$

where F_0 is the cohesion coefficient for dry rock.

Moisture reduces the coefficient of cohesion almost by a factor of two and can thus significantly affect the stability of wetted rocks. The angle of internal friction ρ has the same physical meaning as coefficient A in equation (2.54). As with the coefficient A, the angle of internal friction is independent of moisture content but is dependent on strain rate.

Experimental relationships between the angle of internal friction and strain rate for quartzitic sandstone and limestone are shown in Fig. 2.52 (a). The parallelism of the experimental lines for these two rocks should be considered a special case, typical for the types of rocks tested. Figure 2.52 (b) and (c) shows the relationships of parameters A and B to strain rate for

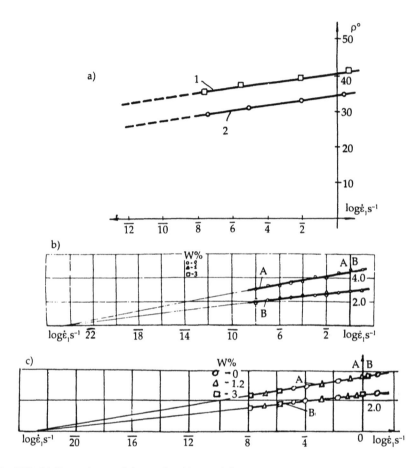

Fig. 2.52: (a) Dependence of the angle of internal friction of (1) quartzitic sandstone and (2) limestone on rate strain. (b) and (c) Relationship between parameters *A* and *B* and strain rate for various levels of moisture; (b) quartzitic sandstone and (c) limestone.

various levels of moisture for quartzitic sandstone and limestone respectively. As can be seen from the Figure, the values of these parameters are not influenced by moisture. These parameters have a meaning similar to that of the angle of internal friction.

2.6.1 Effect of Moisture on the Plastic Characteristics of Rocks

Plastic deformation in rocks is accompanied by strain hardening, which is indicated by the modulus of plasticity S, which may be determined from the general curve of plastic deformation plotted in $(\Delta\tau - \Delta\varepsilon_1)$ coordinates. Here,

$$\Delta\tau = [\tau(W) - \tau_{el}(W)].$$

Figures 2.53 and 2.54 show general curves of plastic deformation in limestone and quartzitic sandstone respectively. The moisture levels and strain rates pertaining to these specimens are indicated on the graphs. The tangent of the slope angle to the general curve is taken to be the plasticity modulus.

It can be seen from these diagrams that the plasticity modulus at the origin (S_0) decreases with increase in moisture and with reduction in strain rate. This result is clearly demonstrated in Fig. 2.55. For each level of moisture, with variation in strain rate, a relationship is obtained in the form of

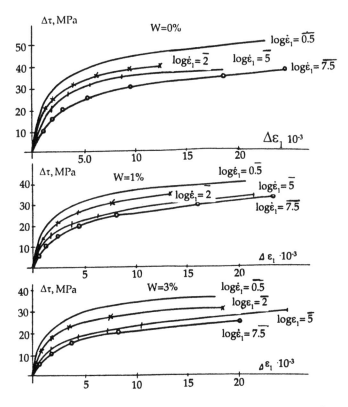

Fig. 2.53: General curves of plastic deformation for limestone at various levels of moisture and strain rates.

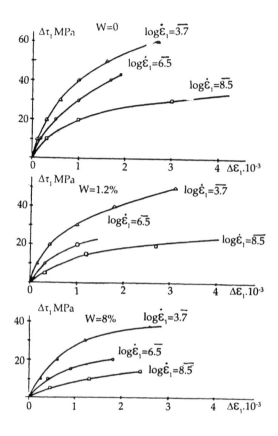

Fig. 2.54: General curves of plastic deformation for quartzitic sandstone at various levels of moisture and strain rates.

a ray from an origin that is common to all rays. At a given constant strain rate, the relationship between the plasticity modulus and moisture can be expressed by the equation:

$$S'_0 (W) = S'_0 \exp^{(-KW)}, \qquad (2.62)$$

where S'_0 is the plasticity modulus in a single plane for rock in which the moisture content $W = 0$; K is a constant determined experimentally.

In addition to its effect on the plasticity modulus, moisture considerably increases the plasticity of rocks per se, i.e., it increases the ultimate residual deformation $\Delta\varepsilon_{1us}$ at any constant rate of deformation. Thus, in dry quartzitic sandstone, at a strain rate log $\dot{\varepsilon}_1 = \overline{3.7}$ at a stress level $\Delta\tau = 40$ MPa, the deformation was $\Delta\varepsilon_1 = 0.9 \times 10^{-3}$. Under similar conditions but

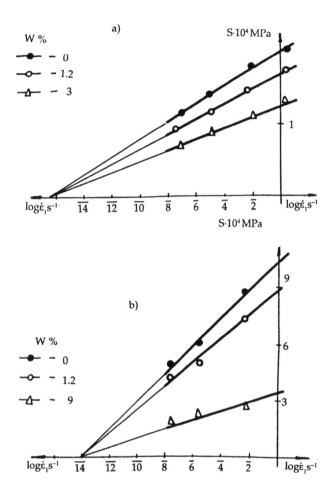

Fig. 2.55: Dependence of the plasticity modulus of (a) limestone and (b) quartzitic sandstone on strain rate for various levels of moisture W.

with a moisture content $W = 8\%$, the deformation reached a value of $\Delta\varepsilon_1 = 2.8 \times 10^{-3}$, i.e., three times greater. A similar picture was also observed in the case of limestone. At a stress level $\Delta\tau = 35$ MPa, specimens of dry limestone deformed to an extent of $\Delta\varepsilon_1 = 3 \times 10^{-3}$, while specimens with moisture content $W = 3\%$ deformed to an extent of $\Delta\varepsilon_1 = 17 \times 10^{-3}$, i.e., the magnitude of deformation was five times greater for the wetted specimen. The strain rate in both cases was equal ($\log \dot{\varepsilon}_1 = \overline{0.5}$).

2.7 CREEP IN ROCKS

The topic discussed here [76, 83, 72] is an addition to the results of investigations mentioned in the previous section. Tests were conducted on spring-based presses designed for long-term testing. The design of these machines is illustrated in Fig. 2.5. Uniaxial compression testing was done under various levels of constant load, which was a specific percentage of the strength obtained for a *standard rate* of deformation equal to 10^{-5} to 10^{-4} s^{-1}. Specimens were prismatic in shape, with dimensions $150 \times 150 \times 300$ mm. The configuration of the specimen is shown in Fig. 2.7. Specimens were covered with a cloth soaked in paraffin to exclude the effect of external atmosphere. Specimens insulated in this manner as well as specimens without insulation were tested. The latter will be discussed later.

Creep curves of insulated specimens of sylvinite from the Verkhnekamsk deposit are shown in Fig. 2.56. Peak strength in the standard regime was $\sigma_1^{us} = 29.1$ MPa. Time is indicated in the graphs in hours and days. Volumetric strains were dilatant; only at the test load of 30% of the Peak strength the volumetric strain indicate contraction. At a load of 85%, progressive creep was observed; at loads of 70, 60 and 50%, creep was stable. At 30% load, the creep rate decayed progressively throughout the test.

The relationships between the dilatant volumetric strains in sylvinite specimens from the Verkhnekamsk deposit and the stress level for different test durations are shown in Fig. 2.57. Time is indicated in hours and days in the graphs.

Creep curves of natural rock salt from Artemovsk (Donbass), loaded at 70% of the peak strength, equal to $\sigma_1^{us} = 29.1$ MPa, are illustrated in Fig. 2.58.

Figure 2.59 shows creep curves for the same Artemovsk rock salt, recorded at loads of 50% and 30% of the peak strength.

Dilatant volumetric strains, at different loads and test durations for the same Artemovsk rock salt, are given in Fig. 2.60 (a). All test specimens were insulated from the external atmosphere. Although the dilatant strain reached a level of 20%, the specimens did not fail. Their structure was intensely damaged, particularly along grain boundaries. Grain sizes reached the centimetre level. This damage is clearly evident from photographs of rock salt specimens prior to and after testing (Fig. 2.60b).

The main reason for this fairly high stability of specimens of rock salt even with an intensely damaged structure was the reliability of the specimen insulation, which precluded penetration of atmospheric water vapour into cracks that formed in the specimen. This was confirmed by tests on uninsulated specimens of sylvinite from the Verkhnekamsk deposit [73]. Results of these tests are given in Fig. 2.61. Specimens were tested for 800 days at a load of 50% of the peak strength. Months of the year are

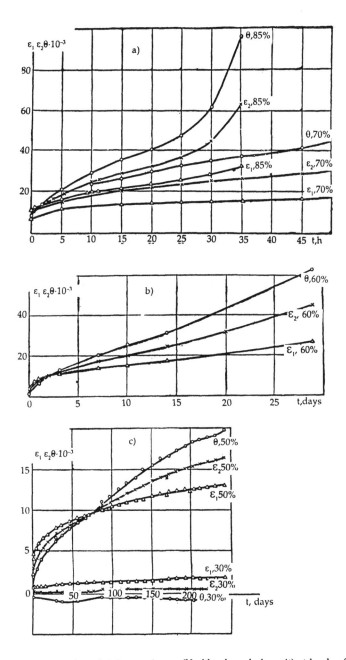

Fig. 2.56: Creep curves for sylvinite specimens (Verkhnekamsk deposit) at loads of (a) 85 and 70%, (b) 60%, (c) 50 and 30% of peak strength. ε_1—axial, ε_2—lateral, θ—volumetric strains.

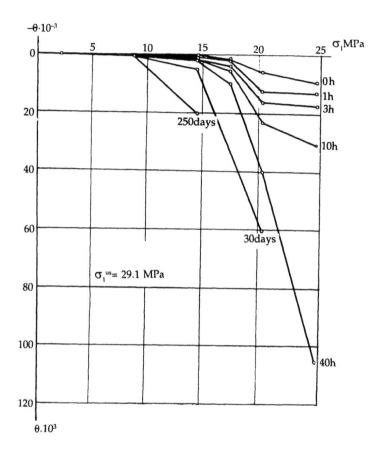

Fig. 2.57: Volumetric strain versus stress curves of insulated specimens of sylvinite for various test durations.

indicated on the time axis by Roman numerals. Creep curves contain sections exhibiting an increased rate of creep. Sections with increased creep rate were observed at the end of June (VI), July (VII), August (VIII) and partly in September (IX), i.e., at that time of the year when the room-heating season ceases in St. Petersburg. During this period the relative humidity in laboratory premises, in the conditions of St. Petersburg, reaches 90–95%. At a load of 50% of the peak strength, intense fracturing of the specimen was observed. The volume of the specimen increased from 5 to 20%, which allowed water vapours to penetrate cracks and interact with the crystalline grid of rock salt in the cracks. This interaction weakened interatomic bonds,

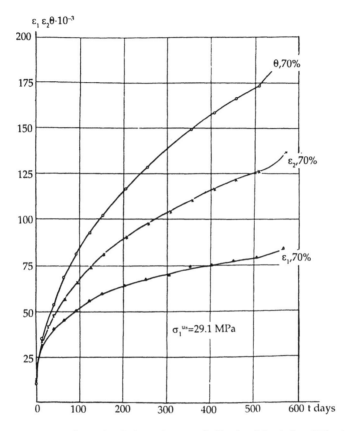

Fig. 2.58: Creep curves for rock salt from Artemovsk (Donbass) loaded at 70% of the peak strength. ε_1—axial, ε_2—lateral, θ—volumetric strains.

promoted diffusion processes and led to intensification of permanent deformations.

Curves of creep induced at a load of 30% of the peak strength are also plotted in the same diagram. Creep in this case exhibited a dampening trend; dilatant volumetric strains were almost nil, which is also the case in Fig. 2.56 at 30% load on insulated specimens of the same sylvinite.

Curves of creep for insulated specimens of lignite from the Shurabsk deposit in Tadzhikistan are plotted in Fig. 2.62. The characteristic feature of these curves is that the axial deformation ε_1 remained almost independent of test duration after 50 days of testing. Lateral and volumetric deformations developed intensely. As coal is highly porous, porosity amounting to 33%, volumetric deformations began changing sign from con-

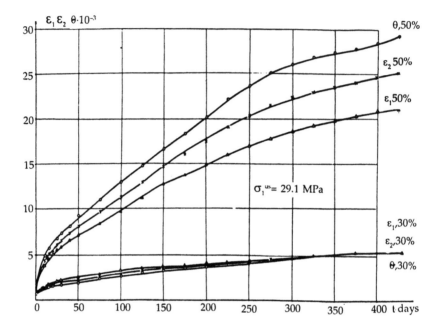

Fig. 2.59: Creep curves for rock salt from Artemovsk (Donbass) at loads of 50 and 30% of the peak (ultimate) strength. ε_1—axial, ε_2—lateral, θ—volumetric strains.

traction to dilatation only at a stress $\sigma_1 = 2.2$ MPa (when peak strength of coal was 3.15 MPa).

Volumetric deformations in specimens of lignite, depending on load and time, are plotted in Fig. 2.63. The maximum is distinctly seen in the curves of volumetric creep.

Cambrian clay, occurring under St. Petersburg, through which tunnels for the city metro line were driven at a depth of 70–75 metres, was also tested. Cambrian clay is very moist, albeit a reliable protection from the inrush of water. Prismatic clay specimens of the dimensions indicated earlier were subjected to creep tests. The specimens were completely encased in an insulating cover.

Curves of creep in clay at various levels of load are plotted in Fig. 2.64. Peak strength of cambrian clay was 9.1 MPa. It should be noted that at all load levels, lateral deformations were less than the corresponding axial deformations, though volumetric deformations were contractile at all load levels, and the coefficient of lateral deformation was less than 0.5.

During the tests, conducted in conditions of uniaxial compression, moisture from the specimen was squeezed out and stored in the gap between the impermeable insulating jacket and specimen body. When the jacket was

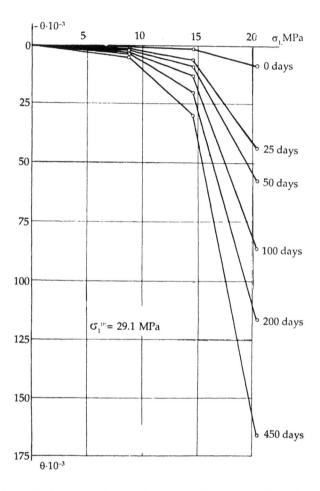

Fig. 2.60(a): Relationship between dilatant volumetric strains in rock salt specimens from Artemovsk (Donbass) for different test durations.

opened, water of 100 cm³ and more in volume poured out. This process of water squeezing occurs over time and depends on the level of applied load: the higher the load, the more intensive the squeezing process. With loss of moisture, the mechanical properties of clay change considerably: it becomes stronger and begins to acquire the properties of a brittle material.

Volumetric deformations in Cambrian clay are plotted in Fig. 2.65. Standard tests for clay strength were conducted in a universal press and yielded the graph shown by the dashed line in Fig. 2.65. Embrittlement of the clay due to loss of moisture, as in any other brittle material, leads to the forma-

Fig. 2.60(b): Photographs of unstrained and strained specimens of rock salt.

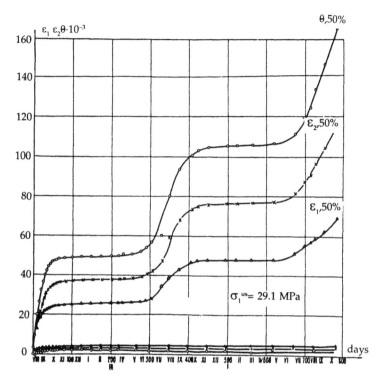

Fig. 2.61: Curves showing seasonal creep in sylvinite specimens (from the Verkhnekamsk deposit) not covered by a protective layer (i.e., open to atmospheric effects), at loads of 50% and 30% of the peak (ultimate) strength. ε_1—axial, ε_2 —lateral, θ—volumetric strains.

tion of cracks and voids in the clay. The maxima in Fig. 2.65 can be explained thusly: two processes occur simultaneously in clay—contraction, associated with closure of pores and voids and expulsion of water, and opening of cracks, which leads to dilatation and increase in volume. The total volumetric strain is the algebraic sum of the two dilatation processes of different sign (i.e., dilatation/contraction).

The results of tests on the creep and durability of rock salt can be useful for assessing the longevity and stability of underground workings driven through salt masses [68].

The kinetic equations (2.1) and (2.2) may be simplified as follows:

$$t = t_0 \exp(\alpha_t \sigma); \tag{2.2a}$$

$$\dot{\varepsilon} = \dot{\varepsilon}_0 \exp(-\alpha_{\dot{\varepsilon}} \sigma). \tag{2.1a}$$

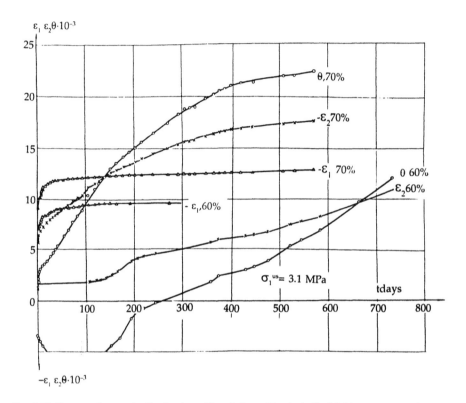

Fig. 2.62: Curves of creep for lignite from Shurabsk coal basin in Tadzhikistan. ε_1—axial, ε_2—lateral, θ—volumetric strains.

Here t, t_0, $\dot{\varepsilon}$, $\dot{\varepsilon}_0$ have the same values and the same meaning as in equations (2.1) and (2.2); σ is the normal stress under uniaxial compression. The coefficients α_t and $\alpha_{\dot{\varepsilon}}$ were obtained through transformations of the kinetic equations; they are equal in absolute magnitude but have opposite signs, i.e.:

$$|\alpha_t| = |-\alpha_{\dot{\varepsilon}}|. \tag{2.63}$$

Figure 2.66 shows graphs for ($\log \dot{\varepsilon}_1 - \sigma$) and ($\log t - \sigma$), plotted on the basis of experimental data on creep of two types of sylvinite, where (1) refers to sylvinite from the Verkhnekamsk deposit and (2) refers to sylvinite from the Starobinsk deposit in Byelorussia. Coefficients α_t and $\alpha_{\dot{\varepsilon}}$ determine the slope of the straight lines shown in Fig. 2.66. Values of these coefficients for the sylvinite from Verkhnekamsk are:

$$\alpha_t = 0.047 \text{ and } \alpha_{\dot{\varepsilon}_1} = 0.045.$$

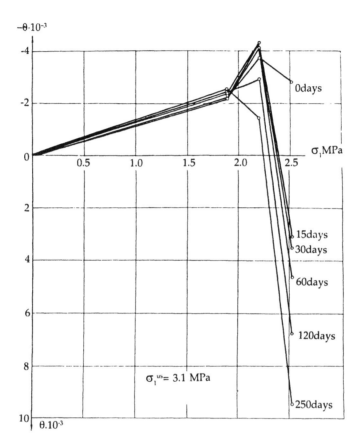

Fig. 2.63: Volumetric deformations in lignite from Shurabsk coal basin at different loads and test durations.

The values of the frequency factors are

$$t_0 = 3.15 \times 10^{-13} \ s^{-1} \text{ and } \dot{\varepsilon}_0 = 6.15 \times 10^9 \ s^{-1}.$$

The coefficients obtained for both types of sylvinite were very similar in absolute magnitude, as is evident from the graphs.

Special experiments were designed to test larger specimens of sylvinite (500 × 500 × 800 mm) directly in a mine in the Verkhnekamsk deposit, to establish their strength and strain properties. Specimens were cut from a pillar and tested in special presses (designed for mining).

The results of these tests revealed complete agreement between the data obtained under mine conditions and that obtained by testing specimens of

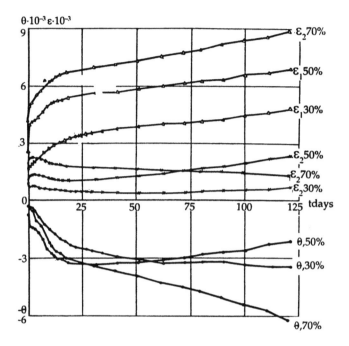

Fig. 2.64: Creep curves of St. Petersburg Cambrian clay. ε_1—axial, ε_2—lateral, θ—volumetric strains.

considerably smaller size in the laboratory. The no scale effect was detected. This fact meant that most of the tests could be conducted in the laboratory, which significantly simplified investigations.

The task of determining the stresses acting on the pillar was achieved by studying the creep rate in the pillar. This rate could be established by erecting datum marks in the mine and taking measurements. Based on these the time before onset of pillar failure could be assessed.

The technique of evaluating stress in a pillar, based on data obtained from sylvinite samples from the Verkhnekamsk deposit, is shown in Fig. 2.67. Lines 1 and 2 in this Figure were obtained through laboratory testing. The strain rate $\dot{\varepsilon}_1$ of the pillar was determined in the mine itself.

Starting with the observed deformation rate $\dot{\varepsilon}_1$, we extend a line horizontally to intersect line 2. We then extend a line vertically to intersect the y-axis to determine the magnitude of acting stress σ_g. Extending this line vertically upwards, we obtain the time before the onset of pillar failure t_1. To plot the longevity line (1), one point determined under high stresses

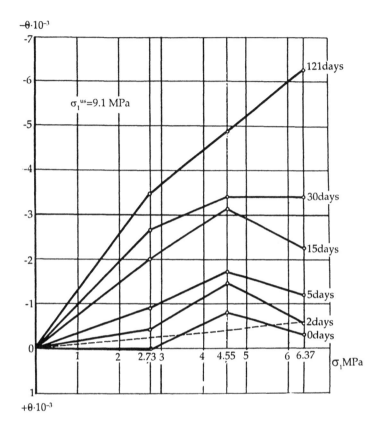

Fig. 2.65: Relationship between volumetric strain of contraction in Cambrian clay subjected to creep and level of load and test duration.

suffices (and saves time). The slope of the longevity line (1) is determined by the relationship (2.63). The corresponding coefficient α_t is obtained through laboratory tests on creep.

The method described above has adequate validity for salt rocks. This is not so for other rocks due to the strong influence of the scale effect.

2.8 AFTER-EFFECTS INDUCED IN ROCK BY PRIOR PERMANENT DEFORMATION

Permanent and plastic deformations in all solids result in the development of internal residual stresses in the solid upon removal of the external

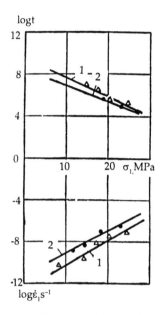

Fig. 2.66: Relationship between longevity log t and the steady-state creep rate log $\dot{\varepsilon}_1$ for sylvinite from the Verkhnekamsk deposit (line 1) and sylvinite from the Starobinsk deposit (line 2).

deforming loads. These residual stresses induce subsequent strain in the bodies, termed *consequent deformations*. Physically, these deformations are creep deformations which arise as the level of residual stresses decreases with time.

Consequent deformations are typical of a number of different materials and metals when subjected to stresses that may be even below the elastic limit. These phenomena are generally attributed to elastic imperfections in the solids caused by a non-uniformity of structure. Rocks are no exception. The magnitudes of consequent deformations depend on the level of permanent (irreversible) deformations and hydrostatic pressures.

The authors tested the following rocks [86]: (1) white marble from the Urals, (2) NBP sandstone from Donbass, (3) limestone from Estanoslanets basin and (4) lignite from Shurabsk.

Core specimens 30 mm in diameter and 80 mm in length were subjected to confined pressure tests using the method described earlier. They were insulated from the hydrostatic fluid by a polyethylene jacket. Depending on the type of experiment, hydrostatic pressure was developed in the range of a few MPa to 600 MPa, and the magnitudes of irreversible axial deformations $\Delta\varepsilon_1$ varied from one or two per cent to 25%. Specimens were never loaded to failure.

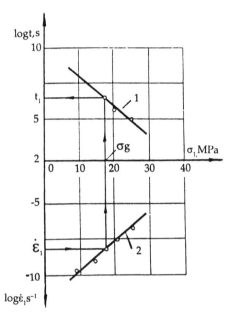

Fig. 2.67: Example of the laboratory technique for evaluating stress magnitude and longevity of a pillar, based on the pillar strain rate $\dot{\varepsilon}_1$ as determined directly in the mine.

After reaching a specific deformation, at a given value of hydrostatic pressure specimens were unloaded (decompressed), removed from the cell to study the after-effects, and placed in a glass cell under vacuum. Specimens could be tested under various gaseous atmospheres.

The specimen was placed in a device within the cell to measure deformations (Fig. 2.68a). Axial and lateral strains were measured by a set of dial gauges (4) and (3) to an accuracy of 0.01 and 0.001 mm. Lateral strains were measured in two mutually perpendicular directions. In Fig. 2.68 (a) only two lateral indicators (3) are shown. After placing the specimen (1) in a metal frame (2), a vacuum of 10^{-2} to 10^{-3} mm mercury column was created in the cell. The insulating polyethylene jacket was then opened. In this case, the experiment was conducted directly under vacuum. Gauge readings were observed visually through the glass cap in the vacuum chamber at specific times.

In addition to vacuum, an atmosphere of different moisture levels was created in the cell by vaporisation of a specific quantity of distilled water. The moisture level was recorded by a moisture meter placed in the cell. Two levels of relative humidity were taken as reference levels—30% and 100%.

Various gaseous atmospheres were also created in the cell: nitrogen, oxygen, hydrogen, carbon dioxide and a 2:1 mixture of oxygen with

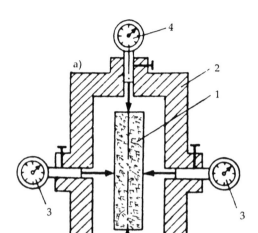
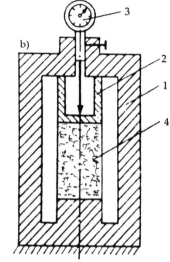

Fig. 2.68: Device for investigating (a) consequent deformations and (b) unlocking the associated forces.

hydrogen. The cell was initially placed under vacuum, after which a certain quantum of gas was pumped in, developing a pressure of 300 to 400 mm in the mercury column. The jacket of the specimen was removed in the cell after achieving the desired atmosphere. The cell temperature was maintained consistently between 20–22°C.

A dynamometer was placed in the cell while it was under vacuum to measure the mechanical repulsion (or "expanding") force generated by the deforming specimen (Fig. 2.68b). The specimen (4) was seated in a rigid frame (1) and the repulsion force measured by a stiff elastic dynamometer (2). This force was recorded using a dial type indicator (3) having a resolution of 0.001 mm.

Marble specimens from the Urals were subjected to more comprehensive tests, comprising not only vacuum and moisture, but also all the aforementioned gaseous atmospheres. Figure 2.69 shows the resultant curves for marble specimens initially subjected to a deformation of 18% under hydrostatic pressure levels $\sigma_2 = 100$, 200 and 400 MPa, plotted in coordinate $\Delta\varepsilon_{1cd}$ (consequent axial deformations) and $\Delta\varepsilon_{2cd}$ (consequent lateral deformations) versus time t. Tests were conducted under vacuum and under relative humidity levels of 30 and 100%. As can be seen from the graph, the lowest amount of consequent deformation was obtained under vacuum and the highest under 100% humidity; results for 30% humidity were intermediate.

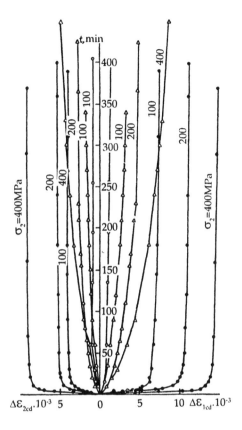

Fig. 2.69: Consequent deformations in marble specimens under vacuum and at two levels of humidity: o—vacuum, Δ—30% humidity, •—100% humidity. The pressure levels σ₂ (in MPa) at which the specimens were pre-strained, are indicated alongside the curves.

A relationship between the magnitudes of consequent deformations and level of hydrostatic pressure σ_2 was established. With rise in σ_2 from 100 to 400 MPa, the magnitude of consequent deformation, in all cases, increased correspondingly by 2–3 times for the same level of preceding irreversible deformation (18%). At the initial stage of the experiment, humidity had a strong effect on the rate of consequent deformation. Thus, at 30% humidity the deformation rate was higher roughly by one order of magnitude than that observed under vacuum, while at 100% humidity this difference was higher by two orders of magnitude. With increased duration of the experiment, the deformation rate dropped sharply under conditions of 100% humidity compared to rates obtained at 30% humidity. Consequent deformations $\Delta\varepsilon_1{}_{cd}$ and $\Delta\varepsilon_2{}_{cd}$ had the same sign and corresponded to expansion of the specimen in both the axial and lateral directions. Other rocks tested had similar signs for consequent deformations.

Figure 2.70 shows the after-effect (consequent) curves of marble from the Urals tested in media of oxygen, nitrogen, hydrogen, carbon dioxide and a mixture of oxygen with hydrogen. The corresponding after-effect curve under vacuum taken from Fig. 2.69 is also shown. In all cases, the specimens were pre-strained to 18% under a hydrostatic pressure $\sigma_2 = 100$ MPa.

As can be seen from the graphs, the minimum value of consequent deformation was obtained under vacuum. This deformation was roughly twice that observed in media containing oxygen, nitrogen, carbon dioxide and hydrogen. In the medium of oxygen mixed with hydrogen, the consequent deformations were roughly three times greater than in vacuum and ap-

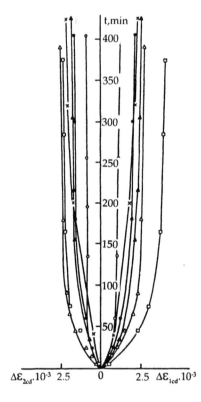

Fig. 2.70: Consequent deformations in marble specimens under vacuum and in different gases: o—vacuum, •—oxygen, Δ—carbon dioxide, ×—nitrogen, ▲—hydrogen, □—mixture of hydrogen and oxygen.

proximately equal to the consequent deformation obtained under 30% humidity (Fig. 2.69). As before, here too, the lateral deformations $\Delta\varepsilon_{2\ cd}$ and axial deformations $\Delta\varepsilon_{1\ cd}$ are dilatant (i.e. expanding)

Curves of consequent deformation in NBP sandstone from Donbass, pre-strained to 15% under $\sigma_2 = 600$ MPa, are shown in Fig. 2.71. Tests were conducted under 100% humidity. The magnitude of consequent deformation $\Delta\varepsilon_{1\ cd}$ was more than 1%. The signs of deformations $\Delta\varepsilon_{1\ cd}$ and $\Delta\varepsilon_{2\ cd}$ correspond to expansion.

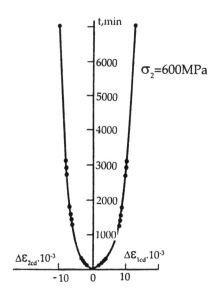

Fig. 2.71: Consequent deformations in NBP sandstone from Donbass under conditions of 100% humidity.

Results obtained for limestone specimens from Estanoslanets basin were similar to the preceding tests. Specimens were pre-strained to 18% under $\sigma_2 = 200$ MPa. Tests for after-effect were conducted under 100% humidity. The experimental curve is shown in Fig. 2.72.

Results of consequent deformations obtained for specimens of lignite from the Shurabsk basin in Tadzhikistan are shown in Fig. 2.73. Lateral deformation was recorded only in those specimens pre-strained at $\sigma_2 = 100$ MPa.

All the results considered above are qualitatively similar, even though obtained on rocks of different composition and origin.

Figure 2.74 shows the observed relationships between consequent deformations $\Delta\varepsilon_{1\ cd}$ and the magnitude of preceding irreversible deformation $\Delta\varepsilon_1$ in tests ranging from 5 to 25% conducted on marble specimens at a

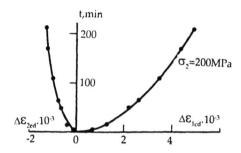

Fig. 2.72: Consequent deformation in limestone under conditions of 100% humidity.

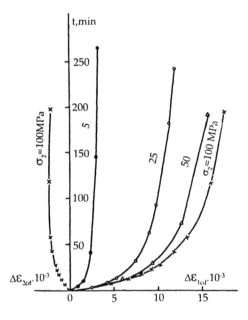

Fig. 2.73: Consequent deformations in lignite specimens under conditions of 30% humidity, pre-strained to 18% under σ_2 = 5, 25, 50 and 100 MPa.

pressure σ_2 = 100 MPa. Tests for determining consequent deformation were conducted under 30% humidity. The test duration was the same in all cases, namely, 2000 min, since beyond this time period consequent deformations were negligible.

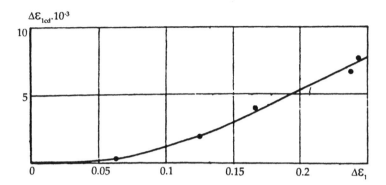

Fig. 2.74: Relationships between consequent deformations in marble specimens at 30% humidity and magnitude of preceding irreversible deformation $\Delta\varepsilon_1$. Observation time 2000 min.

As can be seen from the graph in Fig. 2.74, with an increase in $\Delta\varepsilon_1$ by a factor of 5, magnitude of consequent deformation $\Delta\varepsilon_{1\ cd}$ increased by more than one order of magnitude.

The magnitude of consequent deformation $\Delta\varepsilon_{1\ cd}$ also depended on the level of hydrostatic pressure σ_2 under which the preceding irreversible deformation occurred.

The relationships between consequent deformations and level of confining pressure σ_2 under which all the specimens were pre-strained to 18% are shown in Fig. 2.75. The total duration of testing for consequent deformation was 400 min in specimens of marble and 250 min in lignite specimens. It can be seen that at 100% humidity the consequent deformation in marble (curve 1) considerably exceeded the value obtained under conditions of 30% humidity (curve 2). All other conditions remaining the same. The magnitude of confining pressure σ_2 definitely increased consequent deformation.

Tests on marble specimens, similar to those shown in Fig. 2.69, were repeated and continued for 1500 min. Specimens were pre-strained to 18% under pressures $\sigma_2 = 100$, 200 and 400 MPa. Testing for consequent deformation was done at 30% and 100% humidity. The after-effects for both axial and lateral deformations in an atmosphere of 30% humidity (•) are illustrated in Fig. 2.76 by curves O–A_1, O–A_2, O–A_3 for different levels of σ_2. The after-effects for axial and lateral deformations in an atmosphere of 100% humidity (o) under different pressure levels, are represented by curves O–C_1, O–C_2, O–C_3. These curves are similar to those shown in Fig. 2.69 except for the test duration. Under long-term testing it was possible to record relatively higher magnitudes of consequent defomations.

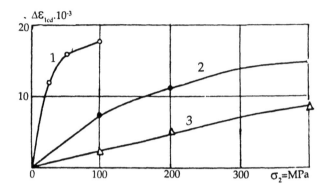

Fig. 2.75: Relationships between consequent deformations in specimens of marble and lignite and magnitude of confining pressure σ_2. Marble at 30% humidity ($\bullet$) and 100% humidity (o) and lignite (Δ) at 30% humidity. All specimens were subjected to the same magnitude of permanent deformation equal to 18%.

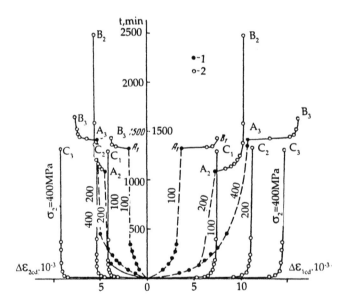

Fig. 2.76: Time-dependent consequent deformations in marble specimens under humidity levels varying from 30% ($\bullet$) to 100% (o).

The series of experiments at 30% humidity involved a novel feature. The humidity was rapidly increased from 30% to 100% in the vacuum cells, after a specific observation time corresponding to points A_1, A_2, A_3. Later, test duration was extended until it corresponded to points B_1, B_2, B_3.

At points A_1, A_2, A_3, after increase in humidity from 30% to 100%, the rate of after-effect increased sharply and attained values typical for an atmosphere of 100%, after which the rate dropped quickly. The overall trend of curves in this section does not differ from the curves of consequent deformations obtained under 100% humidity only. Total deformation observed in the same specimen at 30% and 100% humidity, corresponding to points B_1, B_2, B_3, proved essentially the same or very close to deformation corresponding to points C_1, C_2, C_3 obtained in specimens tested under 100% humidity only. This result indicates that the magnitudes of consequent deformations, independent of atmosphere, asymptotically tend to some ultimate values. These values were attained over different time periods, depending on the surrounding atmosphere.

A series of experiments conducted to study the after-effects in a stiff dynamometric strength measuring device (Fig. 2.67b) showed that permanently pre-strained specimens were capable of creating an repulsion force over time, the magnitudes of which were capable of counteracting external forces. Here, the repulsion force depended on the conditions of preceding permanent deformation in the specimen, magnitude of hydrostatic pressure σ_2 and magnitude of irreversible deformation $\Delta\varepsilon_1$. The higher the pressure σ_2, the greater the repulsion force; at constant σ_2, the higher the $\Delta\varepsilon_1$, the greater the force. The rate of increase in repulsion force, similar to the rate of consequent deformation, depended strongly on the composition of the surrounding medium in which the testing was carried out.

Relationships between the values of repulsion stresses σ, MPa along the specimen axis and time at different levels of humidity for marble specimens pre-strained under pressures $\sigma_2 = 100$ and 200 MPa, are shown in Fig. 2.77. The repulsion stresses recorded were caused by residual stresses in the specimens, but these are not mutually equal. The highest rate of increment in repulsion was obtained in specimens pre-strained at $\sigma_2 = 200$ MPa and tested for after-effects in an atmosphere of 100% humidity (curve 1). In the case of specimens pre-strained at $\sigma_2 = 100$ MPa and likewise tested in 100% humidity (curve 2), the rate of increment in repulsion was less than in the preceding case by more than one order of magnitude. Specimens irreversibly pre-strained at $\sigma_2 = 100$ MPa and tested for consequent deformations in 30% humidity (curve 3) showed the least incremental rate in repulsion.

Curves showing increment in repulsion stresses as a function of time, obtained by testing specimens of NBP sandstone from Donbass, are shown in Fig. 2.78. The highest rate of increment in repulsion stresses and the maximum magnitude of repulsion stress in specimens pre-strained to 15% under $\sigma_2 = 600$ MPa, were obtained under conditions of 100% humidity (curve 1).

Studying the graphs in Figs. 2.77 and 2.78, one can speak of the existence of an ultimate value of repulsion stress, to which the curves tend asymptotically

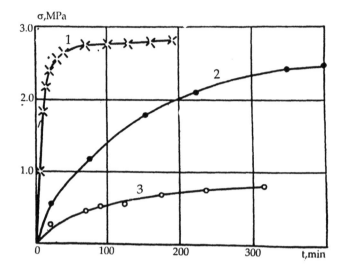

Fig. 2.77: Time-dependent relationship of magnitude of repulsion stress σ developed by specimens of marble which had been permanently pre-strained to 18% under $\sigma_2 = 100$ MPa. Conditions for testing after-effects: •—100% humidity, o—30% humidity, ×—results of tests conducted to establish after-effects at 100% humidity after pre-straining to 18% under $\sigma_2 = 200$ MPa.

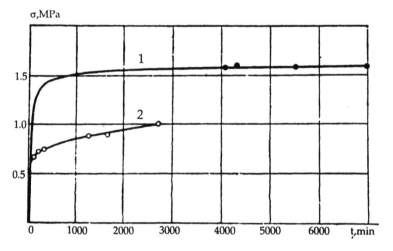

Fig. 2.78: Time-dependent repulsion stresses in specimens of NBP sandstone (Donbass) pre-strained to 15% under $\sigma_2 = 600$ MPa: •—at 100% humidity, o—at 30% humidity.

at various rates. Here the magnitude of ultimate repulsion stress depends on the parameters of preceding irreversible deformation σ_2 and $\Delta\varepsilon_1$.

This assertion is similar to that made regarding invariability (while approaching the asymptote) of consequent deformation $\Delta\varepsilon_1{}_{cd}$ for the given values of σ_2 and $\Delta\varepsilon_1$.

The phenomenon of after-effects has a direct bearing on the problem of a core whose properties vary significantly upon its removal from great depth. The core is subject to consequent deformations.

Effect of Deformation Rate on the 'Post-peak Strength' Properties of Rock and Energy Balance During Dynamic Uncontrolled Fracture

3.1 INTRODUCTION

Any process by which rocks are loaded to collapse culminates in a partial reduction or total elimination of the load-bearing capacity of the material. Study of the properties of rocks in the regime wherein the deformations are continued beyond the peak strength to collapse is hence essential for understanding the rupture process. In particular, it is the properties of rocks in the post-failure regime that determine the energy consumption during rupture, the possible loss of stability of the process and its evolution. Results of investigations into the post-peak properties of rocks for quasi-static deformation rates have been discussed in earlier chapters. Study of the 'post-peak' properties under dynamic conditions is of special interest since these conditions arise often in practical mining and also in earthquakes.

Under mining conditions, dynamic loading may develop in two distinctly different ways.

1) Dynamic loading may be produced by an external effect; for example, may be generated in a rock mass adjacent to some region that has failed in a brittle manner. In this case, rock elsewhere that is in various stages of deformation, both in the pre-peak and post-peak regimes, may be significantly affected by the wave arriving from dynamic event. The wave usually produces an increase in the force acting on the loaded rock and causes a response that will depend on the pre-existing condition of the loaded rock.

2) Dynamic conditions result from a loss of stability in the critically loaded region itself. This can occur only if the loaded region is deformed beyond the peak strength. A fall in the load-bearing capacity is inevitable

for this loading situation. This type of dynamic process always occurs during earthquakes and rockbursts.

The two regimes of dynamic deformation are discussed below.

3.2   EFFECT OF DEFORMATION RATE ON ROCK PROPERTIES IN THE POST-PEAK STRENGTH REGIME

The procedures for conducting dynamic tests on rocks in the post-peak strength regime have been described in Chapter 2.

Studies by the authors and by other researchers [50, 49, 6, 43, 44, 30, 39] reveal that different rocks exhibit different responses to changes in the strain rate. This is true both in the pre-peak and post-peak regimes. This is illustrated by the series of *stress-strain* diagrams (Fig. 3.1) obtained in tests at various rates on several rocks (marble, sulphide ore, granite, two types of sandstone, rock salt [80, 92], lignite and sylvinite [39]). The tests were all conducted in uniaxial compression. The specimen was subjected to a specific constant strain rate from the onset of load until complete collapse. The strain rates are as indicated in the caption to Fig. 3.1.

Similar changes in mechanical behaviour were observed in specimens of marble, granite, ore, sandstones and lignite. Thus both the peak strength and ductility increased with increase in strain rate, the drop modulus M reduced in the post-peak region of the curves, and residual strength increased—such that, overall, the energy consumed in the collapse process was also increased. Specifically, the increase in energy consumed in deforming these rocks in the post-failure strength regime, in the given range of strain rates, was: marble 220%, granite 70%, sulphide ore 80%, NBP sandstone 200%, BP sandstone 120% and lignite 230%. The maximum strength increase was observed for the sulphide ore (20.6%); the minimum increase was observed in BP sandstone (9.4%).

For the sylvinite (KCl), increase in strain rate led to an increase in strength by 75%, but also a marked embrittlement, which led to a reduction of 300% (between curves 1 and 3) and 700% (between curves 2 and 3) in the energy consumed during post-peak deformation.

In the case of rock salt, reductions were observed in both the strength (25.5%) and the energy consumption during post-peak deformation (by 63%).

Analysing the results of the above-mentioned investigations and those conducted by other authors, it is clear that under conditions of uniaxial compression, the properties of most rocks change with increase in strain rate according to the classic kinetic concept in strength of materials, i.e., both the strength of these rocks and the energy required for their rupture increase. The ductility and residual strength of the material were observed

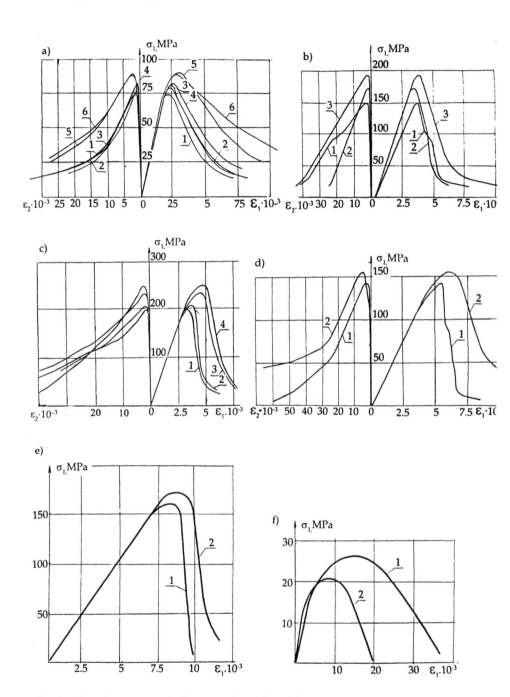

Fig. 3.1: Complete *stress-strain* diagrams obtained at different strain rates for the following rocks.

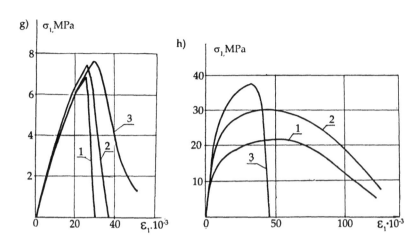

Fig. 3.1: (Contd.)

(a) marble

Curve no.	1	2	3	4	5	6
$\dot{\varepsilon}_1$ s^{-1}	2×10^{-6}	2×10^{-5}	2×10^{-4}	2×10^{-3}	2×10^{-2}	2×10^{-1}

(b) sulphide ore

Curve no.	1	2	3
$\dot{\varepsilon}_1$ s^{-1}	10^{-5}	5×10^{-2}	2×10^{-1}

(c) granite

Curve no.	1	2	3	4
$\dot{\varepsilon}_1$ s^{-1}	10^{-5}	2×10^{-4}	5×10^{-2}	2×10^{-1}

(d) NBP sandstone

Curve no.	1	2
$\dot{\varepsilon}_1$ s^{-1}	3×10^{-6}	10^{-1}

(e) BP sandstone

Curve no.	1	2
$\dot{\varepsilon}_1$ s^{-1}	10^{-5}	2×10^{-1}

(f) rock salt

Curve no.	1	2
$\dot{\varepsilon}_1$ s^{-1}	2×10^{-5}	2.5×10^{0}

(g) lignite

Curve no.	1	2	3
$\dot{\varepsilon}_1$ s^{-2}	10^{-4}	10^{-2}	10^{0}

(h) sylvinite

Curve no.	1	2	3
$\dot{\varepsilon}_1$ s^{-1}	10^{-8}	10^{-6}	4×10^{0}

together with a flattening of the post-peak deformation modulus. The same general trends in property variation were observed under increasing levels of confining pressure, σ_2. The similarity in effect of increased strain rate and confining pressure on rock properties described above was established by a detailed comparative study of the effect of each of these parameters [92, 91]. The results of these studies are outlined below.

  Stress σ_1 versus *axial strain* ε_1 and *lateral strain* ε_2 relationships for marble are shown in Fig. 3.2. Specimens were subjected to uniaxial compression at two strain rates: static ($\dot{\varepsilon}_1 = 2 \times 10^{-6}$ s^{-1}, test duration 1 h, curve 1) and dynamic ($\dot{\varepsilon}_1 = 2 \times 10^{-1}$ s^{-1}, test duration 0.05 s, curve 2), as well as under a static rate with a confining pressure selected such that the strength of the specimen tested statically was the same as for the specimen tested dynamically (curves 3, 4, confining pressure approx. 4 MPa). As can be seen from the Figure, the responses shown in curves 2, 3 and 4 are quite similar.

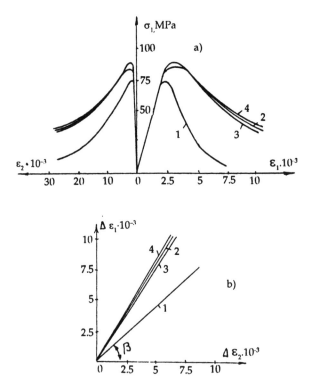

Fig. 3.2: Relationships between (a) $\sigma_1 - \varepsilon_1 - \varepsilon_2$ and (b) $\Delta\varepsilon_1 - \Delta\varepsilon_2$ for marble under the following test conditions:
1—uniaxial compression at a strain rate $\dot{\varepsilon}_1 = 2 \times 10^{-6}$ s^{-1}; 2—uniaxial compression at a strain rate $\dot{\varepsilon}_1 = 2 \times 10^{-1}$ s^{-1}; 3, 4—compression with confining pressure $\sigma_2 = 4$ MPa and strain rate $\dot{\varepsilon}_1 = 2 \times 10^{-6}$ s^{-1}.

The relationships between the residual (irreversible) axial $\Delta\varepsilon_1$ and lateral $\Delta\varepsilon_2$ strains, plotted from processing the results of the curves given in Fig. 3.2 (a) are presented in Fig. 3.2 (b). The tangent of the angle of inclination of these relationships, in the selected coordinate system, is equal to the coefficient of permanent lateral deformation μ. Increase in the strain rate led to a reduction in the value of μ, as in tests conducted under confining pressure. Irreversible changes in the volumetric strain were also the same for both testing methods. In terms of the discussed above statistical model, a reduction in μ and increase in permanent volumetric strains in the pre-failure and post-failure strength regimes indicate the participation of a large number of structural elements and shear planes ω in the deformation process, which should be accompanied by an increased degree of disintegration in the material. This has been confirmed by sieve analysis of specimen fragments obtained after the tests. In the case of specimens subjected to quasi-static uniaxial compression, 25% by weight of the specimen comprised fines less than 0.8 mm in size. In uniaxial dynamic and static confined pressure loading, the quantities of fines were 40% and 45% respectively.

Thus the disintegration of specimens when loaded under (i) static confined pressures and (ii) a high rate of loading was almost the same, and 1.5 times more than the disintegration under static uniaxial compression. It is known that the degree of crushability of a material increases with increase in strain rate. Results similar to those presented in the preceding chapter and have been reported by other investigators—see, for example [40].

The authors later undertook a series of experiments in which the objective was to study the effect of variation in test conditions (confining pressure and strain rate) introduced at different stages of loading, on the development of deformation processes. Curve 1 in Fig. 3.3 (a) was obtained by testing a specimen under uniaxial compression until complete failure at a static strain rate ($\dot{\varepsilon}_1 = 2 \times 10^{-6}$ s^{-1}); curve 2 was obtained under dynamic loading ($\dot{\varepsilon}_1 = 2 \times 10^{-1}$ s^{-1}). Curves 3, 4 and 5 correspond to tests conducted under varying conditions. Points A in the diagrams correspond to the beginning of a sharp rise in strain rate from $\dot{\varepsilon} = 2 \times 10^{-6}$ s^{-1} to $\dot{\varepsilon} = 2 \times 10^{-1}$ s^{-1} at different stages of post-peak deformation.

Curve 1 in Fig. 3.3 (b) was obtained by quasi-static uniaxial loading. Curve 2 indicates the data obtained by testing at the same rate but at a confining pressure $\sigma_2 = 4$ MPa. Curves 3, 4 and 5 were obtained under conditions of variable loading, i.e., uniaxial compression up to point A, and compression under confining pressures of 3, 2, 1 MPa respectively beyond A.

The curves in Fig. 3.3 (a) and (b) are very similar. Application of confining pressure and increase in strain rate produced similar effects throughout the post-peak deformation regime, i.e., hardening of material, increase in plasticity and reduction in drop modulus M.

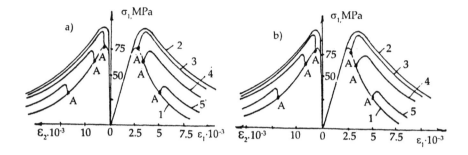

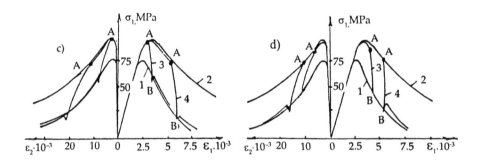

Fig. 3.3: Stress-strain relationships ($\sigma_1 - \epsilon_1 - \epsilon_2$) obtained by testing marble under the following conditions.
(a) 1—uniaxial compression at strain rate $\dot{\epsilon}_1 = 2 \times 10^{-6}$ s^{-1}; 2—uniaxial compression at strain rate $\dot{\epsilon}_1 = 2 \times 10^{-1}$ s^{-1}; 3, 4, 5—uniaxial compression under changing strain rate $\dot{\epsilon}_1 = 2 \times 10^{-6}$ s^{-1} up to point A; $\dot{\epsilon}_1 = 2 \times 10^{-1}$ s^{-1} beyond A. (b) 1—uniaxial compression at strain rate $\dot{\epsilon}_1 = 2 \times 10^{-6}$ s^{-1}; 2—compression under confining pressure at strain rate $\dot{\epsilon}_1 = 2 \times 10^{-6}$ s^{-1}; 3, 4, 5—changing confinement regime (uniaxial compression up to A; compression under confining pressure at strain rate $\dot{\epsilon}_1 = 2 \times 10^{-6}$ s^{-1} beyond A). (c) 1—uniaxial compression at strain rate $\dot{\epsilon}_1 = 2 \times 10^{-6}$ s^{-1}; 2—uniaxial compression at strain rate $\dot{\epsilon}_1 = 2 \times 10^{-1}$ s^{-1}; 3, 4—uniaxial compression at strain rate $\dot{\epsilon}_1 = 2 \times 10^{-1}$ s^{-1} up to point A; strain rate $\dot{\epsilon}_1 = 2 \times 10^{-6}$ s^{-1} beyond A. (d) 1—uniaxial compression at strain rate $\dot{\epsilon}_1 \times 2 \times 10^{-6}$ s^{-1}; 2—compression under confining pressure at strain rate $\dot{\epsilon}_1 = 2 \times 10^{-6}$ s^{-1}; 3, 4—confining pressure $\sigma_2 \times 4$ MPa to point A; confining pressure reduced to atmospheric (i.e., uniaxial compression) beyond A—at strain rate $\dot{\epsilon}_1 = 2 \times 10^{-6}$ s^{-1}.

The results of experiments in which the test conditions were reversed are shown in Fig. 3.3 (c) and (d). Up to point A in Fig. 3.3 (c) the specimen was subjected to dynamic loading at a strain rate of $\dot{\epsilon}_1 = 2 \times 10^{-1}$ s^{-1}, while in Fig. 3.3 (d) the load was applied under a confining pressure $\sigma_2 = 4$ MPa. Beyond points A, in the first case the strain rate was reduced suddenly to the quasi-static level ($\dot{\epsilon}_1 = 2 \times 10^{-6}$ s^{-1}); in the second case, the confining pressure was dropped to atmospheric (i.e., no confining pressure).

A sudden reduction in the strain rate and elimination of the confining pressure led to a reduction in load-bearing ability of the specimen to the same value as that observed for the specimen tested under static uniaxial compression (curve 1) at the same strain rate. In both cases, the load-bearing ability of the specimen was reduced even at very small axial and lateral deformations. The magnitudes of the axial deformation and the drop modulus M in segment AB for the given loading conditions were established by the rate of change in strain rate and the stiffness of the loading system. A rapid change in the loading rate from dynamic to static was accomplished by the use of special limiting arrestors in the testing apparatus described earlier (see Fig. 2.4). When the testing conditions were changed, the modulus of the post-peak characteristic (i.e., drop modulus) varied from $M = 0.16 \times 10^5$ MPa to $M = 5 \times 10^5$ MPa.

Thus, a sharp reduction in the strain rate, or a reduction in confining pressure result both in a considerable reduction in the energy required for rupture, and an increase in the modulus of post-peak deformation. These changes may lead, in turn, to a loss of equilibrium between the rock mass surrounding the volume that is moving towards collapse (i.e., in the post-peak strength regime) and may produce rapid instability and rupture.

A series of tests was conducted under conditions of triaxial compression to determine the residual strength of the various rock specimens. A specimen subjected to confining pressure under stiff loading conditions at a static strain rate $\dot{\varepsilon}_1 \times 10^{-5}$ s^{-1}, was tested until the curve touched the residual strength limit $\Delta\sigma_{1 \text{ res}}$ (point A on the curve in Fig. 3.4) and later, without changing the cell (confining) pressure a dynamic strain rate

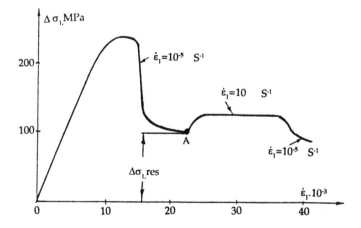

Fig. 3.4: $\Delta\sigma_1 - \varepsilon_1$ curve obtained for sandstone under changing conditions of deformation in the residual strength regime.

$\dot{\varepsilon}_1 \times 10 \text{ s}^{-1}$ was applied. A rise in residual strength was observed with increase in strain rate. After the dynamic effect had passed, the specimen regained its initial residual strength.

Similar experiments were conducted at different levels of confining pressure σ_2. Results of the tests on marble and sandstone are given in Table 3.1.

Table 3.1

	Marble							
σ_2, MPa	0.1	0.1	1	1	10	10	21	21
$\dot{\varepsilon}_1 \text{ s}^{-1} \Delta\sigma_{1\,\text{res}} \text{ MPa}$	10^{-5}	10	10^{-5}	10	10^{-5}	10	10^{-5}	10
	10	15	20	26	56	68	80	102
	Sandstone							
σ_2, MPa	0.1	0.1	5	5	10	10	21	21
$\dot{\varepsilon}_1 \text{ s}^{-1} \Delta\sigma_{1\,\text{res}} \text{ MPa}$	10^{-5}	10	10^{-5}	10	10^{-5}	10	10^{-5}	10
	9	18	44	62	71	94	100	136

These tests showed that the effect of strain rate on the residual strength intensified with increase in confining pressure. For marble, the increment in strength induced by increased strain rate was as follows: at $\sigma_2 = 0.1$ MPa the strength increase was 5 MPa, and at $\sigma_2 = 21$ MPa the strength increase was 22 MPa. For the same values of confining pressure and change in strain rate, the strength increase for sandstone was 9 MPa and 36 MPa respectively.

Based on these test results, as well as those conducted earlier and mentioned in the previous sections, graphs reflecting the relationships between logarithmic values of the peak shear strength (τ_{us}) and the residual shear strength (τ_{res}) with the parameter C ($= \sigma_2/\sigma_1$) for marble and sandstone, are shown in Fig. 3.5 for both static (line 1) and dynamic (line 2) strain rates. It can be seen that the point of intersection (τ_{res}^d and τ_{us}^d) for dynamic loading is shifted to the left, i.e., towards lower values of C compared to the point of intersection for the static loading relationships. This indicates that the conditions at which only plastic deformation takes place (zone of simple shear) in dynamic loading occur at lower levels of confining pressure compared to static loading, i.e., dynamic loading 'substitutes', to some extent, for the effect of confining pressure. This confirms the results of experimental and theoretical studies discussed in Chapter 2.

Thus, based on results presented in this section, it can be concluded that the *speed* (rate) of deformation imposed on rocks of the type considered here, induces changes in the mechanical characteristics and the mechanics of development of deformation processes that are similar to the effect of confining pressure.

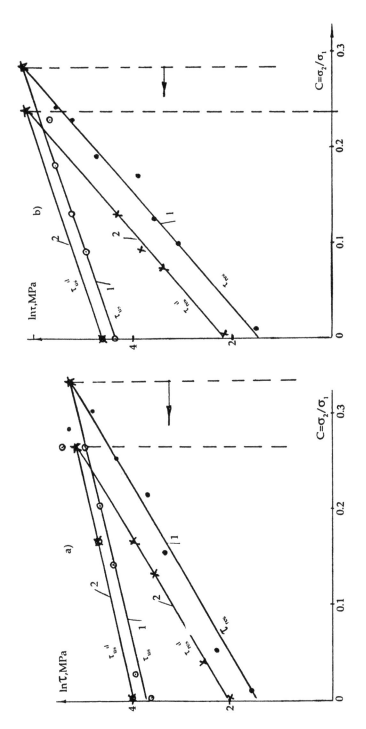

Fig. 3.5: Relationships between the peak strength and residual strength and the parameter C (= σ_2/σ_1) obtained under static (line 1) and dynamic (line 2) strain rates for (a) marble and (b) sandstone.

3.2.1 Simplified Method for Determining the Influence of Strain Rate on the Strength and Energy Consumed in the Rupture of Rocks

The effect of strain rate on the strength and deformation properties as well as on the energy required for rupture is determined by comparing corresponding indices obtained on different specimens of the same rock tested under static and dynamic rates of loading. However, when some specimens of the rock have a non-uniform and jointed composition while others differ sharply in the strain-strength characteristics, it becomes difficult to establish the magnitude of the effect of strain rate on the aforementioned indices. A large number of tests have to be conducted under static and dynamic strain rates and the mean values determined and compared for the various parameters of interest. If there is a wide scatter of experimental data and only a weak effect of strain rate on strain-strength characteristics, it is practically impossible to establish the degree of this effect with any confidence. For example, $\sigma_1 - \varepsilon_1$ diagrams, illustrated in Fig. 3.6, were obtained by testing bituminous coal from (a) Vorkutinsk and (b) Kuzbass deposits. The curves shown by dashed lines were obtained in static loading tests ($\dot{\varepsilon}_1 = 10^{-6}\ \mathrm{s}^{-1}$); curves shown by continuous lines were obtained in dynamic tests ($\dot{\varepsilon}_1 = 10^{-1}\ \mathrm{s}^{-1}$). It is impossible to determine from these test results whether or not dynamic loading has any effect.

A method for deriving and displaying the results of static and dynamic loading tests that allows an assessment to be made of the degree of the dynamic effect on rock behaviour, even under such complex conditions, is discussed below [92, 93, 95].

Let us consider the experimental curves of $\sigma_1 - \varepsilon_1$ in Fig. 3.7 obtained by uniaxial compression tests of (a) marble and (b) granite under various speeds of loading. The mechanical characteristics of these rock specimens show very little scatter (about 5%). This made it possible to assess patterns of variation in the strength and deformation properties as a function of loading rate at various stages of post-peak deformation. These curves were obtained in a manner similar to those shown in Fig. 3.3 (a). Curve 1 corresponds to static loading at a strain rate of $\dot{\varepsilon}_1 = 2 \times 10^{-6}\ \mathrm{s}^{-1}$. Curve 2 corresponds to dynamic loading at a strain rate of $\dot{\varepsilon}_1 = 2 \times 10^{-1}\ \mathrm{s}^{-1}$. Curves 3, 4 and 5 were obtained using a 'two-stage' loading process: static up to point A and dynamic loading beyond A. The ultimate values of the applied loads are indicated in the curves as follows:

σ_{us} — peak strength of specimen in the static tests;

σ_{us}^{d} — peak strength of specimen in dynamic tests;

σ_A — residual strength of a failed specimen—beyond which it was loaded dynamically [i.e., strength corresponding to point A on the curve];

σ_A^{d} — ultimate strength of a failed specimen under dynamic loading.

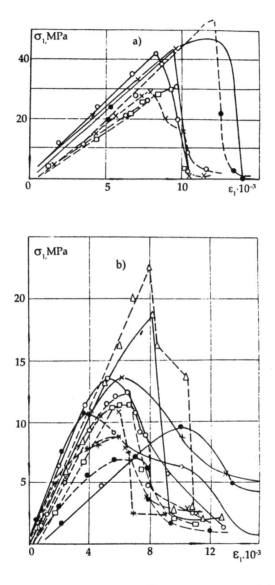

Fig. 3.6: $\sigma_1 - \varepsilon_1$ relationships for bituminous coal from (a) Vorkutinsk and (b) Kuzbass deposits obtained under static (dashed line) and dynamic (continuous line) strain rates.

It is evident from these graphs that, as the degree of rupture of the specimens in the post-failure regime increases, the effect of strain rate on the magnitude of 'absolute dynamic hardening' $(\sigma_A^d - \sigma_A)$ decreases. If we define the 'degree of rupture' in a specimen by the following ratio

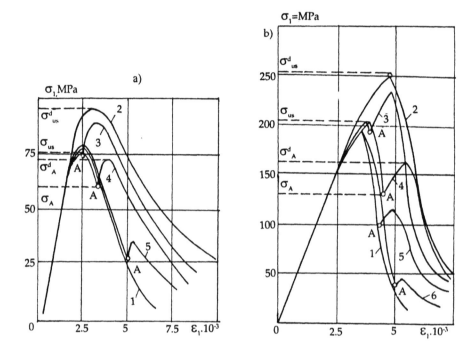

Fig. 3.7: $\sigma_1 - \varepsilon_1$ curves for (a) marble and (b) granite obtained under uniaxial compression at various loading rates.

$$H = (\sigma_{us} - \sigma_A)/\sigma_{us},$$

then the relationship between $(\sigma_A^d - \sigma_A)$ and H is seen to be almost linear (see Fig. 3.8a). Also, the magnitude of 'relative dynamic hardening', defined as $(\sigma_A^d - \sigma_A)/\sigma_A$, is independent of the degree of rupture H (see Fig. 3.8b).

Tests conducted in this way allow the rock strength to be determined for both static and dynamic strain rate using a single specimen. In order to do this, it is necessary to load the specimen statically just up to the peak strength, such that it begines to lose its load-bearing ability. The specimen should then be subjected to dynamic loading. The values of σ_{us}, σ_A and σ_A^d can then be determined directly while the dynamic strength is computed according to the formula

$$\sigma_{us}^d = \sigma_{us} (K + 1),$$

where $K = (\sigma_A^d - \sigma_A)/\sigma_A$.

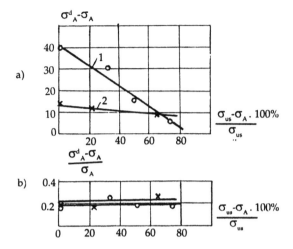

Fig. 3.8: Relationship between (a) the absolute and (b) relative values of dynamic hardening and the degree of rupture in specimens of (1) granite and (2) marble.

If we take into account the fact that the post-peak branches of dynamic loading curves for specimens that are either loaded dynamically from a given static point or tested dynamically over the entire loading cycle, are roughly parallel (see Fig. 3.7), then complete static and dynamic diagrams can be constructed with some confidence. The amount of energy consumed in the failure process can then be determined. Figure 3.9 shows examples of two types of bituminous coal from (a) Vorkutinsk and (b) Kuzbass deposits, as well as (c) for granite. Continious lines here present the diagrams resulting from the tests. Parts of the diagrams are shown by dashed lines have been extrapolated. Ray CD, linking peaks of the static and dynamic diagrams, should be maintained parallel to the elastic section in the diagram. The difference in the energy needed to fracture granite, between the value as calculated using plotted diagram and that observed experimentally (Fig. 3.7b), was about 6%.

3.3 ENERGY BALANCE IN BRITTLE FAILURE PROCESSES IN ROCK

The features of the dynamic investigations discussed in the previous sections are the following. Rock was tested dynamically either over the entire loading cycle or after preceding static deformation. The high strain rate was applied suddenly, as a shock, and was maintained constant upto the end of dynamic process. Under such loading conditions, dynamic hardening

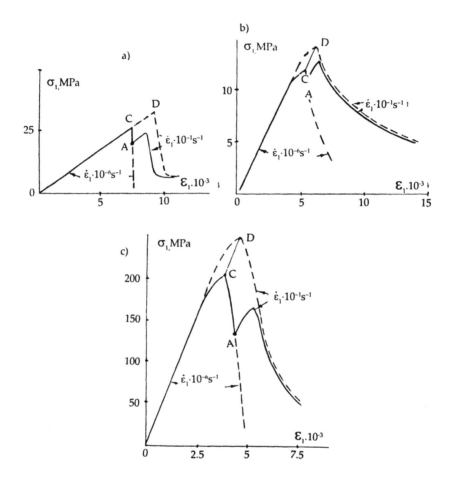

Fig. 3.9: Static and dynamic stress-strain diagrams for specimens of bituminous coal from (a) Vorkutinsk and (b) Kuzbass deposits and (c) granite, as determined from the results of tests conducted under variable (static as well as dynamic) loading conditions.

and the formation of a second maximum in the $\sigma_1 - \varepsilon_1$ diagram were observed, in the post-peak deformation regime. The initiation and development of a number of important dynamic processes in nature (e.g., earthquakes and rockbursts) do not occur in this way. In these situations, the peak loads in the mass undergoing failure develop statically, and the dynamic process results from a loss of stability between the two parts of the system: (i) the *volume of rock under failure* and (ii) the *surrounding rock mass*. In this case, the resisting force from (i) drops with loss of driving force from (ii). The change in strain rate during the dynamic stage of such a process is determined by the energy balance between (i) and (ii) in the system.

Cook [14, 15] and, later, other researchers [44, 5, 61, 62, 32, 16, 51, 46, 47, 38], established that unstable deformation and rupture of rocks occur whenever the amount of energy required for continued deformation is less than that released as a result of the deformation. The excess energy leads to dynamic rupture, which is accompanied by the radiation of acoustic and seismic energy from the source of the instability and may, in some cases, result in flying fragments. This type of instability can occur in large-scale (earthquakes, rockbursts) or in small-scale rupture events.

The mechanics by which the system becomes unstable has been studied intensively by several researchers [14, 15, 44, .5, 61, 52, 32, 16, 51, 46, 47, 38]. It is also necessary to investigate the mechanics operative at the onset of the dynamic process itself. The initial elastic energy, stored in the system prior to failure, is converted into several different forms of energy. What controls the redistribution of this energy into other forms? What are the governing laws and how can the energy balance be influenced in order to reduce the dynamic nature of the failure process? Understanding the mechanics by which dynamic failure of rocks develops due to a loss of stability, is essential in order to effectively combat such phenomena as earthquakes and rockbursts.

The conditions under which dynamic processes occur in nature can differ markedly. In some cases, often seen in earthquakes, the instability develops along pre-existing planes of rupture and is manifest in the form of dynamic slippage along the plane. In other cases, the loss of stability is associated with the rupture of a particular volume in the rock mass. Rupture processes may occur under conditions of confined compression (in deep-seated parts of a rock mass), or under plane-stress conditions (e.g. along the periphery of mine workings), or under conditions of uniaxial compression (in rock pillars). Each of these conditions imposes specific controlling conditions on the redistribution of energy during the rupture process. For example, flying rock fragments are usual when rupture occurs at an exposed face, i.e., under uniaxial or plane-stress conditions. Even so, all rupture processes are the result of a loss of stability and have many features in common.

Laboratory test conditions are simpler than those that arise in fild conditions. However, they can serve to provide useful insights into actual, full-scale beheviour in the field. For example, rupture of rock specimens under uniaxial compression can be considered to be a reasonable analogue of failure of a pillar in a mine working, while the rupture process of a specimen under volumetric compression can be an analogue for processes occurring at depth.

Results of laboratory investigations pertaining to energy balance in the processes of brittle uncontrolled rupture, under conditions of uniaxial or volumetric compression, are given below.

3.3.1 Energy Balance in Brittle Failure when Energy Consumed for Post-failure Deformation is Greater than Zero (Uniaxial Compression)

It is necessary first of all to characterise the types of energy which arise during the process of uncontrolled dynamic rupture and to decide which might need to be determined in experimental studies. For this, we shall analyse the curves shown in Fig. 3.10, plotted in P–Δl coordinates pertaining to load-strain in the specimen (OBE) and stiffness characteristic of the loading system (BC). (Hereafter, '*loading system*' will be abbreviated LS.) Point B on the diagram corresponds to ultimate equilibrium state of the system *specimen-LS*. In a particular case, it might coincide with ultimate strength of the specimen or might be located on the post-peak branch of the diagram. The shaded area in the diagram corresponds to total work done in deforming and rupturing the specimen. This work can be divided into three components:

— amount of work W_0 spent on irreversibly deforming the specimen up to point B (area OBD);

— amount of work W_c done for permanent deformation and rupturing (area of triangle DBA) of the specimen after reaching point B by means of elastic energy stored in the specimen (in this case, the work done is equal to that spent on elastic deformation W_{el} up to point B);

— amount of work W_{pfd} corresponding to energy consumed for deforming and fracturing the specimen in the post-failure strength zone that is contributed by the LS (area of triangle ABE).

Elastic energy of compression is stored in the system under conditions of ultimate equilibrium corresponding to point B in the diagram. This energy

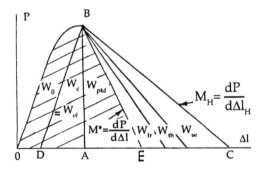

Fig. 3.10: Components of energy balance during loading and fracturing of the specimen.

is concentrated in the volume of the specimen as well as in the LS. The first portion of elastic energy is indicated in the Figure as W_{el}; the second portion, referred to as W_H, corresponds to the area of triangle ABC.

Even a very small deviation in deformation to the right of B in the system leads to violation of the condition for ultimate equilibrium and the system makes a transition to a state of uncontrolled dynamic rupture. Elastic energy $(W_{el} + W_H)$ concentrated in the system prior to loss of stability, is the source of energy for this process. This energy gets transferred on the whole into other forms of energy, as listed below:

1) Part of the energy is spent on fracturing the specimen. The area of triangle DBE in the diagram corresponds to this part of energy. The one fraction of this energy equal to W_C is contributed by the specimen itself, while the second fraction of the energy equal W_{pfd} is contributed by the LS.

2) Part of the energy is transformed into kinetic energy W_{fr} of flying rock fragments of the specimen after failure.

3) Part of the energy is converted into seismic energy W_{se} of the oscillatory processes which arise in the LS after specimen failure.

4) Part of the energy is transformed to heat energy W_{th}, realised in the process of deformation and fracture of the specimen.

Thus, the total energy balance in the process of uncontrolled dynamic fracture can be represented in the following manner:

$$W_{el} + W_H = W_c + W_{pfd} + W_{fr} + W_{se} + W_{th}. \tag{3.1}$$

Experimental investigations have to establish interrelationships among all the aforementioned components of energy balance and ascertain the mechanics governing the conversion of potential elastic energy of compression into other forms of energy associated with the dynamics of the fracture process.

Conditions for loss in stability and energy balance of the system *specimen-LS* depend on properties of the test material. Two variants of the process of uncontrolled rupture are distinguished in the aforementioned publications:

1) under the condition $W_{pfd} > 0$ (class 1, according to Wawersik, see Fig. 3.11);

2) under the condition $W^*_{pfd} < 0$ (class 2, according to Wawersik).

We shall first consider the energy balance in brittle failure when energy consumed W_{pfd} is greater than zero.

Loss of system stability in this case is possible when the condition

$$M^* > M_H \tag{3.2}$$

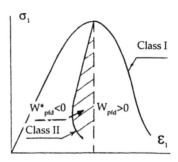

Fig. 3.11: Rock classification according to Wawersik.

is fulfilled, where $M^* = dP/d\Delta l$ characterises the post-peak branch in the curve related to the specimen; $M_H = dP/d\Delta l_H$ is the stiffness characteristic of the LS (see Fig. 3.10).

Numerous tests conducted on rocks belonging to class 1 according to Wawersik's classification showed that under stiff loading (i.e., at $M^* \, M_H$), the process of deformation beyond peak strength occurred in a stable regime. A huge amount of elastic energy W_{el}, stored in the specimen at peak strength, is completely dissipated in internal processes of deformation and crack propagation of the specimen (i.e., it is converted into energy W_c); hence it is not available for dynamic failure. The complete stress-strain diagrams obtained by testing very strong and brittle rocks in a high stiffness-stable deformation system (see Fig. 1.13) are a good example. In these rocks, the elastic energy W_{el} stored at peak strength, exceeded the energy of post-peak deformation W_{pfd} by a factor of 30. Dynamic effects developed during fracture of these rocks only when condition (3.2) was satisfied, i.e., when the energy released from the LS exceeded the energy consumed for the rupture process.

Thus, the excess elastic energy stored in the loading system is the source of energy for the dynamic fracture effects for this type of rock. The excess energy determined by the difference between the energies W_H and W_{pfd}, is used in accelerating the broken rock fragments into flight (energy W_{fr}) and in generating oscillation in the LS after rupture of the specimen (energy W_{se}): Part of the excess energy is eventually converted into heat (energy W_{th}).

The energy balance in such a process is of the form:

$$W_H - W_{pfd} = W_{fr} + W_{se} + W_{th}. \tag{3.3}$$

Experimental studies of the energy balance in the process of brittle uncontrolled failure were conducted on a testing machine specifically developed for this purpose and described below.

3.3.2 Testing Methodology

The testing machine [84, 92, 108] (see Fig. 3.12) consists of a stiff frame (1) and a stiff Wedge (W) drive (2) and (3) used to generate the load; the principle of operation has been described in Chapter 1.

In this version, the machine is extremely stiff and the energy stored in the loading system (LS) is negligibly small. Brittle rocks in the post-peak strength regime can be stably deformed in this machine. An elastic element (4) attached to a wedge (2) is the 'accumulator' for the metred energy in the loading system. This element consists of a thin-walled steel sleeve or ring (see Fig. 3.12a). A set of replaceable elastic elements allows the amount of energy stored in the press during the experiments to be varied over more than three orders of magnitude. The values of stiffness and period of natural oscillations of the elastic elements used in the experiments are given in Table 3.2.

Table 3.2

Form of elastic element	Stiffness, 10^7 N/m	Period of natural oscillation, 10^{-3} s
Without elastic element	1000	
Sleeve	34	0.3
Sleeve	17.3	0.375
Ring	11	0.25
Ring	6.9	0.4
Ring	3.07	0.8
Ring	2.6	1.0
Ring	1.73	1.1
Ring	1.4	1.3
Ring	0.73	2.0
Ring	0.45	2.5
Ring	0.3	4.0

Strain gauges (5) are glued to the housing of the elastic element. These gauges can record deformation in the element, both when the specimen is being loaded and during rupture. A replaceable inertial mass (6) with a piezoelectric accelerometer (7) is attached to the upper end of the elastic element. The weight of this inertial mass was changed over one order of magnitude (0.5 to 5 kg) during the tests. The period of natural oscillations for the acclerometer was 0.4×10^{-4} s (i.e., a natural frequency of 2.5×10^4 Hz). A stiff dynamometer (9) is affixed to the non-moving platen of the loading machine. Gauges (11) to measure axial deformations are attached to the specimen.

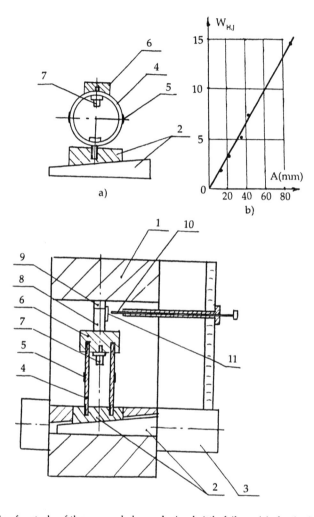

Fig. 3.12: Machine for study of the energy balance during brittle failure: (a) elastic ring element (schematic); (b) typical calibration curve for the accelerometer.

The experiment is conducted as follows. A specimen (8) is placed on the inertial mass (6) and statically loaded to the peak strength using the W-drive. After attaining the peak strength, fracture (failure) of the specimen occurs due to the energy stored in the elastic element (4). The static and dynamic phases of deformation are recorded by the stiff dynamometer (9); strain gauges, both on the specimen (11) and on the elastic element (5); and the accelerometer (7).

During dynamic process, deformation is induced in the specimen by the elastic element and is equal to the deformation in this element, as recorded by the gauges (5). The energy consumed in deforming the specimen in the post-failure strength zone (W_{pfd}) is determined from the *load (P)-deformation* (Δl) curve obtained from readings of gauges (9) and (5) (or 11).

The accelerometer (7) records the acceleration with which the inertial mass (6) moves, both during to and after failure of the specimen. The amplitude of the accelerometer signal (under constant inertial mass and stiffness of the elastic element) is proportional to the energy W_{se} that is not absorbed by the specimen during failure, and is manifested in the form of oscillations of the elastic element and inertial mass.

To record the energy W_{se} released in this manner, the accelerometer is first calibrated as follows. The rock specimen in the testing machine is replaced by a thin-walled glass tube. This glass specimen is compressed to that load which corresponds to the peak strength of the rock specimen. The elastic energy stored in the elastic element in this process is determined from the load applied and the deformation in the element. The glass specimen is then made to fail suddenly by firing a bullet (10) (see Fig. 3.12) against one side. Since the time taken for the glass specimen to disintegrate (5×10^{-5} s) is considerably shorter than the period of natural oscillations of the elastic element with the inertial mass attached (see Table 3.2), almost all of the entire energy of the elastic element is converted into kinetic energy of motion of the elastic element with the inertial mass. The amplitude of the signal recorded on the accelerometer then corresponds to the total energy stored in the loading system prior to failure.

Glass tubes were used in the same way at smaller loads. The dimensions of the tubes are chosen such that the stress developed prior to failure is around 1000 MPa. To achieve such high stress in the tubes, it is necessary to prepare the face ends carefully to ensure strict parallelism, planarity and smoothness of the specimen surfaces. Under the given stress, failure time is minimal and the glass tube disintegrates to a fine dust. A calibration curve presents the relationship between the initial energy (W_H) stored in the elastic element prior to each test, and the amplitude (A) of the signals from the accelerometer. The graph for one such elastic element is shown in Fig. 3.3b. The energy W_H, in Joules, is plotted on the vertical axis, against the amplitude A, in mm, on the horizontal axis.

During testing of a rock specimen, the recorded amplitude of the signal from the accelerometer is compared with the calibration graph. From this it is possible to determine the vibration (or oscillatory) energy (W_{se}) of the loading system.

The kinetic energy W_{fr} of the flying fragments from the disintegrating specimen is established from the formula for the total energy balance (equation 3.3):

$$W_{fr} \times W_H - W_{pfd} - W_{se} - W_{th}$$

The following eavest should be noted with respect to the above relationship for W_{fr}. The energy W_{th} was not determined in the experiments and was not considered in the computations. Experiments (to be discussed) showed that under uniaxial compression, the quantity of energy W_{th} was a negligible component of the total energy and exerted no influence on the general validity of the energy distribution pattern derived for the energy balance during the rupture process.

3.3.3 Results of Experimental Investigations

The rock types used for the dynamic tests, whose results are presented in Fig. 3.1, were also used for the energy distribution tests [92, 84], i.e., marble, granite, sulphide ore, two types of sandstones (BP and NBP), lignite, rock salt and sylvinite. All of these specimens, except marble, were taken from mines known to present a rockburst or outburst hazard.

Examples of the experimental (load-deformation) results $P–\Delta l$ are shown in Fig. 3.13. These diagrams illustrate (i) the post-peak characteristics of rock during the process of uncontrolled dynamic rupture (curve 1); (ii) the post-peak characteristics of the same rock under conditions of static (stable) deformation in a stiff loading system (dashed line, curve 2); and (iii) the stiffness characteristic of the loading system (curve 3). Diagram (a) was obtained on sandstone and diagram (b) on rock salt.

It can be seen from the diagrams that energy consumption for post-failure deformation under dynamic conditions, W^d_{pfd}, differs considerably from that under static conditions, W_{pfd}. Two types of rocks were distinguished in the course of these studies:

A) Rocks that consume more energy for rupture under conditions of uncontrolled dynamic rupture than under static loading (as in the case of sandstone in Fig. 3.13a).

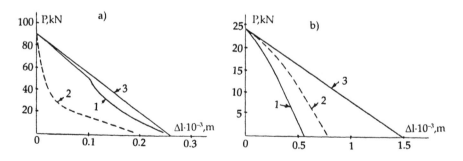

Fig. 3.13: Post-failure branches of diagram $P–\Delta l$ for (a) sandstone and (b) rock salt obtained under dynamic (curve 1) and static (curve 2) rupture. Curve 3—the stiffness characteristic of the LS.

B) Rocks that consume less energy during dynamic rupture (as in the case of rock salt in Fig. 3.13b).

Tests were conducted for several different stiffnesses of the LS, which was varied in order to change the initial elastic energy W_H stored in the LS prior to the start of uncontrolled rupture. As W_H increased, rupture became more dynamic. At high values of W_H, even a relatively plastic rock such as marble failed violently with a ringing sound and produced fly rock.

These experiments established that the quantity of energy consumed in rupturing rocks, W_{pfd}^d, is a function of the initial elastic energy, W_H. Figure 3.14 shows the variation of W_{pfd}^d as a function of W_H for type A and type B rocks. Initial points on the curves, corresponding to the energy consumption for rupture under a static stable regime of rupture (W_{pfd}), are indicated by the letter C.

As seen in Fig. 3.14, at low values of W_H, the change in W_{pfd}^d with increase in energy stored in the LS is quite large. However, this relationship gradually attanuates and starting from a certain value of W_H the two are

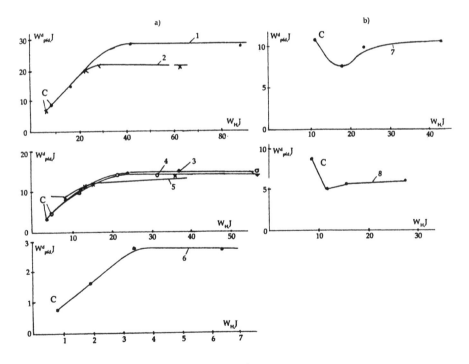

Fig. 3.14: Variation of the energy consumption W_{pfd}^d as a function of the energy W_H stored in the loading complex prior to failure for (a) type A rocks; 1—marble, 2—granite, 3—sandstone NBP, 4—sandstone BP, 5— sulphide ore, 6—lignite; (b) type B rocks: 7—rock salt, 8—sylvinite.

practically independent. Such a relationship between W_{pfd}^d and W_H can be explained by considering the interaction between the LS and the failed specimen beyond the peak strength. We will consider the combined diagram (Fig. 3.15), which shows the post-peak deformation portion (AC) of the specimen failure curve and the stiffness characteristic of the LS (AB).

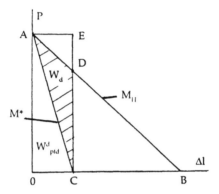

Fig. 3.15: Interaction between the LS and the test specimen.

In segment OC, the specimen is deformed and ruptured due to the elastic energy of the LS. In segment CB the failured material of the specimen is accelerated by the LS. Work W_d (corresponding to the shaded area to triangle ADC) is used in accelerating deformation and rupture of rock specimen and can be determined from the equation

$$W_d = W_{pfd}^d (1 - W_{pfd}^d / W_H) = 0.5 (M* - M_H) \Delta l_{pfd}, \Delta l_H. \qquad (3.4)$$

where Δl_{pfd} is the magnitude of absolute deformation of the specimen in the post-peak zone (segment OC in Fig. 3.15), Δl_H is the magnitude of absolute deformation of the LS (segment OB in Fig. 3.15).

It can be seen from the diagram that when the slope of the stiffness characteristic of the loading complex M_H is close to that of the specimen post-peak characteristic $M*$, even a slight reduction in M_H (accompanied by increase in W_H) leads to a significant increase in the area of the shaded portion of the triangle and consequently to an increase in the energy W_d. When $M_H << M*$, an increase in energy W_H, of the same magnitude as in the preceding condition, produces very little increse in W_d. Figure 3.16 shows the realtionship between W_d and W_H, computed according to formula (3.4) for sandstone.

The energy W_d determines the deformation rate of the specimen. The maximum rate of deformation of the specimen at point C (Fig. 3.15) may be evaluated by equating the kinetic energy of the mass $(m_0 + m_H)$ to the

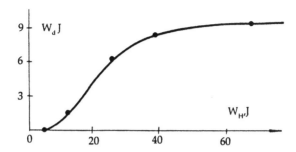

Fig. 3.16: Relationship between energy W_d and energy W_H stored in the loading complex prior to rupture for sandstone specimens.

work done, W_d (where m_0 is the mass of the test specimen, m_H is the inertial mass of the LS):

$$V_{\text{def} \cdot \text{max}} = \sqrt{2W_d/(m_0 + m_H)} =$$

$$= \sqrt{2W_{pfd}^d \, (1 - W_{pfd}^d/W_H)/(m_0 + m_H)}. \qquad (3.5)$$

The maximum strain rate of the specimen may be determined as follows:

$$\dot{\varepsilon}_{1 \, \text{max}} = \sqrt{2W_d/(m_0 + m_H)}/l_0$$

$$= \sqrt{2W_{pfd}^d \, (1 - W_{pfd}^d/W_H)/(m_0 + m_H)}/l_0, \qquad (3.6)$$

where l_0 is the length of the specimen.

The relationship between the strain rate of a sandstone specimen and the energy W_H is illustrated in Fig. 3.17. Calculations were made according to formula (3.6); values of W_d were taken from the graph in Fig. 3.16. The relationship of $\dot{\varepsilon}_1$ versus W_H is essentially similar to the relationships of W_{pfd}^d versus W_H and W_d versus W_H shown in Figs. 3.14 and 3.16.

As shown in Section 3.1, the energy consumption for rupture and deformation of rocks beyond the peak strength is a function of rate of deformation (strain rate). In the case considered here, the strain rate changed only over the initial section of diagram $\dot{\varepsilon}_1$ (W_H) i.e., at values of $W_H < 50$ J. These strain rate variations led to a corresponding change in the energy consumed for rupture W_{pfd}^d with variation in the values of W_H. At values of $W_H > 50$ J, the rate of deformation became independent on W_H, as did the relationship between W_{pfd}^d and W_H, as evident in Fig. 3.14.

Analysis of the variation in strain rate of the specimen volume undergoing rupture as a function of the elastic energy W_H (Fig. 3.17), enables us

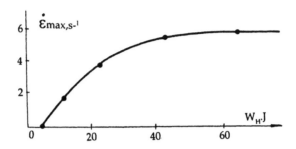

Fig. 3.17: Specimen strain rate versus the energy W_H (sandstone).

to infer that the dynamic processes of rupture that result from loss of stability due to the elastic energy W_H accumulated in the LS prior to failure, are constrained such that the strain rate cannot attain high values. For example, the maximum strain rate in the sandstone tested, did not exceed 6 s^{-1} (see Fig. 3.17).

The maximum strain rate depends not only on the energy W_H, but also on the energy consumed during rupture W_{pfd}^d (see equation 3.6). Thus, the maximum rate may differ for various rocks even though the initial values of W_H are the same. This is seen in Fig. 3.18 for a given case, plotted with values derived using equation (3.6) for the condition $W_H = \text{const} = 10$ J. At values of W_{pfd}^d equal to 0 and W_H, the strain rate becomes zero. The strain rate reaches maximum when the ratio $W_{pfd}^d/W_H = 0.5$.

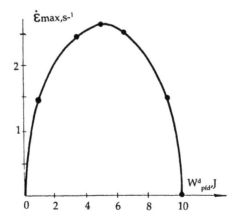

Fig. 3.18: Strain rate of different rock versus the energy required for rupture W_{pfd}^d when $W_H = \text{const}$.

Figure 3.19 illustrates the variation in the principal components of energy balance W^d_{pfd} W_{fr}, W_{se} as a function of W_H, for all of the rocks tested. Curves numbered from 1 to 6 (shown in the figure by label a) refer to rocks of type A for which an increase in energy consumption for rupture was observed with increase in deformation rate. Curves numbered 7 and 8 (shown in the Figure by label b) correspond to rocks of type B, for which the energy consumption on rupture decreases with increase in deformatin rate.

In rocks of type A (Fig. 3.19a), energies W_{fr} and W_{se} are seen to increase slowly with a substantial increase in energy consumption for rupture W^d_{pfd}, (in section CD of the diagram). Since these two energy components are used to create dynamic effects of rupture (fly rock and vibrating processes in the LS after rupture), then manifestation of dynamic effects in such rock types will be attenuated by the increasing energy consumed in rupture.

The opposite trend is observed in rocks of type B (see Fig. 3.19b). Here the energy consumed in rupture W^d_{pfd} in segment CD decreases as the energies W_{fr} and W_{se} rise sharply. For these rocks, loss of stability in the system *volume under rupture-LS*, induces the violent dynamic rupture even for a small excess of W_H over the static energy consumed in rupture W_{pfd}. This effect was observed by the authors during experiments.

It is necessary to consider these relationships between energy consumption for rupture of rocks and the elastic energy reserves in the system in practical calculations to predict the dynamic aspects due to rupture. This is very important since as seen, rocks may be of two types. Figure 3.19 illustrates the difference in the dynamics of the processes of rupture of type A and B rocks under similar initial conditions, i.e., for the same increment of elastic energy W_H over the static energy W_{pfd} at the onset of uncontrolled rupture. Let us consider the sandstone (4) and rock salt (7).

In the diagrams pertaining to these rocks, a band of 7 J is indicated by two broken lines. This portion indicates the difference between the energies W_H and W_{pfd}. When a specimen of sandstone ruptures due to a large increase in the energy consumed for rupture W^d_{pfd} compared to static energy W_{pfd}, the energy of the vibratory process W_{se} is 1.5 J and the energy of fly rock W_{fr} is about 0.15 J. The comparable values in the case of rock salt are: W_{se} about 10 J, W_{fr} about 1 J.

It is seen that

1) the values of dynamic energy components for salt are almost 7 times larger than those for sandstone;

2) the value of the dynamic components of energy released as a result of salt rupture exceeds the initial excess energy contained in the system ($W_{se} + W_{fr} = 11$ J $> W_H - W_{pfd} = 7$ J).

The interrelationships between components of energy balance, illustrated in Fig. 3.19, were obtained for a constant ratio between the mass of the test specimen (m_0) and the inertial mass of the loading complex (m_H) involved in the vibratory process after rupture of the specimen. Let us now consider

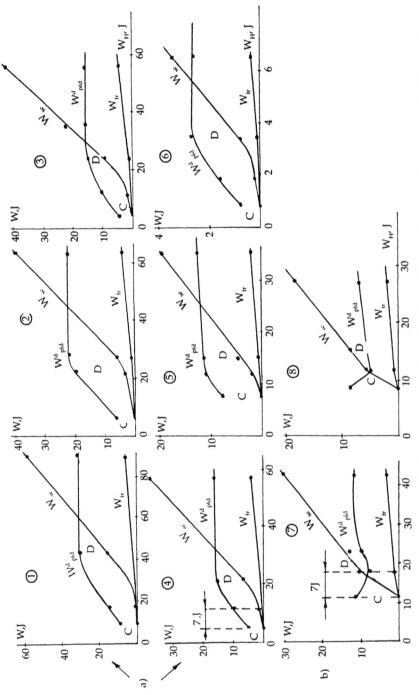

Fig. 3.19: Relationship between the energy balance components, W_{pfd}, W_{fr}, W_{se}, and the elastic energy W_H stored in the LS prior to onset of uncontrolled rupture of the specimen for (a) rocks of type A: 1—marble, 2—granite, 3—NBP sandstone, 4—BP sandstone, 5—sulphide ore, 6—lignite; (b) rocks of type B: 7—rock salt and 8—sylvinite.

the changes in the components of energy balance W_{fr} and W_{se} when the ratio between masses m_H and m_0 is varied. This evaluation is based on the following premises. Increase in the rate of deformation and rupture of the specimen (i.e., segment OC of the diagram in Fig. 3.15) and further acceleration of the broken rock material (segment CB) result from acceleration of the inertial mass (6) of the LS. In the machine shown in Fig. 3.12, this acceleration occurs when the inertial mass (6) is driven by the elastic energy in element (4). The combined movement of the inertial mass of the LS and the mass of broken rock occurs until the inertial mass of the LS reaches a maximum velocity. Beyond this point, fragments of failed specimen continue to move at the maximum velocity while the inertial mass of the LS decelerates and begins to vibrate. The maximum velocities of the inertial mass of the LS and the mass of broken rock at the moment of separation are determined from the expressions

$$V_{Hmax} = \sqrt{2W_{se}/m_H} \tag{3.7}$$

and

$$V_{0max} = \sqrt{2W_{fr}/m_0}, \tag{3.8}$$

where W_{se} is the energy of the vibratory process measured in the experiment; m_H the inertial mass of the LS; W_{fr} the kinetic energy of fly rock; m_0 the mass of the failed specimen. Equating equations (3.7) and (3.8), we get the ratio between energies W_{se} and W_{fr}:

$$W_{se}/W_{fr} = m_H/m_0. \tag{3.9}$$

Values of the energy W_{fr} in the graphs shown in Fig. 3.19 were determined from the expression $W_{fr} = W_H - W_{pfd}^d - W_{se}$. The ratios of W_{se} to W_{fr} were close to the ratio between the masses m_H and m_0, which in these experiments was maintained constant, approximately equal to 10 for all values of W_H and for all rocks. When the ratio of masses was altered in other experiments, the ratio between the energies also varied almost proportionally. This result confirms the validity of the analysis of interaction between the LS and the test specimen conducted above and also explains the mechanics of redistribution of elastic energy W_H between the components of energy balance W_{se} and W_{fr}. The transfer of energy from the LS to the volume of ruptured material is similar to the action of catapult, as illustrated in Fig. 3.20 (a). Here the potential energy of elastic compression, accumulated only in the LS can be converted into dynamic forms of energy (kinetic energy due to motion of mass m_0 and energy of vibratory processes of the LS), after a sudden withdrawal of the compressing force P.

The relationship obtained as a result of the foregoing investigations is given by

$$W_H = W_{pdf}^d + W_{fr} + W_{se} \tag{3.10}$$

which enables us to conclude that the energy related to thermal losses W_{th} under conditions of uniaxial rupture, compared to other types of energies, is very low; it does not exceed in magnitude the probable error in determination of the components of energy balance.

Thus, it can be concluded that for rocks for which energy consumption in post-failure deformation $W_{pdf} > 0$, the relatinships between the components of energy balance in the process of uncontrolled dynamic rupture may be determined from equations (3.4) to (3.10). The interrelationships between the components of energy balance, arising from these equations, can be expressed as follows:

$$W_{se} = m_H(W_H - W_{pfd}^d)/(m_H + m_0) \tag{3.11}$$

$$W_{fr} = m_0(W_H - W_{pfd}^d)/(m_H + m_0) \tag{3.12}$$

$$W_{pfd}^d = W_{pfd}^d(\dot{\varepsilon}_{max}). \tag{3.13}$$

The velocity of flying fragments can be determined by

$$V_{fr} = \sqrt{2(W_H - W_{pfd}^d)/(m_H + m_0)}. \tag{3.14}$$

3.3.4 Energy Balance in Brittle Failure of Rock for which the Energy Required for 'Post-peak' Deformation is Negative (Uniaxial Compression)

Between the situations in which energy is released by the specimen during rupture (i.e. $W_{pfd}^* < 0$) and that in which additional energy is consumed during rupture (i.e. $W_{pfd} > 0$), we have an intermediate case in which $W_{pfd} = 0$ (see Fig. 3.20a). In this case the amount of elastic energy contained in the specimen is sufficient to disintegrate the specimen, even under conditions of absolutely stiff loading. In this case, the rupture process is not dynamic in nature, since no energy is available to accelerate the fragments. If the loading system is not stiff, then the entire energy W_H stored in the LS is spent completely on creating dynamic effects (fly rock, vibratory processes).

The case of $W_{pfd}^* < 0$ is shown in Fig. 3.20 (b). It should be noted that in this case, the energy W_{pfd}^* differs in some respects compared to the cases considered earlier. In the preceding cases, the energy W_{pfd}^* is the energy consumed in deforming and rupturing a specimen in the post-peak zone of deformation; in the present case, W_{pfd}^* is the free energy stored in a specimen prior to the initiation of uncontrolled rupture.

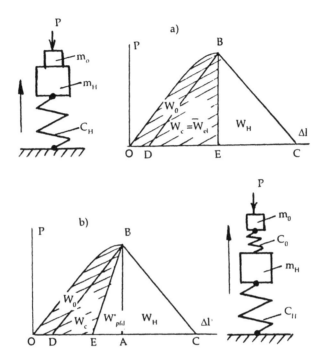

Fig. 3.20: Kinematic loading schemes for the situations $W_{pfd} \geq 0$ and $W_{pfd}^* < 0$, and the corresponding *load-deformation* diagrams for the specimen (OBE) and stiffness characteristic (BC) of the LS, together with the corresponding forms of energy.

The excess elastic energy W_{pfd}^* contained in the specimen behaves in essentially the same manner as the elastic energy W_H stored in the LS, helping to develop the dynamic conditions for rupture. The energy balance for such a system (as in the case of any other system) is determined by the manner in which the potential elastic energy of compression is transformed into other types of energy. For the case in which $W_{pfd}^* < 0$, more than one of the possible mechanics occurs. The particular mechanics that occurs is determined by the ratio of inertia masses of the two key elements in the system (i.e., the LS and the specimen), the duration of the rupture process, the acoustic transparency of the interface between the specimen and the LS for elastic waves generated during the failure stage. If we neglect the possibility of acoustic transmission of energy acrosss the barrier separating the two elements of the system, which determines the possibility of energy loss from the specimen volume undergoing rupture, three different energy balance situations can be distinguished.

1) When the inertias of both elements of the system are comparable in magnitude, then the relation between the energies W_{se} and W_{fr} (as in the case of $W_{pfd} > 0$) is determined by the ratio of the mass of the LS to the mass of the specimen, and the total amount of energy contributing to the development of dynamic effects is equal to the sum $W_{pfd}^* + W_H$. In this case, the energy balance for the dynamic process may be written

$$W_{pfd}^* + W_H = W_{fr} + W_{se}; \ W_{se}/W_{fr} = m_H/m_0.$$

2) When the inertia of the LS is much larger than the inertia of the test specimen, the excess energy concentrated in the test specimen is converted into kinetic energy of the fragments, which are accelerated to high velocity so rapidly that the external system LS is unable to react to this event and hence unable to transmit its energy to the disintegrating specimen. The energy balance for such a system may be written

$$W_{pfd}^* + W_H = W_{fr} + W_{se}; \ W_{se} = W_H; \ W_{fr} = W_{pfd}^*.$$

3) The third case is intermediate between the first and second. Here the total energy balance is the same, i.e., the entire elastic energy released from the system is converted into kinetic energy of flying fragments and into energy of vibration of the LS ($W_{pfd}^* + W_H = W_{fr} + W_{se}$) but the ratio between the energies W_{se} and W_{fr} is determined by whatever fraction of the energy W_H of the LS is able to be transferred to the test specimen (or vice versa) until the instant when contact between the LS and the failed specimen is lost.

In addition to the ratio between the inertias of the LS and test specimen, the acoustic transparency (impedance) of the boundaries between the LS and the test specimen is a very important characteristic of the system. This characteristic of the boundary conditions can produce significant changes in the ratios between the energy components derived above. The results of experimental studies on this topic are discussed below.

The energy balance during brittle failure was studied for the case $W_{pfd}^* < 0$ by testing specimens of soda-lime glass [67]. This glass is a typical representative in Wawersik's class II materials. It was selected because specimens of this glass exhibit very uniform mechanical properties. This is important in tests intended to identify how the energy balance of components change when the initial conditions of rupture are changed. Moreover, the tensile strength of the glass is 70–80 times less than the uniaxial compressive strength (i.e., much greater than most rocks, for which this difference does not exceed 20). Thus the glass brittleness is markedly higher, which creates conditions favourable to the generation of large energy flows, 'carried away' by the decompression wave that is generated upon the sudden rupture of the specimen. Large variations in energy ratios can be recorded easily during tests, leading to more reliable results. It was not possible to find any rock with such characteristics.

Since the dynamic effects of the rupture process are governed by the ability of the elements of the system (i.e., the LS and the test specimen) to exchange the potential energy of elastic compression stored in them before rupture, it is very important to study the conditions for this energy exchange. A special device was designed for this purpose. The elements of the original apparatus used are shown in Fig. 3.21 [67].

Tubular soda-lime glass specimens (1), internal diameter 30 mm, thickness 1 mm and length 200 mm, were tested in this device. The tubes were annealed at 500°C before testing to remove residual stresses. The ends of the specimens were carefully ground, then encased in steel bushings (2) and a liquid glue poured into the gap (3) between the bushings and the specimen. The design of the thrust bearings for load transfer from the press to the specimen differed in cases (a) and (b), as shown in Fig. 3.21. Different

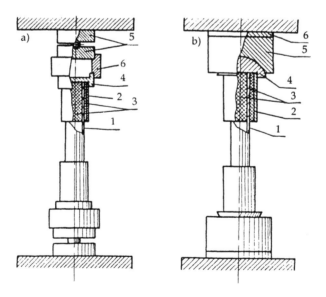

Fig. 3.21: Apparatus for attaching glass tube specimens to thrust bearings.

designs were used to create different conditions at the contacts, namely, case (a) in which the flow of elastic energy from specimen to LS or vice versa, was inhibited; case (b) in which conditions more favourable for energy transformation in the form of elastic unloading waves, which develop at the moment of specimen rupture, were provided.

In the case (a), the end of the glass tube passes through a steel part (4) and rests against a ring (6). Such thrust bearings hinder the passage of elastic waves through them because, firstly, the *steel-glass* interface is less transparent for elastic waves due to the large difference in acoustic impedance between glass and steel. Secondly, an additional air barrier is related between parts (4) and (5). The coefficient of reflection R of elastic waves at the *glass-steel* interface is more than 30% and is determined by the expression

$$R = (Z_1 - Z_2/Z_1 + Z_2)^2 \cdot 100 \qquad (3.15)$$

where Z_1 and Z_2 are values of acoustic impedance of the glass and the steel.

The acoustic impedance is defined as the product, $Z = \rho C$, where ρ is the density of the material; C is the velocity of longitudinal sound waves (**P** wave velocity) through the material.

When $Z_1 = Z_2$, the coefficient of reflection is equal to zero and elastic waves pass through the interface without reflection. This situation is encountered with the thrust bearing design used in case (b). Here the parts (4) and (5) in contact with the glass are made from aluminium. The acoustic impedance of glass and aluminium differs slightly so that the coefficient R is less than 2%. This situation makes the interface very 'transparent' to the passage of elastic waves generated during rupture of the specimen. The aluminium part (5) is massive; its volume is more than 100 times greater than that of the glass specimen. In both cases, the lower nodes, in contact with the specimen, are similar in construction to the upper nodes.

The sequence of rupturing the glass specimen in the tests was as follows. Rupture begins with growth of the first major crack, leading to rapid unloading of the specimen and formation of an elastic unloading wave. The velocity of propagation of a rupture crack through glass is very high, about one-third the velocity of sound in glass [36]. The unloading (relief) elastic wave, is a compression wave. It traverses the specimen and upon reaching the boundaries (including the ends of the specimen), is reflected, changing from compression to tension. Rupture occurs under the effect of the tension wave, forming new fragments in which alternating processes of *tension-compression* progressively break up the material. These processes continue until the entire reserve of elastic energy remaining in the specimen after initiation of rupture has been dissipated.

In tests conducted on machines equipped with thrust bearings of type (a), the elastic energy contained in the specimen prior to the onset of rupture remains trapped within the specimen, creating a huge crushing effect in glass, in contrast to tests conducted on machines equipped with thrust bearings of type (b).

Tests were conducted for various values of specific elastic energy Q_{el} stored in the specimen prior to the initiation of rupture. The value of Q_{el} was varied over a range of more than 30:1. The quantity of energy was controlled by varying the compressive load on the specimen. The maximum amount of energy was transmitted to the specimen when the stresses reached the compressive strength. This was followed by automatic self-disintegration of the specimen. When lower loads were applied, rupture of the specimen was initiated artificially by a slight blow from a diamond, or hard alloy bullet on the lateral surface of the glass tube. The microcrack created by the blow led to rupture of the tube at stresses of as low as one-tenth of the uniaxial compressive strength. The amount of energy conveyed to the specimens by the bullet impact was less than 0.001 of the minimal value of elastic energy developed in the tube prior to the initiation of rupture in the specimen.

Other methods of initiation of rupture were also used, such as applying the pre-assigned compressive load at a given eccentricity, or holding a hot wire close to the surface of the specimen. Results obtained with the various methods of initiating rupture were similar.

The parameters measured in the tests included the specific elastic energy Q_{el} stored in the specimen prior to failure and the quantity of fragments obtained after specimen failure.

The amount of energy Q_{el} was computed by the formula;

$$Q_{el} = \sigma^2/2E, \tag{3.16}$$

where σ is the applied axial compressive stress and E is Young's modulus for the glass.

It was proven experimentally that equation (3.16) is sufficiently valid to be applied to glass specimens. Tests showed Young's modulus and the Poisson ratio of glass to be constant over the entire range of loading up to the ultimate strength under uniaxial compression.

The quantity of fragments was determined by sieve analysis and weighing the mass of crushed material obtained after failure.

The relationship between the number of fragments and the specific elastic energy Q_{el} stored in the specimen prior to failure, is illustrated for two series of tests (a) and (b) in Fig. 3.22. The relationship is plotted on semilogarithmic coordinates.

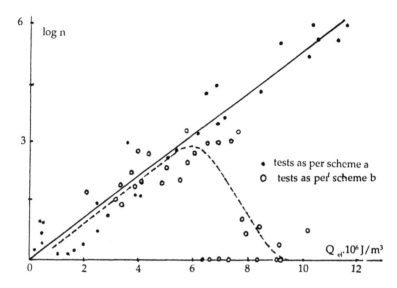

Fig. 3.22: Relationship between the number of fragments n into which the specimen disintegrated and the specific elastic energy Q_{el} stored in the specimen prior to failure for two series of tests (a) and (b).

Results obtained from the series (a) tests are indicated by black dots and approximated by a continuous line. The number of fragments observed at maximum specific energy $Q_{el} = 12 \times 10^6$ J/m^3 was almost one million per cu. cm in these tests. Specimen failure was accompanied by a loud blast-like noise and rapid projection of fragments.

Results of the series (b) tests are shown in the Figure by open circles and approximated by a dashed line. Up to the energy level $Q_{el} = 6 \times 10^6$ J/m^3, the results of the two series of tests coincided within the limits of scatter of the data. Divergence of the dashed line from the straight ray began at rather high values of elastic energy. Also, the number of fragments dropped sharply. In some tests the specimens were not crushed at all, but disintegrated into several large pieces. In such cases, the fly-fragments phenomenon was not observed. Failure was accompanied by a sound resembling the dull thud of a lead hammer against an anvil.

The sharp drop in the degree of disintegration of the specimen material is due to the transfer of part of the energy of the unloading wave into the large aluminium end caps (5), which are 100 times larger in volume than the specimen. The unloading wave can propagate readily into these caps where the energy density of the wave drops sharply due to their large volume. Reverse transmission of energy from the caps back into the specimen was inhibited because of the small contact area between the aluminium caps and the ends of the specimen. The aluminium caps may be considered to act as a perfect black body (or energy sink) in the acoustical sense, in which the specimen/aluminium cap interface acts as a slot through which the unloading wave is discharged into the volume of the total 'black body'.

The similarity of results obtained with different test schemes at low elastic energies, is due to the imperfect contact between the ends of the glass tube and the caps at the low normal stresses across the interfaces. When the contact is not 'perfect', the interface becomes less 'transparent' to the unloading wave (i.e., some reflection occurs at the interface).

Measurement of the kinetic energy of fly rock W_{fr} was attempted in a series of tests [67], using the set-up shown schematically in Fig. 3.23. The method of attaching a tubular specimen is the same as that described earlier. The ends of the specimen (1) were encased in steel sockets (2) by means of an epoxy glue poured into the gap (3) between the two sockets. A load P was applied along the axis of the specimen whose ends rested against steel or aluminium end caps (7), providing different acoustic impedance (transmissivity) across the boundary between the specimen and the LS. Tests

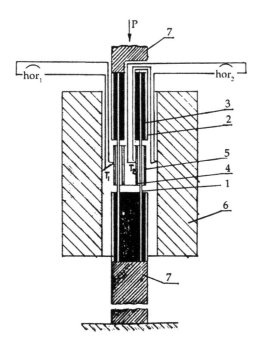

Fig. 3.23: Set-up for calorimetric tests to measure the kinetic energy of flying fragments of a specimen.

were also conducted under conditions of a fully transparent interface (i.e., no reflection of energy) between the test specimen and the LS. Glass tubes 400 mm long were used in these tests and each end attached with epoxy glue to a long steel socket. The long ends of the tubes served to absorb the energy of the relief (unloading) wave. The volume of these ends was 10 times larger than the disintegrated part of the specimen.

The elastic energy Q_{el} was varied by applying the load P eccentrically relative to the axis of the specimen. In the absence of eccentricity, rupture occurred at maximum load, corresponding to the peak strength of the glass under uniaxial compression. Increase in eccentricity reduced the breaking load. The range of variation in breaking stress was found to be the same as that encountered in tests in which rupture was initiated by the impact of a diamond bullet.

The kinetic energy of flying fragments was measured using a calorimeter placed around the specimen. The calorimeter consisted of internal (4) and external (5) bushings of pure copper. When the specimen ruptured, fly rock struck the copper bushings and heated them, since the kinetic energy was converted into heat. The quantity of heat was determined by measuring the temperature of the calorimeter using differential copper-constantan thermocouples T_1 and T_2. Hot junctions were placed on the walls of the calorimeter and cold junctions on the large external shield (6) made of aluminium.

The maximum increase in temperature of the calorimeter, due to impact of the fragments, did not exceed 1°C. The sensitivity of the instruments used in the tests allowed temperature variation to be measured with an accuracy up to 0.01° C. Conversion of the recorded temperature into a quantity of heat which, in turn, was equivalent to the original kinetic energy, was effected using a calibration graph. Calibration was carried out using microheaters consisting of electrically resistant constantan wire extensometers 30 microns in diameter. Constantan gauges were glued to the surface of the bushings of the calorimeter, through which a correctly measured amount of electrical energy was supplied. The temperatures of the calorimeter bushings were recorded using the method described earlier.

Results of tests conducted to determine the kinetic energy of the fly rock are shown in Fig. 3.24—with the specific kinetic energy of flight of fragments Q_{fr} plotted as a function of the potential elastic energy Q_{el}. Results of tests with acoustically 'less transparent' contacts between the steel end caps and the glass specimen are indicated by black circles, approximated by curve 1. Results of tests conducted with 'acoustically transparent' contacts (i.e., non-reflective interface) between the aluminium/glass contacts and glass/glass contacts are indicated by black squares and triangles respectively. The two transparent contact systems gave similar results within the limits of scatter and are approximated by the single curve 2.

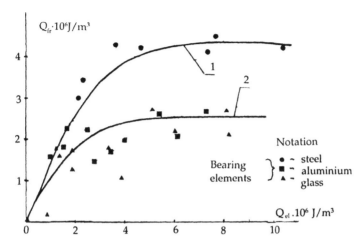

Fig. 3.24: Specific kinetic energy of fly fragments Q_{fr} as a function of the specific elastic energy stored in the system Q_{el} before failure, for different cases of acoustic transparency of the test specimen/LS interface.

The kinetic energy of the fragments was found to be higher in the case of the less transparent contacts since in such cases, more of the elastic energy of the relief (unloading) wave was 'trapped' within the specimen volume. The apparent independence of Q_{fr} and Q_{el} at values of $Q_{el} > 5 \times 10^6$ J/m^3 may be due to errors in the measurement of the energy Q_{fr}. Additional studies using modern, more sensitive methods for recording kinetic energy of the fly fragments will be required to resolve this question.

The duration of the brittle failure process is important as the rate of energy release determines the power of the process. To investigate this question, tests were conducted on a machine schematically similar to that shown in Fig. 3.23 except that the calorimeter was replaced by piezoelectric transducers positioned around the specimen. Collision of rock fragments with the piezoelectric transducers, generated electrical signals that were amplified and transmitted to a cathode ray oscillograph with a triggered sweep. The time interval between the impact of the first and last fragment was taken as the indicator of the duration of the failure process. Results of these experiments are shown in Fig. 3.25.

The duration t is plotted on the vertical axis as a function of the specific elastic energy Q_{el} stored in the specimen prior to the initiation of rupture. The time duration was determined with an accuracy of approximately 10^{-4} s, with a scatter in the range of $\pm 10^{-4}$ s; hence the experimental results lie in three rows. Results indicate that within the limits of accuracy of

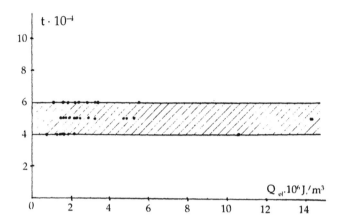

Fig. 3.25: Relationship between duration of the brittle rupture process for glass specimens and the specific elastic energy Q_{el} stored in the specimen before the initiation of rupture.

determination, the duration of the failure process is independent of the specific elastic energy Q_{el} stored in the test specimen prior to failure. The average duration of the failure process and flight of rock fragments was 5×10^{-4} s.

We shall now consider the total energy balance of the system [i.e., the *volume under rupture (the test specimen) and LS*]. Each component of the energy balance (in absolute units) is determined as a function of the total quantity of elastic energy that is stored in the system prior to failure—and able to create dynamic failure (similar to the relationships shown in Fig. 3.19).

The energy W_{el} is easily determined from the values of specific elastic energy in the specimen prior to failure, as indicated in Figs. 3.22 and 3.24, and from the dimensions of the specimen (the volume of the specimen was 7.5×10^{-6} m^3).

The energy W_c was not determined in the tests but can be calculated. Assuming a specific surface energy of 1000 erg/cm^2 [59] and the maximum 'degree of crushability' during rupture (from the graph shown in Fig. 3.22), we obtain a value of W_c equiavalent to about 0.3 J.

The energy W_{pfd}^* is the free energy contained in the specimen prior to the initiation of rupture, and is determined from the relationship $W_{pfd}^* = W_{el} - W_c$. Since the value of energy W_c in the tests was less than 1% of the energy W_{el}, we can take $W_{pfd}^* = W_{el}$ without significant error.

The value of the energy W_H is limited to the energy stored before initiation of failure in those parts of the glass tubes (encased in sockets) that were not ruptured during the test. This approximate estimate of W_H suffices for a qualitative evaluation of the energy transfer between elements of the system *specimen-LS*.

Figure 3.26 shows the relationship between the dynamic components W_{fr} and W_{se} of the energy balance during the rupture process of glass as a fraction of the total reserve of free elastic energy in the *specimen-LS* prior to initiation of uncontrolled rupture ($W_{pfd}^* + W_H$). Two experimental conditions are shown: (a) less transparent contacts and (b) transparent contacts. The dashed line shows the free energy W_{pfd}^* stored in the test specimen before initiation of failure, as a fraction of the total energy balance. Comparison of the relationships obtained for the two cases shows that (apart from the difference in values of energy W_{fr} highlighted earlier) in case (a), over the lower range of energies ($W_{pfd}^* + W_H$), the kinetic energy of flying fragments exceeds the free elastic energy W_{pfd}^*. This range (in which $W_{fr} > W_{pfd}^*$) is shaded in the diagram. This result is indicative of the fact that in the case of the less transparent contact, the kinetic energy of the fragments is increased due to the energy W_H stored in the loading system before failure. In tests with transparent contacts, the energy W_{fr} did not exceed energy W_{pfd}^* for any test. For the case wherein the initial reserve of elastic

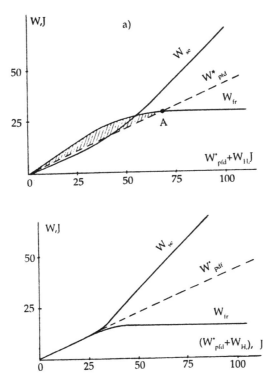

Fig. 3.26: Relationship between the dynamic components W_{fr} and W_{se} of energy balance in the rupture process of glass and the total reserve of free elastic energy ($W_H + W_{pfd}^*$) in the *specimen-LS* before initiation of uncontrolled rupture, for the two test conditions: (a) less transparent contacts and (b) transparent contacts.

energy ($W_H + W_{pfd}^*$) in the system was large, the fraction of energy W_{fr} in the total balance decreased while the proportion of W_{se} increased—for both the transparent and the less transparent interface situations. As mentioned above, the apparent independence of W_{fr} from ($W_{pfd}^* + W_{fr}$) for large values of the initial elastic energy in the system, could well be a result of experimental errors in measurement of energy Q_{fr}.

The foregoing studies show conclusively that the degree of acoustic transparency of the boundaries between a test specimen and the loading system exerts a strong influence on the redistribution of initial elastic energy of compression (i.e., that stored in the system prior to failure) to other types of energy into which it is converted. Reduction in acoustic transmissivity of the boundaries hinders the escape of energy from the rupture zone in the form of a relief or unloading wave, with the following consequences:

— an increase in the energy of rupture W_c—manifested in increased fragmentation;
— reduction in the seismic energy fraction W_{se};
— increase in kinetic energy of flying fragments W_{fr}.

These results on the influence of acoustic transparency of boundaries between test specimen and LS have been used as the basis for consideration of this factor in analysis of the energy balance in natural dynamic phenomena [7].

3.3.5 Energy Balance in Triaxial Compression

Before directly embarking on the study of energy balance in the process of dynamic uncontrolled rupture under triaxial (volumetric) compression, it is essential to recognise the influence of confining pressures σ_2 on the character of deformation and rupture of rock specimens. The difference in character of defermotion affects the process of energy exchange between the loading system and the test specimen [94].

In general, increase in confining pressure on a rock leads to an increase in plastic behaviour, i.e., to an increase in the residual deformation in the pre-peak and post-peak strength regimes and a reduction in the post-peak drop modulus ($M = d\Delta\sigma_1/d\varepsilon_1$). However, in many rocks this pattern of variation in the modulus M may change over a particular range of confining pressures σ_2. For example, let us consider the relationship between *stress* $\Delta\sigma_1$—*relative axial* ε_1 and *lateral deformation* ε_2 of the specimen shown in Fig. 3.27 for various confining pressures σ_2 in tests on (a) granite, (b) sandstone and (c) lignite. The confining pressure σ_2 for each test is indicated on the curves. It is evident that over a specific (critical) range of σ_2, the rock becomes more brittle as the confining pressure is increased, i.e., the post-peak branch of the diagram $\Delta\sigma_1 - \varepsilon_1$ becomes steeper, and the value of the modulus M increases. The pressure at which this effect was greatest was

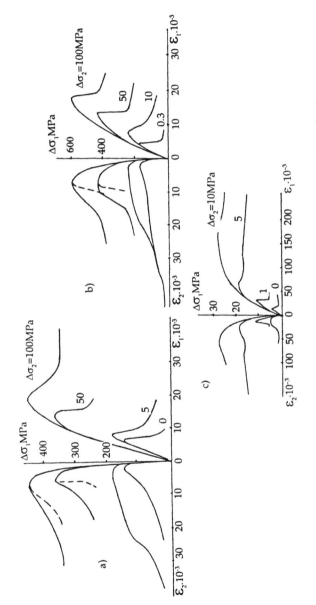

Fig. 3.27: Diagrams of $\Delta\sigma_1$ versus $\varepsilon_1 - \varepsilon_2$ for (a) sandstone, (b) granite and (c) lignite

50 MPa for granite and sandstone and 1 MPa for lignite. In the sandstone, the modulus M increased by a factor of three, compared to its value at low levels of confining pressure; in granite, the increase was a factor of 6 and in lignite 6.5. Graphs of M versus σ_2 are shown for these rocks in Fig. 3.28.

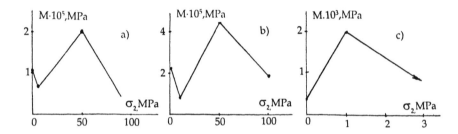

Fig. 3.28: Relationship between the post-peak drop modulus M and confining pressure σ_2 for (a) sandstone, (b) granite and (c) lignite

This behaviour is explained by a change in the mechanics of the deformation processes over a given range of confining pressures σ_2. Under low confining pressures, the deformation process involves the growth of numerous microcracks, which are distributed almost evenly over the entire volume of material, and the specimen dilates considerably, as seen by the high values of lateral deformation ε_2 in the graphs of Fig. 3.27. The process of multiple crack formation requires very large energy inputs, as indicated by the large area under the post-failure deformation $\Delta\sigma_1 - \varepsilon_1$ curve, and a relatively low value of M.

With increase in confining pressure, at a particular moment this uniformity of deformation over the entire volume is violated and the post-peak strength deformation process localises along a single shear plane. The lateral deformation drops sharply and is no longer distributed evenly over the specimen volume.

In conducting these tests, the lateral deformation of the specimen was measured by several extensometers at different locations around the specimen. Graphs of $\Delta\sigma_1 - \varepsilon_2$, based on the extensometer readings, are shown in (Fig. 3.27) for both the highest (continuous line) and the least (dashed line) change in lateral deformation. In the direction coincident with the rupture plane, deformation was almost absent and, in individual cases, even became negative due to elastic contractions as the load on the specimen decreased. In the perpendicular direction, deformation was produced by movement along the rupture plane. Very little volumetric disintegration occurred during the plane formation. The energy consumption during this process was considerably less than in the preceding case (i.e., under low confining pressure) and, correspondingly, the area under the $\Delta\sigma_1 - \varepsilon_1$ curve in the post-peak regime decreased, thereby increasing the value of M sharply.

This difference in the mechanics of development of the deformation process in the post-peak strength regime for various confining pressures, leads to the following important feature of energy distribution.

Under conditions of uniform development of the deformation process throughout the entire volume of the specimen, the change in specimen length does not significantly affect the ratio of energies W_{el}/W_{pfd} since, with change in length, there is a concomitant proportional increase in the elastic energy stored in the specimen and in the energy dissipated in permanent complete rupture of the specimen.

The situation is different when testing rocks under conditions in which rupture develops along a single shear plane. In this case, the energy exchange process varies considerably. This is explained by the fact that the energy content for the development of rupture plane remains constant, irrespective of the specimen length, while the quantity of elastic energy stored in the specimen varies in proportion to its length. Thus, with increase in specimen length, a large quantity of energy is supplied to the rupture zone from the test specimen itself and less energy (equal to W_{pfd}) is needed from the loading system. Diagrams of the *axial load P* versus *axial deformation* Δl of the specimens, obtained for specimens of different lengths, exhibit different steepness of the post-peak portion of the diagram; moreover, the longer the specimen, the steeper the post-peak branch. When the specimen length exceeds a specific critical length (l_{crit}), the rupture process proceeds only due to the stored elastic energy of the material of the specimen and is uncontrollably dynamic in nature even under conditions of perfectly stiff loading.

Typical diagrams of $P - \Delta l$ (OBEK) for specimens of different lenghs, are shown in Fig. 3.29. The middle diagram (b) corresponds to a specimen of critical length $l = l_{crit}$, in the diagram (a) $l < l_{crit}$ and in the diagram, (c) $l > l_{crit}$. The load P is the product of the differential axial stress in the specimen and its cross-sectional area.

Line BC reflects the stiffness characteristic of the LS. The cross-hatched areas under the curves refer to specific types of energy.

Area OBEKH represents the total deformation and the rupture energy of the specimen. This includes:

— irreversible work W_0 used to deform the specimen to point B;
— work done in permanent deformation and rupture W_c of the specimen after reaching point B, derived from the elastic energy stored in the specimen;
— work W_{pfd}, which in the case of (a) is the energy of deformation and rupture of the specimen beyond the peak strength, derived from the LS;

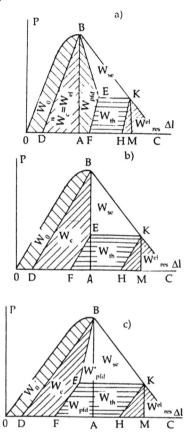

Fig. 3.29: Typical $P - \Delta l$ curves for specimens of different length, indicating the main energy components in the energy balance during uncontrolled rupture under triaxial compression.

— work W_{pfd}^*, which, in the case of (c), is the reversible energy stored in the specimen before the onset of uncontrolled rupture;

— work used to deform the specimen to its residual strength W_{th}, which is dissipated in frictional movement along the rupture plane and is converted to heat.

Under peak load equilibrium conditions corresponding to point B on the diagram, the elastic energy of compression is stored in the system and is concentrated over the volume of the specimen (W_{el}) and in the LS (W_H), where W_{el} corresponds to area DBA and W_H to area ABC.

When the rupture process is complete, some elastic energy W_{res}^{el} (area *HKC*) remains in the system due to the residual strength of the specimen. Part of the energy is stored in the specimen (area HKM) and part in the LS (area MKC).

The energy released during rupture is converted into seismic (vibratory) energy. As can be seen in the diagrams, increase in length of specimen increases the area representing this free energy (unshaded portion) and, consequently, the rupture process becomes 'more dynamic'. The free energy for cases (a) and (c) is determined from the following expressions:

(a) $W_{se} = W_{el} + W_{H} - W_{c} - W_{pfd} - W_{th} - W_{res}^{el}$;

(c) $W_{se} + W_{pfd}^{*} = W_{el} + W_{H} - W_{c} - W_{th} - W_{res}^{el}$.

All components in the energy balance can be calculated readily from these diagrams.

Thus, the energy balance during dynamic uncontrolled rupture under triaxial compression has its own particular characteristics that differ from those of the energy balance under uniaxial compression. These characteristics are:

1) Failure is not accompanied by the flight of fragments; hence this part of the energy is absent in the energy balance.

2) Due to the residual strength, governed by frictional forces between parts of the failed material, a major portion of the energy is converted into heat. This component may amount more than 50% of the total energy of the process.

3) After failure, some elastic energy remains in the *specimen-LS* due to the residual strength of the specimen.

4) Under triaxial conditions, if the fracture process occurs along a single shear plane, the extent to which instability develops (and associated dynamic effects) depends on the length of the test specimen.

The mentioned above change in the mechanics of the deformation process in the post-peak strength regime, in which the energy consumed in fracture decreases and the post-peak (or 'drop') modulus M increases over a certain range of confining pressures, can have serious consequences in the form of dynamic fracturing in the vicinity of mine workings. We shall consider the equilibrium of the system (i) *volume undergoing fracture* and (ii) *the surrounding rock mass* at different distances from the surface of a mine excavation.

Figure 3.30 shows a typical characteristic of the stress state of a part of the rock mass in the region of stress concentration near the excavation. Points 1 and 2 in this diagram represent a zone deformed beyond the peak strength, point 3 in the region of peak strength and point 4 in the pre-peak bearing-pressure regime. The magnitude of the minimum principal stress σ_3 increases as the point of interest moves farther from the excavation and deeper into the rock mass. For the four selected points, the minimum principal stress changes such that: $\sigma_3^1 = 0 < \sigma_3^2 < \sigma_3^3 < \sigma_3^4$. A typical $P - \Delta l$ diagram, representing the load deformation properties of the rock for the

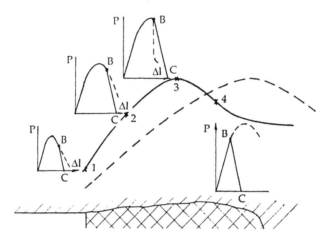

Fig. 3.30: Diagrams illustrating the conditions of equilibrium in the system *volume subjected to fracture-surrounding rock mass* in the immediate vicinity of a mine working

corresponding values of σ_3, is shown for each point. B in each diagram corresponds to the (load) stress intensity at the particular point. Left of point 3, the rock is in the post-failure deformation regime and right of point 3, in a zone of pre-failure deformation. That part of the $P - \Delta l$ diagram lying beyond point B is shown by a dashed line. The stiffness characteristic of the loading system, represented by the rock mass outside the failure region, is assumed to be the same in all cases and is indicated by the segment BC.

It is seen that at points 1, 2 and 4, the system is in stable equilibrium. At point 3, due to a change in the mechanics of deformation in the post-peak strength region at a critical value of σ_3 and a corresponding decrease in the energy consumed in fracture, the system develops a condition of instability. The zone of instability is situated in the region of high stress. As a result, dynamic character of the fracture increases. Moreover, loss of support in the deep-seated parts of the rock mass, carrying a major portion of the total load, can result in failure of the entire region around the excavation. A typical bearing-pressure curve after failure is shown as a dashed line. The failure zone in the region near the excavation is shaded. The resultant increased area of failure means a large increase in volume of failed rocks, which eventually will 'burst' into the excavation.

4

Mechanics of Deformation and Fracture of Saturated Rocks

4.1 INTRODUCTION

Under *in-situ* conditions, rocks are usually in contact with liquids or gases; naturally the pores, void and cracks in them become filled with fluid. The presence of a liquid or gas in rock interstices leads to changes in the properties of the rock mass. The interaction between the rock and the fluid(s) will vary with such factors as the state of stress in the rock, fluid pressure and temperature. These interactions affect the response of the rock and hence are important for a variety of practical geotechnical problems, including mine excavations, oil and gas exploitations, stability of deep boreholes, outbursts of rock and gas into mine workings, construction of underground storage chambers for liquid and gaseous fuel etc.

This chapter examines the effects of 1) liquid and gas flows in rocks and 2) pore pressure (liquid and gaseous) on the mechanical properties of rocks.

Systematic laboratory studies of these issues have been carried out by the authors since the early 1960s. The results of earlier studies were first published in 1968 [67]. In this chapter, interesting experimental procedures and results of studies are described, which reflect the extensive experience of the authors in dealing with these aspects.

4.2 METHODS USED TO STUDY PORE PRESSURE AND FLUID FLOW PROCESSES IN ROCKS

Testing machines were modified by the authors to enable conduction of tests over a wide range of experimental conditions, and analysis of the processes of fluid flow in rocks and the effect of pore pressure on their mechanical properties. The principle of operation of one such machine is described, together with details of components of other machines. These machines are intended for specific experiments—for example: study of the flow (percolation) properties of low-permeability rocks; study of brittle rocks in the post-peak strength regime; determination of rock porosity under *in-situ* conditions by saturating the rock with liquid or gaseous fluid; determination of the recoverability of fluid from rock when pore pressure is reduced; and study of fluid percolation into the walls of a borehole under conditions simulating hydraulic fracture or when using drilling mud to maintain borehole stability.

Figure 4.1 illustrates the basic details of a machine for testing rock specimens under axial compression at different levels of hydrostatic pressure [82]. The main elements in this machine are a hydraulic press, capacity 1000 tons, and a confining cell designed for a maximum hydrostatic pressure of 300 MPa. The press consists of a robust frame made of two massive crosspieces (1), four columns (2) and a hydraulic ram (3) fixed to the upper crosspiece. The ram is connected to a pressure source (5) by tubing through the pressure accumulator (4). The accumulator helps to hold the load on the specimen constant during long-term experiments. The hydraulic ram has a return cell in which gas pressure is maintained at the required level by means of an accumulator (20) to ensure return of the ram piston to its original position upon completion of the test.

The working cell consists of a thick outer cylinder (6), loading ram (14) and upper cover (19). The shoulders of the loading ram and cover form a pressure compensation cell (9) linked by a groove (21) to the working area of the cell. The (horizontal) cross-sectional areas of the working and compensating chambers are equal, in order to ensure that the pressure in the working cell remains constant when the loading ram is displaced during a test. A pump (7) generates the confining pressure in the cell. An additional accumulator (8) keeps this pressure constant while conducting long-term tests.

The specimen (11), placed in a tightly waterproof polyethylene jacket and metal thrust bearings (15), is attached to the loading rod (14). The specimen is centred in the cell by means of spherical surfaces on the rod (14) and on the dynamometer (10). Electrical wires from axial and transverse strain gauges attached to the specimen, and also on the dynamometer, are fed through the cell via the electric cable inlet (13).

The flow properties of the specimen can be measured in both the axial and transverse directions. In the first case, the fluid that is to flow through the specimen is fed to the ends of it through the thrust bearings (14). In the second case, the fluid is fed to special attachments (12) that are hermetically sealed to the lateral surface of the specimen. Fluid supply lines are led from the cell through the loading ram. The incoming supply is connected to a pressure source (17), and the outgoing lines to flow meters (16). When the effect of pore pressure is under study, the outgoing line is closed. A vacuum pump (18), provided to suction air from the specimen before the start of a flow experiment, is included in the fluid percolation circuit.

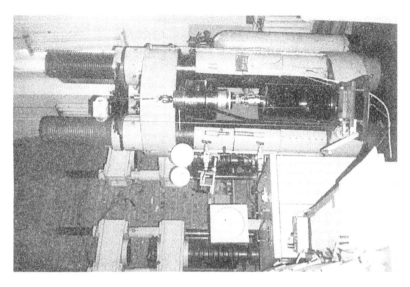

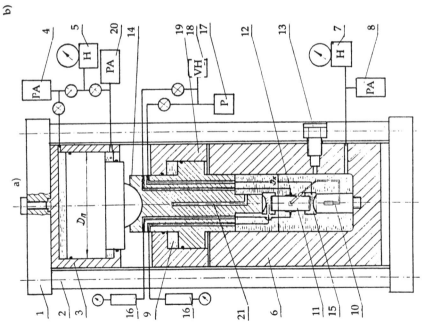

Fig. 4.1: (a) Schematic diagram of a machine for studying fluid flow in rocks; (b) Photograph: general view of the machine.

The internal diameter of the working cell is 160 mm and hence able to accommodate test specimens up to 100 mm in diameter and 200 mm in length. Specimens of such dimensions were used in tests to simulate conditions in deep boreholes, for study of the mechanics of core disking at depth and also to determine parameters for use in hydraulic fracturing tests.

The flow properties of brittle rock in the post-peak zone were studied by testing in a stiff machine, schematically illustrated in Fig. 4.2 [92, 106].

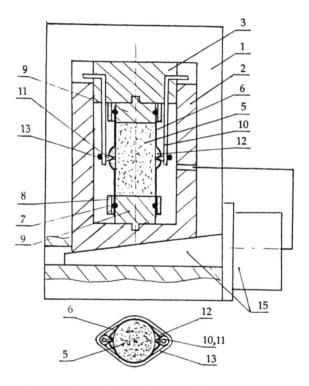

Fig. 4.2: Schematic diagram of a stiff press for study of fluid percolation in the post-peak strength regime of rock deformation.

The machine consists of a stiff monolithic steel frame (1), a W-type hydromechanical drive (15) and a high-pressure cell containing a cylinder (2) and an upper cross-head (3). The principle of operation of such a stiff press has been described in Chapter 1. Here we shall concentrate on the method of preparation of a specimen for study of fluid flow properties. The machine has channels to allow flow both in the axial and transverse directions through a specimen. For axial flow the fluid is supplied and drained through end thrust bearings (9), the lower bearing serving simultaneously as a dynamometer. The scheme of connecting flow channels for studying percolation in the transverse direction only is shown in the Figure; a simple and reliable method of assembly and hermetic sealing of the test specimen was used and is described below.

A specimen (5) is placed in a thin-walled polyethylene tube (6). Metal heels (12) are installed between the specimen and this tube. The heels contain a series of grooves along one surface in order to supply fluid to the specimen surface. The ends of the specimen are sealed by thick metal 'end-pieces' (9) with rubber O-rings (7) pressed tightly to the polyethylene tube by sleeves (8). The specimen is then attached to the cross-head (3). The heels (12) are connected to inlet and outlet tubes (10) by conical draught holes drilled from the upper surface of the heels, into which conical flanges, fixed to the ends of the tubes (10), are inserted. Conical flanges are pressed into the holes along with the polyethylene film of the tube (6) in which holes have been previously drilled for free inflow (and outflow) of the fluid from the tube to the specimen. The conical flange is sealed to prevent the reduced section from being covered by the inserted polyethylene film. Preliminary pressure to ensure hermetic sealing of the flange is effected by rubber plaits (bands) (13). During a test the pressure in the working cell intensifies the force pressing on the cones, thereby ensuring better sealing of the flange. Experience has shown that this conical connection design is very reliable in operation.

This stiff flow-testing machine allows investigations to be carried out at high confining pressures in the cell and at fluid pressures up to 250 MPa. The stiffness of the machine is 10 MN/m.

In cases in which it is necessary to record low volumes of fluid flow with high accuracy, a special machine is used, shown schematically in Fig. 4.3 [102]. A machine of this type was used in particular to study flow in rocks of low permeability; to determine open (exposed) porosity in *in-situ* rocks, as well as to measure the recoverability of fluid from pore space upon release of the pore pressure; and to study the flow properties of borehole walls while simulating sections of boreholes at large depths and while conducting hydraulic fracturing studies.

The machine is stiff. As shown in the schematic diagram in Fig. 4.3 (a), the high-pressure cell is equipped with a device for flow (injector) which is placed in the stiff loading press during experiments. A general view of the cell with injector, and a view of the machine per se are shown in the photographs in Fig. 4.3. It is essential, in order to ensure high accuracy and sensitivity in measuring the volume of fluid being fed into the test specimen and flowing through it, to *reduce to an absolute minimum the parasitic volume which comprises the volume of all voids and flow channels feeding into the specimen and those draining from it.* To satisfy this requirement, all inlet channels must have a minimum cross-sectional area and machine parts in contact with the specimen should be as short as possible. Imposing these requirements on

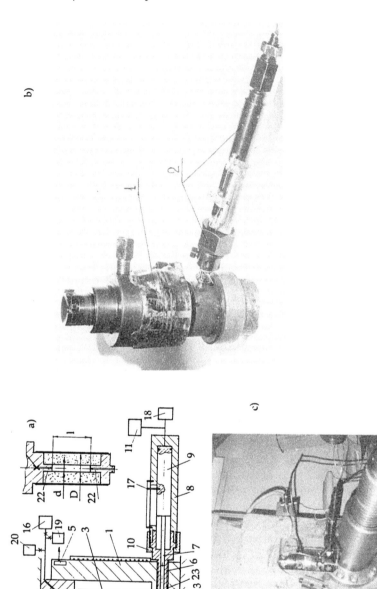

Fig. 4.3: (a) Schematic diagram of a high-pressure cell with an injector for measurement of low-volume flows. Photographs: (b) general view of high-pressure cell with injector and (c) the stiff flow testing machine.

the construction provides high stiffness in the loading system. The specimen (3) attached to the loading rod (2) is encased in a waterproof jacket. This arrangement does not differ in principale from similar constructions described above. The cell is equipped with an external electric heater (4) and a cooling system (5) situated in the zone of the hermetic seals, to allow tests at high temperature to be done with the machine.

A system for pumping fluid into the specimen, in the form of an injector, is the key unit in this machine. The injector has two pressure channels: one for high, the other for low pressure. The high-pressure channel consists of a cylinder (6) of small internal diameter and a piston (7). The cylinder (6) is attached to the bottom of the cell (1). The assembled pair—*cylinder (6)-piston (7)*—is a detachable element in the system. Selection of such pairs of different working diameters allows the volume of fluid being fed into the specimen to be recorded accurately over a wide range of volumes.

The low-pressure channel consists of a cylinder (8) of large internal diameter and a piston (9) and is linked to the high-pressure channel by a coupling screw (10). The low-pressure chamber is connected to pressure source (11) and the high-pressure chamber to pressure source (12). The machine is also equipped with a vacuum pump (16), source of back-pressure (19), flow meter (20) at the exit point from the celll and a system of valves. The volume of pressurised fluid supplied to the specimen from the chamber of the high-pressure injector is measured on the basis of the motion of pistons (7) and (9) by resistance gauges (17). Pore pressure is controlled using a strain gauge (23) cemented to a cylinder (6). The method of preparing specimens in order to conduct hydraulic fracturing tests is illustrated in Fig. 4.3 (a).

The machine is designed to operate at a maximum fluid pressure of 250 MPa. Accuracy in measurement of the volume of fluid transmitted to the specimen from the injector is about 0.005 mm^3.

The sequence of conducting tests using this machine depends on the objectives of the tests and is described later in the appropriate sections where test results are discussed.

In studying the processes of crack-pore space formation in rocks induced by permanent deformation, the varying volume of voids in the specimen was recorded by a U-shaped manometer (see Fig. 4.4). In this case the hermetically sealed capsule in which the specimen (1) was located, was connected to the pressure tube (2) with a manometer (3) with its free end exposed to the atmosphere. Any variation in the volume of void space induced by deformation in the specimen, created a corresponding change in the levels of the U-shaped tubes of the manometer. Special tests have revealed that crack formation processes develop very intensely mid-length of the specimen. Hence, to provide a dependable link between the manometer and the zone of maximum disintegration, a capillary axial hole (4) was drilled through the specimen. Naturally, such a method of measuring

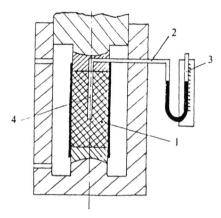

Fig. 4.4: System for recording volumetric strains in a specimen using a U-shaped manometer.

volumetric strain can be used only when the pore pressure in the specimen is atmospheric. For these cases, this method is highly preferable to the use of an injector because it is simple to construct and operate, and errors associated with friction between the piston and seal in the injector are excluded.

4.3 EXPERIMENTAL INVESTIGATION OF FLUID PERCOLATION PROPERTIES OF ROCKS

This section first discusses results providing insights into the nature of variation of fluid flow properties with change in stress state and during the deformation processes. This is followed by a discussion of the results of some complex comparative studies of two types of sandstones with very different characteristics. The first sandstone (burst-prone) belongs to the category of rocks classified as hazardous in relation to dynamic collapse behaviour such as outbursts of rock and gas; the second sandstone is classified as a non-burst-prone rock.

The testing machines described above allowed the fluid percolation properties (permeability) of the rock specimen to be measured in two directions: 1) along the specimen axis, which coincides with the direction of action of the maximum principal stress σ_1; and 2) across the specimen, i.e., in the direction of action of the minimum principal stress $\sigma_2 = \sigma_3$. A difference in stress level in different directions in the specimen influences the permeability measured in the different directions. As permanent deformation increases, this difference in permeability should also increase, since microcracks which develop at this point have a definite orientation, determined by the stress field. This fluid flow behaviour measured in different directions provides complementary information about the structural variations in the material induced by variations in stress state or deformation.

In the case wherein the volumetric compression in isotropic rocks is uniform, it seems likely that the fluid flow properties measured in different directions will be equal. However, experience revealed a significant difference in values of permeability coefficient computed by the same methodology for cases of fluid flow along and across the specimen. The relationship between the permeability K_{ft} and level of hydrostatic compression $\sigma_1 = \sigma_2 = \sigma_3$ pertaining to (a) marble and (b) lignite is shown in Fig. 4.5. Flow percolation was implemented by nitrogen under a pressure of 1.5 MPa at the inlet for marble and under 0.5 MPa for coal at a pressure equal to atmospheric at the outlet. The value of K_{ft} was calculated using the well-known formula:

$$K_{ft} = Vl \, \eta / St \, (P_{il}^2 - P_{ol}^2), \qquad (4.1)$$

where V is the volume of gas passing through the specimen; l is the base length of flow; η the viscosity of the gas; S the cross-sectional area of the specimen; t the duration of flow; P_{il} the gas pressure at the inlet; P_{ol} the gas pressure at the outlet.

Graphs were plotted in semi-log coordinates. The upper curves (1) correspond to flow across the specimen, while the lower curves (2) were obtained for the case of flow along the specimen. The general trends of the

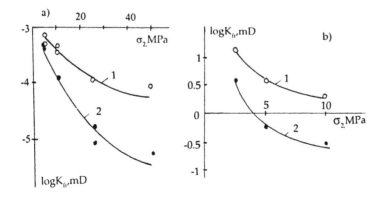

Fig. 4.5: Relationship between the permeability coefficient K_{ft} and hydrostatic pressure $\sigma_1 = \sigma_2 = \sigma_3$ for (a) marble and (b) lignite. Curves 1 were obtained for flow across the specimen and curves 2 for flow along the length of the specimen.

upper and lower curves are similar but the difference in numerical values of K_{ft} is quite large. Further, this difference increases with increment in the hydrostratic pressure since, with rise in σ_2, the value of K_{ft} for percolation along the specimen decreases more rapidly. For marble at $\sigma_2 = 5$ MPa the difference in values of K_{ft} may be considered to be the usual limits of uncertainty in evaluation of specific values; at $\sigma_2 = 50$ MPa the difference is one order of magnitude. For lignite, over the given range of pressures, the difference in values of K_{ft} varies by a factor of 4 to 8. Such a difference in coefficients of permeability is probably attributable to a scale effect. In the tests the base length of the specimen was 80 mm and the base width across the specimen was 25 mm. Formula (4.1) for computing K_{ft} assumes linearity between the value of K_{ft} and the base length of flow l. Most probably, at low values of l the relationship is not linear. To obtain comparable results of evaluations obtained by the two methods described above, it is necessary to make some corrections. Both methods can be used for qualitative evaluation of the effect of hydrostatic pressure on permeability.

Change in stress state and the associated elastic deformation of rock, and at higher stresses rock disintegration and dilatation as well as permanent deformation, exert considerable influence on the fluid flow properties of rock.

Graphs reflecting the variation in permeability during the deformation of marble specimens [35, 83] are shown in Fig. 4.6. The graphs were plotted in semi-log coordinates: the values of permeability K_{ft} in millidarcies on the vertical axis, with values of the differential axial stress $\Delta\sigma_1$ in the specimen shown on the horizontal axis. Flow occurred in the specimen in two directions—axially (Fig. 4.6a) and transversely (Fig. 4.6b). Tests were

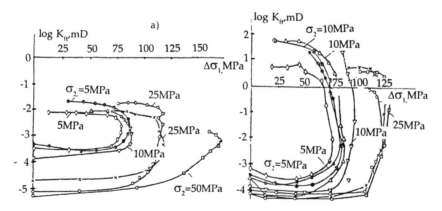

Fig. 4.6: Relationship between the permeability K_{ft} of marble and axial load $\Delta\sigma_1$ during pre-failure and post-failure deformation, obtained at various confining pressures σ_2 for cases of fluid flow (a) along the specimen and (b) across the specimen.

conducted at various confining pressures, the values of which are indicated on each curve. Nitrogen was used as the fluid flowing through the rock. Pressure at the inlet was 1.5–2.0 MPa and at the outlet, atmospheric. Pre-peak and post-peak deformations were investigated. In the zone of elastic deformation, the decrease in K_{ft} was associated with closure of pores and pre-existing cracks in the specimen. With the onset of permanent deformations, accompanied by expansion of crack-pore space in the material, the value of K_{ft} also increased. Near to the peak strength, since a large number of shear planes ω participate in the deformation process (see the deformation model in Section 1.4.3), leading to the rapid development of crack-pore space, a sharp rise in K_{ft} is observed. During post-peak deformation the end of the rapid rise in K_{ft} is related to localisation of the deformation process along one or a small number of shear planes ω—along which fracture occurs.

The huge difference in variation of permeability measured in axial and transverse directions is worth noting. For example, at confining pressures $\sigma_2 = 5$ and 10 MPa, in the first case K_{ft} increased by 1.0–1.5 orders of magnitude compared to 4.5–5.0 orders in the second case. At $\sigma = 25$ MPa, this difference was less: for flow along the specimen K_{ft} increased by 3 orders of magnitude versus 4.5 orders across the specimen. Two factors account for this difference: 1) those associated with test procedure and 2) the nature of the increase in microfracturing of the material during deformation.

1) In the case of axial flow, the end parts of the specimen play a role in the test. The stress distribution in the specimen is distorted by the effect of friction between the ends of the specimen and the loading platen of the press applying the load. The deformation process at the ends of the specimen is complicated and increases rapidly at mid-height of the specimen where disintegration reaches maximum. Due to uneven deformation along its length, the specimen eventually resembles a barrel. That part of the specimen abutting the ends remains less deformed and hence flow properties at these two places are changed only slightly compared to the initial material. For transverse flow the effect of the specimen ends on the value of K_{ft} is excluded.

2) The second reason for differences in permeability coefficient is the feature of the mechanism of the permanent deformation development. This aspect is detailed later (see Fig. 4.12).

Thus, transverse flow is obviously more sensitive to structural changes taking place in the specimen during the growth of deformation processes. Measurement of fluid flow across the specimen is preferable even when testing materials that hardly loosen as a result of deformation since, as shown by tests (see Fig. 4.5), the transverse permeability of a specimen is quite high, which makes it easy to record small rates of fluid flow during an experiment. Relationships between K_{ft} and axial stress $\Delta\sigma_1$ for lignite, obtained when specimens were subjected to strain under various confining pressures, are shown as an example in Fig. 4.7. Values of σ_2 are indicated on the graphs. Relationships were obtained while measuring flow (a) along the specimen and (b) across it. Absolute values of K_{ft} for case (b) were relatively high, while the magnitude of variation of K_{ft} in the deformation process was almost the same in both cases. In the post-peak zone, at $\sigma_2 = 2.5$ and 5 MPa, a slight increase in the fluid flow properties of the rock was observed. At $\sigma_2 = 10$ MPa, the entire process of deformation was accompanied by closure of pores and reduction in percolation properties.

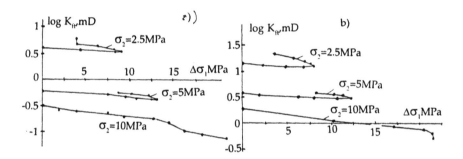

Fig. 4.7: Relationships between permeability K_{ft} for lignite as a function of the axial load $\Delta\sigma_1$ in the pre-peak and post-peak strength regimes obtained at various confining pressures σ_2 for cases of percolation (a) along the specimen and (b) across the specimen.

The deformation processes discussed above and the percolation flow associated with them occurred in a zone of relatively low pressures corresponding to the situation of rock at small depths. Deformation processes under the conditions that obtain at greater depths possess certain specific features, which affect the flow and reservoir properties of rocks. We shall examine these features using the example of marble [102].

A diagram of the mechanical state for marble, discussed earlier in Sec. 1.4.3, is shown in Fig. 4.8. This Figure shows the effect of confining pressure σ_2 on the peak strength τ_{us} and elastic limit τ_{el} of the rock. A line of 'critical state', obtained from experimental results, is also depicted in the diagram. This line divides the diagram into two zones: zone A, lying below the critical line and zone B (striped), lying above the critical line. In zone A, permanent deformation is typically not accompanied by increased disintegration or 'damage' to the rock structure nor an associated increase in volume of material (i.e., dilatation). In zone B, permanent deformation induces damage and growth in crack-pore space. The existence of these two zones with quite different behavioural patterns in rocks is indicative

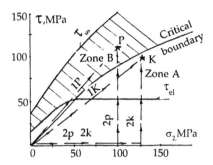

Fig. 4.8: Diagram of mechanical state for marble showing the trajectories of the simple (1) and complex (2) loading paths.

of the strong dependence of their structural characteristics as well as flow percolation properties on the history of variation in the stress state to which they have been subjected during the course of attaining their final state. We shall examine variations in indices of open porosity and permeability when specimens of marble are loaded through different stress paths to reach the same final stress state.

We shall consider two situations: in the first, the final stress state will be located in zone A (indicated by point K in the diagram); in the second, the final stress state is located in zone B (point P in the diagram). We shall also consider each of the two final stress states corresponding to points K and P, which are reached in the experiment through two paths: one by proportional loading and the other by a complex change in principal stresses. The proportional loading path is indicated in the diagram by numbers 1_K and 1_P and the paths of complex loading by numbers 2_K and 2_P. The following changes in rock structure can be expected when the final points are reached through these different paths. Since both paths of loading leading to point K pass through zone A, no disintegration ('damage') should occur in either path. A different picture is observed when point P is reached through the paths indicated. The path of proportional loading 1_P enters zone B immediately after passing the boundary of the elastic limit and travels in this zone up to point P, which should lead to severe disintegration of the rock. A large portion of the path of complex loading 2_P passes through zone A after reaching the elastic limit (without disintegration) and enters zone B only at the end of the path after crossing the critical line; consequently, the degree of disintegration will be relatively insignificant.

Figure 4.9 shows the relationships obtained in the test of shear stress τ in the material and relative changes in volume of open porosity θ versus the magnitude of axial deformation ε_1 when point K was reached through

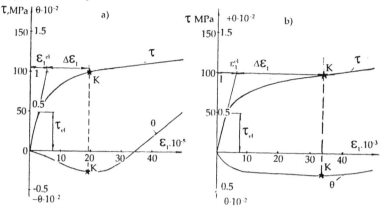

Fig. 4.9: Diagrams showing the relationships for marble of shear stress τ and relative volume of open porosity θ versus the magnitude of axial deformation ε_1 for two paths of loading up to point K: (a) complex path 2_K and (b) proportional path 1_K (see Fig. 4.8).

the complex loading path 2_K (diagram a) and through simple (i.e., proportional) loading 1_K (diagram b). Loading path 2_K was realised under the condition $\sigma_2 = 125$ MPa = const, and loading path 1_K under the condition $\sigma_2/\sigma_1 = 0.385$ = const. The final situation attributed to point K is indicated in the two diagrams by the letter 'K'. As can be seen from the graphs, the loading process was accompanied in both cases by development of quite large permanent deformations $\Delta\varepsilon_1$ (ε_1^{el} in the graphs corresponds to the amount of elastic strain at point K). For the complex loading path at point K, $\Delta\varepsilon_1 = 14 \times 10^{-3}$, and for proportional loading $\Delta\varepsilon_1 = 28 \times 10^{-3}$. In both cases, the change in the volume θ of open porosity in the material has the sign of reduction (i.e., a net volumetric contraction). Curves of θ were obtained by testing in a machine fitted with a U-shaped manometer (see Fig. 4.4). The permeability at point K was 10^{-6} millidarcies (mD).

In both cases the deformation process was continued beyond point K. As the point K was almost situated on the critical line, with increase of stress in material the zone B was reached. The parts of diagrams θ–ε_1 corresponding to zone B, show increase in the volume.

Figure 4.10 shows the relationships of shear stress τ and permeability K_{ft} versus the magnitude of axial deformation ε_1 when pont P was reached through the comx loading path $2p$ (diagram a) and through proportional loading path $1p$ (diagram b). Loading path $2p$ was realized under the condition $\sigma_2 = 100$ MPa = const, and loading path $1p$ under the condition $\sigma_2/\sigma_1 = 0,31$ = const. The diagrams K_{ft} - ε_1 present only parts reflecting the increase of K_{ft}. The situation attributed to point P is indicated in the diagrams by letter 'P'.

It can be seen from these graphs that in the case of complex loading the permeability began to increase only after the shear stress in the material exceeded a critical value and the stress state began to conform to zone B

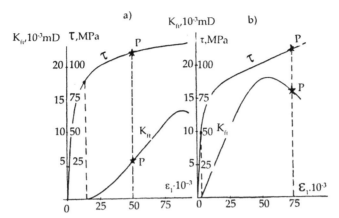

Fig. 4.10: Diagrams showing the relationships for marble of shear stress τ and permeability K_{ft} versus the magnitude of axial deformation ε_1 for two path of loading up to point P: (a) complex path 2p and (b) proportional path 1p. (see Fig. 4.8).

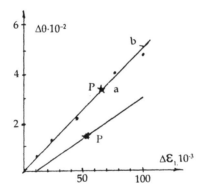

Fig. 4.11: Permanent volume $\Delta\theta$ of open porosity in marble versus magnitude of permanent deformation $\Delta\varepsilon_1$ for (a) complex 2P and (b) proportional 1P paths of loading.

($\tau > 85$ MPa). At the initial moment of rise in K_{ft}, permanent deformation in the body was $\Delta\varepsilon_1 = 8 \times 10^{-3}$ and at point P reached 43×10^{-3}. The permeability rose concomitantly to a value of 5×10^{-3} mD. In the case of proportional loading, increase in filtration activity was observed right after passing through the elastic limit, i.e., starting from $\tau = 50$ MPa, while the intensity of the rise in K_{ft} (i.e. $\Delta k_{ft}/d\varepsilon_1$) was almost 2.5 times higher than that in the preceding case. At point P the magnitude of permanent deformation was 69×10^{-3} while the permeability increased to 16×10^{-3} mD. A sharper rise in K_{ft} in the case of proportional loading is explained by a sharp expansion of void space in the material. Graphs indicating the relationships of permanent volume of open porosity $\Delta\theta$ developed in the marble specimens for deformation along complex (a) and proportional (b) loading paths, are shown in Fig. 4.11. The same loading paths were followed as in the earlier flow test. The values of $\Delta\theta$ and $\Delta\varepsilon_1$ for the stress state which is of interest to us, are marked by points P on these graphs. The volume of empty space ($\Delta\theta$) thus formed in the material when a stress state conforming to point P was attained under conditions of proportional loading (i.e., $\Delta\theta \approx 3.4 \times 10^{-2}$) was approximately 2.25 times greater than under the complex loading path ($\Delta\theta \approx 1.5 \times 10^{-2}$).

It is interesting to note that the volume of empty space $\Delta\theta$ and magnitude of permanent (irreversible) deformation $\Delta\varepsilon_1$ exhibited a linear relationship throughout the entire test while the relationship between K_{ft} and $\Delta\varepsilon_1$ was complex in nature see Fig 4.10. At the initial stage of deformation K_{ft} increased roughly in direct proportion to the value of $\Delta\varepsilon_1$, but then for subsequent deformation the rate of increase in K_{ft} decreased and, finally, the permeability reached a maximum value, after which it began to decrease. Such changes in the fluid flow properties of rocks with change in deformation can be explained as follows. In the given tests flow gauges were positioned on the lateral surface of the specimen (see Fig. 4.12). During the initial stage of deformation, accompanied by loosening and formation of shear planes ω, flow channels formed which traversed the entire specimen. With further deformation, the dimensions of these channels increased and flow along them increased correspondingly. However, as the growth of ω planes occurred simultaneously in all directions, the system of planes situated across the direction of percolation (see Fig. 4.12b) changed the connectivity of the initially straight channels between the flow gauges, thereby making passage through the channels more difficult. At high levels of deformation

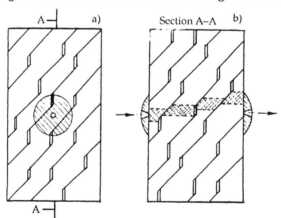

Fig. 4.12: Diagram explaining the reduction in flow properties of rocks at large deformations.

the overall connectivity of the channels or pathways can be significantly reduced as part of the pathways become converted into isolated pores.

The following practical significant conclusions may be drawn from the results of the experiments discussed above:

1) By making use of the mechanical state diagram for a specific rock, based on analysis of the history of variation in stresses to which the rock has been subjected, the evolution of crack-pore space can be assessed. Using the diagram, one can also plan a path of stress change in such a manner that, according to the objective, either maximum 'loosening' of the material is achieved or expansion of additional crack-pore space eliminated.

2) As 'loosening' of rock induced by deformation processes is not necessarily accompanied by increase in its flow properties, the use of rock fracturing methods in order to increase permeability, should be undertaken by taking into account the complex relationship between the deformation and flow properties of the rock described above.

4.3.1 Fluid Flow Properties of Burst-prone (BP) and Non-burst-prone (NBP) Sandstones of Donbass

The results of complex comparative studies of two types of sandstone from the Donbass coal mining basin are presented. One sandstone was taken from a zone classified as hazardous (BP) in relation to the manifestation of dynamic phenomena, such as sudden outbursts of rock and gas during mining operations, and the other from an NBP zone [85]. Experiments were set up with two main objectives—first, to establish significant indices and characteristics whereby the two sandstones could be reliably distinguished and, second, to determine the mechanics of the dynamic phenomena occurring in the form of outbursts. The search for criteria by which to distinguish the sandstones involved studies of the following:

1) behaviour of the mechanical properties of the rocks for various stress states;
2) initial structural state of the rocks and the nature of their variation under loading and increase in permanent deformations;
3) fluid flow properties of the rocks under conditions of hydrostatic pressure and also under increasing levels of permanent deformation right up to fracture for different levels of deviatoric triaxial compression.

Cylindrical specimens 30 mm in diameter and 70 mm in length, which appeared macroscopically to be without cracks, were tested in a stiff machine. This machine allowed loads on the specimen to be controlled during deformation in the post-peak zone.

Investigations revealed that these sandstones have similar characteristics with respect to strength and deformability and, based on these features, it

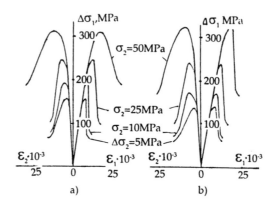

Fig. 4.13: Stress-strain relationships $\Delta\sigma_1$ versus ($\varepsilon_1 - \varepsilon_2$) for (a) NBP and (b) BP sandstones obtained at various confining pressures σ_2.

Table 4.1

Rock type	σ_2, MPa	$\Delta\sigma_1^{us}$, MPa	ε_1^{us}, $\times 10^{-3}$	ε_2^{us}, $\times 10^{-3}$	$\Delta\sigma_1^{res}$, MPa	ε_1, $\times 10^{-3}$	ε_2, $\times 10^{-3}$	$\Delta\theta$, $\times 10^{-3}$
	0	142	5	2	10	6	20	30
	5	160	8.5	4.7	70	9	17	25
NBP	10	185	9.9	6.9	80	10.5	16.5	22.5
	25	250	12	7.9	105	17.5	17.5	17.5
	50	310	20	13	205	27.5	27	26.5
	0	148	7	6	8	9	38	42
	5	156	9.5	4.4	64	10.8	15	19.2
BP	10	173	11.6	8.5	73.5	11.8	15.8	19.8
	25	240	13.2	8.9	95	16.7	16	15.3
	50	323	19	11.3	165	22	25	28

is very difficult to distinguish between the two types of sandstone. A glance at Fig. 4.13 confirms this. Diagrams of $\Delta\sigma$ versus $\varepsilon_1 - \varepsilon_2$ for (a) NBP sandstone and (b) BP sandstone are shown in the Figure. These curves were obtained by testing specimens under various confining pressures σ_2. Values of σ_2 are indicated for each curve. Some strength and deformation characteristics of these sandstones are given in Table 4.1. These are average values obtained from testing several specimens. The following notations are used in the Table: σ_2: confining pressure; $\Delta\sigma_1^{us}$: differential stress at peak strength; ε_1^{us}: axial deformation at peak strength; ε_2^{us}: lateral deformation at peak strength; $\Delta\sigma_1^{res}$: residual strength; ε_1: total axial deformation in the specimen; ε_2: total lateral deformation in the specimen; $\Delta\theta$: total increase in volume of the specimen.

Despite the similarity in the mechanical characteristics of BP and NBP sandstones, significant differences were observed while studying their structural and flow properties. First, we shall consider specific structural features of the initial unstrained rock material and later record typical changes in the structure induced by the deformation processes developing in the material, and finally conclude the study by analysis of the fluid flow properties of the rocks under various loading conditions.

BP sandstones, in a mineralogical sense, belong to the feldspar-quartz variety mixed with rock fragments and allothogenic mica. The terrigenous portion (75–80%) contains fragmented and regenerated quartz (35–65%), feldspars (plagioclase, orthoclase and microcline 15–25%), fragments of rocks (silica, microquartzite, clayey-siliceous schists 5–30%) and mica (mainly muscovite, rarely biotite 1–5%). The cement (15–25%) of the porous and contact-porous type consists of dispersed clayey mineral-hydromica formations with mixed layers of the type hydromica-montmorillonite, with a small quantity of swollen components, mixed with kaolinite and chlorite and also finely dispersed quartz and carbonates. Cementless jointing of grains is marked by sections.

Key parameters of the cracks were established from microsections of nonstandard dimensions using the methodology of VNIGRI. These included permeability of cracks (K_{ck}) 1.0 md, porosity 0.002–0.2%, volumetric density of open pores (T_0) 5–10 1/m. Open and bituminous cracks were developed mainly along the stratification, which very rarely intersected the cracks. Mineralised (ineffective) cracks were oriented differently, predominantly in a steeply inclined manner. According to the classification of E. M. Smekhov, these sandstones can be classified as a porous type of collector (i.e., BP is one with relatively high gas storage capability and relatively high permeability).

NBP sandstones differ from BP in mineralogical composition, namely: low content of fragmented and regenerated quartz and mica minerals; lower average dimension of grains and smaller width of grain-grain contacts. Total composition: fragmented portion 50–70%, cement 30–50%. Cement of basalt type and basaltic-porous; composition: clayey-carbonate mixed with kaolinite, chlorite and silica. NBP sandstones have low reservoir and fluid flow properties due to the high proportion of carbonate in the cementing material. Generally, fine granular carbonate behaves aggressively, corroding and replacing both cement and lumps of various composition. In some places, the calcite grows large polycrystals, tightly juxtaposed. NBP sandstones according to the classification of E. M. Smekhov can be considered as cover rocks. (i.e., NBP has relatively low storage capacity and relatively low permeability).

The petrographic features of BP and NBP sandstones are given in Table 4.2.

Table 4.2

Mineralogical content of terrigenous material, %					
Sandstone type	Quartz	Feldspars	Total content, %	Rock fragments composition	Mica
BP	35–65	15–25	5–30	Siliceous, clayey-siliceous, quartzitic, clayey-micaceous	1–5
NBP	40–55	20–30	10–30	Siliceous, clayey-siliceous, rarely quartzitic	1–2

Cement and its composition, %						
Sandstone type	Total	Clayey (silt)	Clayey	Quartzitic regenerated	Carbonate	
					Sideritic	Calcitic
BP	15–25	3–15	2–20	3–10	3–35	2–5
NBP	30–45	up to 3	—	—	up to 5	30–45

The sandstones under study were katagenetically transformed. Deep-seated katagenetic processes considerably reduce the rock storativity by sealing pores in them, thereby increasing rock strength. Thus a content of authigenic quartz to an extent of 1–7% in BP sandstones results in loss of plasticity and formation of dense brittle merging units. Large seam thickness combined with silicitisation and the presence of lighter crude in the primary micropores, which seeped through tectonic cracks and localised in the fractured and unsealed zone of the Mushketovsk and Koksov thrust faults in the stage of late katagenesis, allowed the rock to retain a fairly high granular porosity (4.5–8.7%).

Microscopic studies of thin sections of BP and NBP sandstones deformed under various confining pressures, allowed two generations of cracks to be distinguished during development of residual deformations in both rock varieties. Intercrystalline cracks, which are older, belong to the first generation, while younger, open and intergranular cracks belong to the second. Cracks of the first generation were narrow, short, intermittent and did not extend beyond the grain boundaries. These cracks grew within the quartz grains along gas-liquid inclusions, along twin and junction planes in grains of plagioclases, mica and calcite. Usually they were filled with a finely ground substance of siliceous, feldspar or carbonate-clay composition. Cracks of the second generation were extended, open, slightly twisted and intersected the entire microsections.

The relative proportion of cracks of these two generations differed under various confining pressures and their influence on the reservoir and flow properties of rocks under loading also varied. Results of a study of the microjointing features in thin sections of original and deformed specimens

Table 4.3

Test conditions σ_2, MPA	Mineralised cracks		Open cracks		K_{ck}, md
	b, μ	T_m, 1/m	b, μ	T_0, 1/m	
Non-burst-prone sandstones					
Original specimen	—	—	2–5	1	0.01
5	5	20–30	15–18	23	2.1
10	5	30–40	5–10	1–2	1.8
25	5–7	40–50	10	2–4	3.4
50	5–10	50–60	10–12	5	4.2
Burst-prone sandstones					
Original specimen	5	5–10	5–7	5–10	1.0
5	5	5–10	18–20	30	3.5
10	5	10–15	5–10	5–7	2.1
25	5–7	15–20	10–15	10	4.3
50	5–10	20–25	15–18	10–12	6.4

of sandstones, according to the VNIGRI method, are given in Table 4.3: opening (or width) of cracks (b, μ), volumetric density of mineralised, flat, ineffective (T_m, 1/m) and open, effective (T_0, 1/m) cracks as well as the crack permeability (K_{ck}, md).

Analysing the data given in Table 4.3, it can be said that the residual deformation in NBP sandstones resulted mainly from growth of intercrystalline cracks and, to a lesser extent, growth of intragranular cracks. In BP sandstones, the opposite was seen due to the growth of intragranular cracks. As a result of the indicated processes, the crack permeability for BP sandstones increased 6 times compared to the initial value, from 1 to 6.4 md. The relative permeability for NBP sandstones increased by two orders of magnitude compared to the initial value, from 0.01 to 4.2 md.

Thus, under conditions of deviatoric triaxial stresses in BP and NBP sandstones, the growth of residual deformations was observed to occur by different mechanisms. Residual deformation in BP sandstones is accompanied by reorientation of clayey minerals in cement and fragmented material due to the rotation, sometimes bending, of individual mica plates. Reorientation helps in the formation and growth of intergranular cracks that detach grains from the cementing substance, thus opening up cracks and leading to rupture in the direction of the maximum principal stress.

Residual deformation in NBP sandstones begins with twinning of calcite crystals, forming the cement of these rocks, and with growth of intracrystalline cracks in rock-forming grains. As a result of twinning of the cement grains, channels, pores and ruptures form in them. Intracrystalline fine cracks generally occur along the twinned seams (joints) at an angle to them although some develop chaotically. Usually the microcracks are short, hair-

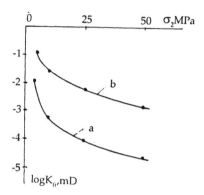

Fig. 4.14: Dependence of permeability K_{ft} on the level of hydrostatic compression σ_2 for (a) NBP and (b) BP sandstones.

line and filled during deformation with a finely dispersed substance. Further growth and opening up of intracrystalline cracks led in the present case to the development of intergranular cracks and cracks intersecting the grains, which, in particular, increased the crack permeability by two orders of magnitude compared to the original rock.

We shall now turn our attention to the fluid flow properties of the two sandstones [85, 92]. Flow was made to occur across the specimen in these tests. Inert gas, such as nitrogen, was used as the flow fluid. In all tests, the pressure of the gas at the inlet was maintained constant at 2.5 MPa and equal to atmospheric pressure at the outlet. Relationships between the permeability K_{ft} and level of hydrostatic compression σ_2 are shown in Fig. 4.14. In the graph, the values of σ_2 are plotted on the abscissa and log K_{ft} on the ordinate. Letter (a) indicates the curve for NBP sandstone and (b) the curve for BP sandstone. A marked difference in the flow properties of these two types of sandstone can be seen. Under hydrostatic pressure $\sigma_2 = 5$ MPa, the permeability of BP sandstone exceeds that of NBP sandstone by one order of magnitude. With increase in σ_2, this difference increases and at a pressure $\sigma_2 = 50$ MPa, it is almost of two orders of magnitude greater.

To understand the mechanics of formation of outbursts under conditions existing in mine excavations, it was necessary to investigate the nature of variation in the fluid flow properties of both types of sandstone during deformation. The specimen was therefore subjected to axial deformation. To reflect the effect of rock pressure in the massif on the fluid flow properties of sandstones, the level of confining pressure in the test chamber was varied from 5 to 50 MPa. It is well known that rock pressure in the massif

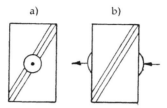

Fig. 4.15: Positioning of fluid flow gauges with respect to rupture plane.

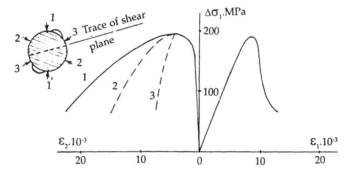

Fig. 4.16: Positioning of three transverse strain gauges in a specimen with respect to the fluid flow gauges and a typical curves of $\Delta\sigma_1$ versus $\varepsilon_1 - \varepsilon_2$, showing the relationships among the transverse deformations recorded by the three gauges.

is a function of the depth and distance from the surface of the mine excavation. Experience has shown that in the given range of confining pressure, the deformation process in the post-peak zone becomes localised along one or a few parallel planes of shear, along which the entire rupture occurs. Here the fluid flow properties being determined depend greatly on the orientation of the flow gauges relative to the already formed rupture planes. Flow gauges were therefore oriented in a specific direction with respect to the rupture plane to enable comparison of the results obtained from different tests, i.e., either along the plane or across it (see Fig. 4.15). This was accomplished practically as follows. Three transverse strain gauges were attached to the specimen in a specific orientation with respect to the flow gauges (see Fig. 4.16) and the signals from these gauges recorded. The deformation process by and large developed evenly in all directions up to pre-peak strength, and all the gauges gave essentially the same readings. In the post-peak strength zone, due to formation of the rupture plane, uniformity in the process of transverse deformation was violated and gauge readings differed. The data thus obtained made it possible to ascertain at an early stage of post-peak deformation how the flow gauges were oriented with respect to the developed rupture plane. If the location of the gauges did not coincide with even just one of the adopted directions of fluid flow, testing was stopped, the specimen removed and the flow gauges repositioned. Testing was then resumed.

Based on the results of the aforesaid tests, relationships reflecting the variation in permeability with loading and deformation of both types of sandstone for various confining pressures were obtained. These relationships provided information not only about the largest variations in the fluid flow properties of the rocks, which arose during flow along the developed planes of rupture, but also about the smallest variation during flow across the rupture planes. Examples of such relationships for (a) NBP and (b) BP sandstones, obtained under deformation at confining pressure $\sigma_2 = 25$ MPa, are shown in Fig. 4.17. These relationships are plotted in semi-logarithmic coordinates. Flow properties of the rocks along the developed rupture plane are indicated by dashed lines and those across the rupture plane by continuous lines. Beyond the post-peak zone, the curves are seen to bifurcate. The mechanics of formation of the major rupture planes differ only very slightly for the two types of sandstone. That is why even the fluid flow characteristics determined along the rupture planes at residual strength (end point on the curves) have almost the same values for similar levels of confining pressure.

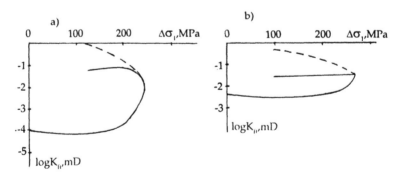

Fig. 4.17: Curves showing the variation in permeability K_{ft} due to the loading and deformation of specimens of (a) NBP and (b) BP sandstones at a confining pressure $\sigma_2 = 25$ MPa. Beyond the post-peak strength, the dashed line indicates flow along the developed rupture plane and the continuous line across the rupture plane.

The relationship K_{ft} versus $\Delta\sigma_1$ obtained in the case of flow across the formed rupture plane, is of major importance. In such a case, flow properties are established through structural variations taking place throughout

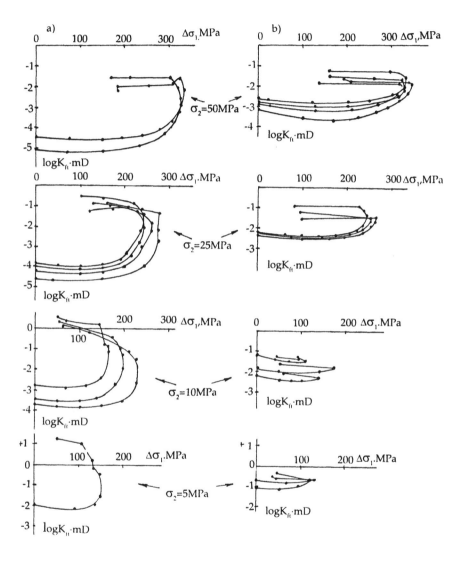

Fig. 4.18: Relationships K_{ft} versus $\Delta\sigma_1$ for (a) NBP and (b) BP sandstones for flow across the rupture plane, obtained for various confining pressures σ_2.

the entire volume of the specimen and not in the local zones coming under the direct influence of the rupture planes. A series of such relationships obtained for (a) NBP and (b) BP sandstones at different pressure levels, is given in Fig. 4.18. Several specimens were tested at each level of σ_2 and fairly good reproducibility of the parameters under study was obtained.

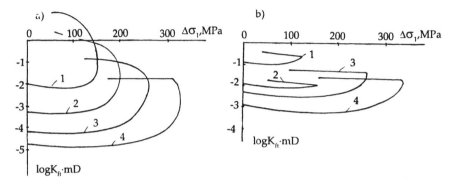

Fig. 4.19: General relationships K_{ft} versus $\Delta\sigma_1$ for (a) NBP and (b) BP sandstones in the case of flow across the rupture plane, obtained at various confining pressures σ_2.

The sharp difference between the relationships indicated in columns (a) and (b) is due to the change in flow characteristics during deformation. The permeability for NBP sandstone increased at all levels of confining pressure by three and more orders of magnitude. For BP sandstone, the largest increments in K_{ft}, under confining pressures 25 and 50 MPa, amounted to one order of magnitude; at low confining pressures, the increase in K_{ft} was less than a factor of two. Such a difference is explained by the peculiarities in growth of the crack system during deformation, as discussed earlier.

For a more visual comparative analysis of variation in flow properties of both types of sandstone under deformation at various levels of confining pressure, general curves were plotted, as shown in Fig. 4.19; these were obtained by averaging the results of a number of similar tests. Relationships (a) depict NBP sandstone while relationships (b) represent BP sandstone. Curves 1 were obtained at $\sigma_2 = 5$ MPa, curves 2 at $\sigma_2 = 10$ MPa, curves 3 at $\sigma_2 = 25$ MPa and curves 4 at $\sigma_2 = 50$ MPa. The following quantities extracted from these curves are given in Table 4.4: 1) values of the permeability determined at the beginning of the test under hydrostatic compression for each level of pressure $\sigma_2(K_{ft}^{int})$; 2) values of the permeability determined at the end of the test after attaining residual strength (K_{ft}^{res}); and 3) the ratio of the change in permeability due to the test ($K_{ft}^{res}/K_{ft}^{int}$).

The observed difference between the initial and final values of permeability obtained for the NBP sandstone from the experiments was higher than that indicated by petrographic studies and is three orders of magnitude. In the case of BP sandstone, the two methods yielded comparable results: the change in permeability varied from 1.7 to 13.3 times depending on the confining pressure.

Table 4.4

Type of sandstone	σ_2, MPa	K_{ft}^{int} $\times 10^{-3}$ md	K_{ft}^{res} $\times 10^{-3}$ md	$K_{ft}^{res}/K_{ft}^{int}$
NBP	5	10	16000	1600
	10	0.64	2000	3125
	25	0.063	120	1904
	50	0.015	12	800
BP	5	120	390	3.25
	10	12	20	1.7
	25	6.4	50	7.8
	50	1.5	20	13.3

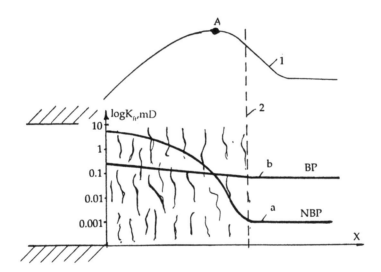

Fig. 4.20: Schematic explanation of the mechanics of outbursts of rock and gas into a mine working.

Based on the results, the following mechanics can be suggested to explain the dynamic outbursts of rocks and gas into a mine working. Figure 4.20 shows the region in the vicinity of an excavation and the associated stress distribution. Curve 1 indicates the vertical stress concentration. Rocks to the left of the vertical line 2 were subjected to permanent deformation with corresponding crack growth, which reached maximum in the regime left of point A, where the rock was in a state of post-peak deformation. The ratio of the change in permeability in the zone induced by the stress concentration is shown in $K_{ft} - x$ coordinates for (a) NBP and (b) BP sandstones.

In NBP sandstones, due to the low initial permeability of the (unstrained) rock and the high permeability of the fractured rock, resulting in a change of three orders of magnitude, gas discharged from the rock mass could escape into the excavation easily and so did not accumulate to create gas pressure in the fully developed crack-pore space. In the case of the BP sandstone, the rate of gas discharge into the fractured zone and the rate of release from it into the mined-out area, were more comparable in magnitude, so that gas was stored under pressure in the fractured zone. The pressures and volumes of gas in the crack-pore space and the resistance offered by the fractured rock to gas flow are such that the gas could accelerate the rupture to cause outbursts of the broken rock into the mined-out area.

4.4 COMBINED EFFECT OF PORE AND CONFINING PRESSURE ON THE MECHANICAL BEHAVIOUR OF ROCK

In this section we shall discuss the effect of reduction in strength as the rock is deformed due to pore pressure developed by liquid or gaseous fluids. The chemical or 'wedging' effect of the fluid on the mechanical properties of rocks, such as described by Rehbinder, is not considered. (Any such effect by reducing the effective surface energy of the cracks would tend to exacerbate the instability of the crack growth process.) To avoid this complication, a neutral gas such as nitrogen or a liquid that did not induce a significant change in the mechanical properties of rocks in the absence of pore pressure was used in the tests.

Figure 4.21 shows the results of investigations into the effect of pore pressure on the strength of three types of rocks (a—lignite, b—marble [83] and c—quartzitic sandstone). The strength values of rocks (τ_{us}) are plotted

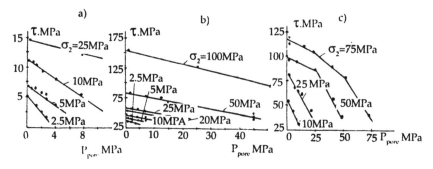

Fig. 4.21: Ultimate strength τ_{us} versus pore pressure P_{pore} for (a) lignite, (b) marble and (c) sandstone at various confining pressures.

on the vertical-axis and pore pressures (P_{pore}) on the horizontal-axis. Tests were conducted at different confining pressures σ_2 as is indicated on the various curves. Pore pressure was generated by nitrogen introduced into the specimen after it had been compressed by confining pressure. For all values of σ_2, increase in pore pressure resulted in a reduction in rock strength. However, the nature of the relationship τ_{us} versus P_{pore} differed for each rock tested. The relationships for lignite are straight lines in which the gradient decreases with increase in the confining pressure. In the case of marble, the relationships are also straight-lined but the change in slope with increase in σ_2 is much less than for lignite. For sandstone, under high confining pressures $\sigma_2 = 50$ and 75 MPa, the relationship between τ_{us} and P_{pore} is non-linear.

The mechanics of these effects of pore pressure on rock strength can be explained by the model of a heterogeneous solid discussed in Chapters 1 and 2 [83, 101]. It was shown that the hardening of such a body with increase in confining pressure σ_2 was induced by the effect of σ_2 on a rupture area **b** contained in a shear plane ω formed during the rupture (see Fig. 4.22). The load strength of the material at peak strength is determined by

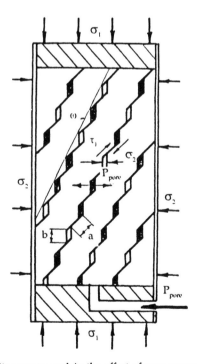

Fig. 4.22: Schematic diagram to explain the effect of pore pressure on rock strength.

the resistance* of all the elements of shear and rupture within the weakest shear plane ω:

$$\tau_{us\omega} = \left(\sum_1^M \tau_{li} + K\sigma_2\chi_{0.7}\right)/M \qquad (4.2)$$

where τ_{li} is shear strength in the microarea of shear **a**; **M** is the number of microareas of shear contained in the macroscopic plane of shear ω; **K** is the number of elementary areas of rupture **b** in the plane ω; $\chi = $ **b/a**.

The pressure P_{pore} of a liquid or gas developed in the rupture cracks **b** counteracts the pressure σ_2 and reduces its hardening ability. When the permeability of the material reaches its maximum, i.e., when each microcrack **b** in the plane ω is fully accessible to the fluid pressure, the strength of the material is determined by the effective stress $\sigma_2^{eff} = \sigma_2 - P_{pore}$:

$$\tau_{us\omega} = \left[\sum_1^M \tau_{li} + K\left(\sigma_2 - P_{pore}\right)\chi\, 0.7\right]/M \qquad (4.3)$$

In this case, at $\sigma_2 = P_{pore}$, the strength of the material is equal to the uniaxial compressive strength.

In those cases wherein the material is only partially permeable or completely impermeable to the fluid, the change in the strength characteristic is different. In Fig. 4.22, pores that are filled by fluid are shown in black. In these conditions, shear resistance along the ω planes would depend on the proportion of rupture cracks **b** contained in the plane ω that are accessible to percolation of fluid into them, such that pore pressure P_{pore} can develop in these cracks:

$$\tau_{us\omega} = \left[\sum_1^M \tau_{li} + (K\sigma_2 - zP_{pore})\chi\, 0.7\right]/M, \qquad (4.4)$$

where **z** is the number of pores formed over the rupture areas into which the pore pressure P_{pore} can penetrate.

Considering the results shown in Fig. 4.21, in terms of this model the decrease in the slope of the relationships τ_{us} versus P_{pore} with increase in confining pressure can be explained as follows: rock permeability decreases with increase in σ_2, and the number of pores accessible to the fluid decreases also. The discontinuities in the slopes observed in the case of sandstone, is explained by the fact that at low values of pore pressure P_{pore}, a portion of the pores becomes inaccessible to the fluid; as the pore pressure P_{pore} increases, the number of filled pores increases correspondingly. These arguments are confirmed by experimental results, which will be.presented later.

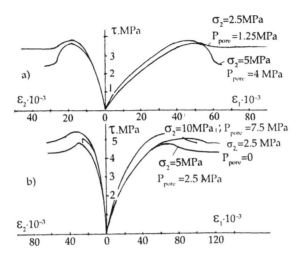

Fig. 4.23: Complete diagrams of τ versus $\varepsilon_1 - \varepsilon_2$ for lignite for various ratios of confining pressure σ_2 to pore pressure P_{pore}.

Figure 4.23 [83] shows complete shear stress/shear strain diagrams [τ versus $\varepsilon_1 - \varepsilon_2$] for lignite, obtained for various ratios of confining pressure σ_2 to pore pressure P_{pore}. The values of these pressures are indicated on the curves. The effective pressure $\sigma_2^{eff} = \sigma_2 - P_{pore}$ was maintained almost constant in the tests. For curves (a) σ_2^{eff} was 1 and 1.25 MPa; for curves (b) σ_2^{eff} = 2.5 MPa. Confining pressures varied from 2.5 to 10 MPa. Over this range of confining pressures, lignite exhibited a fairly high permeability; hence the experimental curves obtained are close to each other. However, the slight change in slope of the relationships τ_{us} versus P_{pore} (Fig. 4.21), obtained at σ_2 = 2.5 MPa and σ_2 = 10 MPa, indicates that in the second case accessibility of a part of the pores to the fluid was reduced. The result is seen in the curve obtained at σ_2 = 10 MPa and P_{pore} = 7.5 MPa (Fig. 4.23). Here, certain enhanced strength characteristics exhibited by the specimens in the pre-peak strength regime, were followed by a rapid reduction in these characteristics beyond peak strength. This was caused by increased percolation in the post-peak zone and filling of pores which until then were inaccessible to fluid flow. In the given case this effect is very small and possibly non-existent given the natural scatter in the results obtained on different specimens. Figure 4.24 [83] shows the corresponding results for marble. The effect is more clearly evident. Complete diagrams, obtained at (a) σ_2 = 5 MPa and (b) σ_2 = 10 MPa in the absence of pore pressure, are shown. These may be compared with the curves obtained for various combinations of confining and pore pressure on specimens having the same values of peak strength. It should be noted that these curves were

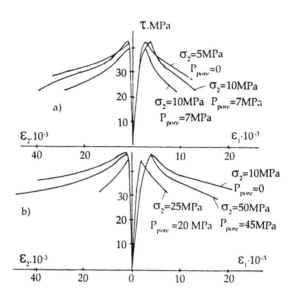

Fig. 4.24: Complete diagrams of τ versus $\varepsilon_1 - \varepsilon_2$ for marble for various ratios of confining pressure σ_2 to pore pressure P_{pore}.

obtained for lower effective pressures $\sigma_2^{eff} = \sigma_2 - P_{pore}$ than the initial curves: for case (a) $\sigma_2^{eff} = 10 - 7 = 3$ MPa; for case (b) $\sigma_2^{eff} = 50 - 45 = 5$ MPa and $\sigma_2^{eff} = 25 - 20 = 5$ MPa. After reaching the ultimate (peak) strength, a more rapid drop in the strength characteristic is observed for these curves. In the pre-peak zone, due to the inaccessibility of a portion of the pores to fluid flow, it was necessary to increase the pore pressure in order to obtain the same peak strength, whereas after passing through the peak strength regime, where intensive cracking processes were initiated, the permeability of the rock increased and pore pressure developed in previously inaccessible pores, and hence reduced the 'clamping' (confining) effect of the confining pressure.

The influence of the permeability of a rock on its strength and deformability can also be illustrated by the test results shown in Fig. 4.25 [89]. Complete load-deformation diagrams τ versus $\varepsilon_1 - \varepsilon_2$ for (a) sandstone, (b) limestone and (c) amphibolite are shown. Three series of tests were conducted on each of these rocks. Curves 1 in the diagrams were obtained on jacketed specimens (i.e., made impermeable to confining pressure $P = \sigma_2$ fluid on the specimen), subjected to loading and deformation. Curves 2 were obtained at the same fluid pressure ($P = \sigma_2$) in the cell but, contrary to the first series, the specimen was not made impermeable to the fluid (non-jacketed); hence confining pressure developed within the internal pore spaces of the rock.

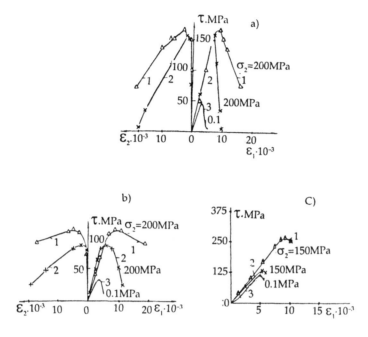

Fig. 4.25: Complete load-deformation curves τ versus ε₁ – ε₂ for (a) sandstone, (b) limestone and (c) amphibolite, obtained under different types of loading: curve 1—jacketed specimen subjected to deformation at confining pressure σ₂; curve 2—non-jacketed specimen subjected to deformation at the same pressure σ₂; curve 3—specimen loaded in uniaxial compression.

Curves 3 were obtained under uniaxial compression. The tests in series 2 were carried out using an organic oil, non-chemically active, for the rocks studied. The confining pressure $\sigma_2 = P$ is indicated on each curve.

It is evident from these test results that sandstone, seen to be almost impermeable to the fluid when tested under loading condition 2, behaved similarly in the tests in series 1 with waterproofing. In both cases, the full confining pressure effect $P = \sigma_2$ led to intense hardening of the specimens compared to the strength observed under uniaxial compression. The degree of hardening obtained under regime 2, compared to regime 1, was about 91%. Beyond the elastic limit, crack formation processes were initiated in the specimen and the pressurised fluid entered the pores formed, resulting in a sharp drop in the specimen strength. Limestone was partially permeable with respect to the confining fluid. Thus, the hardening in series 2 was about 68% of that observed in series 1. By contrast, the strength of amphibolite in test series 2 (i.e., non-jacketed specimens) was close to the uniaxial compressive strength (as in series 3). This indicates that highly permeable rock which allows complete penetration of the fluid under pressure, does not harden—although hardening should occur when the rock is subjected to external triaxial compression.

The following conclusions can be drawn from analysis of the experimental data.

1) Pore pressure leads to reduction in the strength and deformation properties of rocks. This is due to the build-up of fluid pressure inside the pores. This counteracts the externally applied static pressure, thereby reducing the triaxial compression of the rock. The effectiveness of this counteraction depends on the accessibility of the pores in the specimen to percolation of the fluid under pressure. If the rock is very permeable, this effectiveness is maximum and the triaxial (volumetric) compression of the rock mass is determined by the effective stress (σ_{geo}^{eff}), which is the difference between the geostatic stress and the pore pressures ($\sigma_{geo}^{eff} = \sigma_{geo} - P_{pore}$). For rocks that are less permeable, reduction in the (triaxial) volumetric compression due to pore pressure depends on the degree of filling of the pore space by the pressurised fluid and, in this case, the assumption that σ_{geo}^{eff} is the difference ($\sigma_{geo} - P_{pore}$) is incorrect, i.e., it does not reflect the actual state of stress in the rock. The magnitude of rock permeability should also be considered when plotting experimental relationships of strength versus σ_2^{eff}.

2) Liquid and gaseous fluid pressures on rock can be used to create a state of triaxial (volumetric) compression of the rock to achieve rock hardening. The maximum effectiveness of these pressures is obtained when the rock is completely impermeable to the fluid. The 'clamping' effect decreases with moderate permeability and is absent in 'completely permeable' rock. To clarify, the mechanics of this clamping effect of fluid pressure on a porous solid can be illustrated by the following example. If a bundle of crushed wire is placed in a pressurised liquid or gaseous medium (for example, at the bottom of a water reservoir), the pressure of the fluid medium causes no change in the configuration of the bundle because all the 'pores' (spaces between the wires) are subjected to pressure. If the bundle is initially placed in a waterproof jacket, the medium pressing against it reduces these pores (i.e., clamps the wires against each other) and creates a volumetric compression in the wires. The effect of rock clamping due to liquid or gaseous fluid pressure, resulting in hardening (strengthening) of the rock is widely used in mining practice, for example to ensure the stability of the walls of deep boreholes. The effectiveness of this method for ensuring borehole stability is discussed in the next section.

4.4.1 Effect of Pore Pressure and Permeability on Well Stability in Deep Boreholes

Boreholes which penetrate a zone of high-level compressive stress change the stress state in the rock mass surrounding the borehole. This stress state is defined by the principal stresses: $\sigma_1 > \sigma_2 > \sigma_3$. Values of the principal compressive stresses around a borehole can be determined, for example, by the equations formulated by S. G. Lekhnistsky [52]:

$$\sigma_1 = \sigma_\theta = -(\lambda P_{geo} - P_{bh})R_{bh}^2/R^2 - P_{geo}, \tag{4.5}$$

$$\sigma_2 = \sigma_z = -P_{geo}, \tag{4.6}$$

$$\sigma_3 = \sigma_r = (\lambda P_{geo} - P_{bh})R_{bh}^2/R^2 - \lambda P_{geo}, \tag{4.7}$$

where P_{geo} is the magnitude of the lithostatic or geostatic pressure; P_{bh} is the hydrostatic pressure of the column of drilling fluid exerted on the borehole wall; λ is the coefficient of lateral outward thrust; R_{bh} is the borehole radius; R is the distance from borehole axis to the point under study in the rock mass.

The various stresses and graphs indicating the stress distribution around a borehole are shown in Fig. 4.26. Graphs were plotted for the case $\lambda = 1$ and $\rho_{df} = 0.5\,\rho_{rock}$, where ρ_{df} and ρ_{rock} are the specific gravity of the drilling fluid and rock respectively. In this computation scheme, Archimedian forces were not taken into account since they are not of major significance in the investigations to be discussed later.

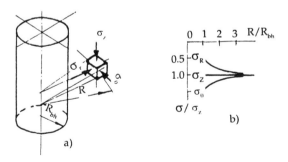

Fig. 4.26: (a) Notation of stresses around a vertical borehole and (b) graphs showing the distribution of elastic stresses along a horizontal drawn from the borehole axis.

The rock immediately adjacent to the borehole wall is subjected to the most extreme conditions since tangential stresses σ_θ reach their maximum values here, while the radial stresses σ_r, which determine the magnitude of triaxial confining pressure on the rock and hence its strength are minimum. The latter assertion follows from the results of studies [11, 53, 41] in which it was experimentally established that under conditions of nonuniform triaxial (volumetric) compression of the type $\sigma_1 > \sigma_2 > \sigma_3$, the resistance of rock to loading is determined by the value of the minimum principal stress σ_3 and depends only slightly on the intermediate stress σ_2. Where pressure is not exerted by a drilling mud on the borehole wall, however, i.e., under $P_{bh} = 0$, the radial stresses $\sigma_r = 0$, the rock in the borehole wall is subjected to conditions of plane stress and the rock strength is close to the uniaxial compressive strength. One function of the drilling mud is to increase the stability of the borehole wall. The pressure exerted by the mud on the borehole wall acts, on the one hand, to increase the value of σ_r to a value determined by the weight of the overlying fluid column and, on the other, to reduce the value of σ_θ. Experience in drilling deep boreholes has shown that drilling mud does not always fulfil this stabilisation role. Increase in the specific gravity of the fluid (or 'mud-weight') very often produces the opposite effect, i.e., a reduced stability of the borehole wall. The causes for this and other effects are discussed below [89, 101].

The method of investigation was dictated by the following considerations. Since drilling mud in a borehole acts directly on the rock and is intended to compress and harden the rock by generating a state of triaxial compression in it, tests were required to verify how effectively the drilling mud performs this function.

A specimen was placed in a high-pressure cell without a waterproof jacket and the drilling mud or other liquid or gas of interest used as the pressurising medium. After generating the necessary pressure in the cell, the specimen was deformed axially until complete failure, and the strength and strain characteristics were measured. Such a regime of testing has been designated above as regime 2 (or as series 2 tests). To evaluate the effectiveness of specimen hardening as a result of the direct pressure exerted on it by the drilling mud—or any other fluid—tests on similar specimens were conducted under regime 1 (series 1 tests), i.e., with pressure exerted on the specimen through a waterproof jacket. The results of testing different fluids and drilling muds in one type of rock (granite) are shown in Fig. 4.27. The values of tangential strength τ are plotted on the vertical axis in the graphs and the fluid pressure in the cell P (or level of σ_2 for curves indicated by the letter A) on the horizontal. In all four diagrams (a, b, c, d) the relationships obtained in testing regime 1 are indicated by the letter 'A'. The remaining relationships were obtained by testing specimens under regime 2. Cylindrical specimens, 30 mm in diameter and 70 mm in length, were used in these tests.

In these tests (results given in Fig. 4.27a), a gas (nitrogen), water and three types of drilling mud were used: No. 1—clay solution in water, No.

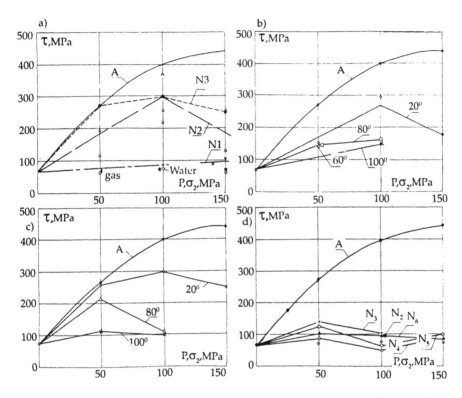

Fig. 4.27: Relationship between strength τ of granite and confining pressure P (σ_2) obtained under testing regime 1 (curve A) and regime 2: (a) for three types of drilling muds, water and gas at a temperature of 20°C; (b) in drilling mud No. 2 under several different temperatures; (c) in drilling mud No. 3 under several different temperatures; (d) in different drilling muds at a temperature above 100°C.

2—organic oil-based solution, No. 3—glycerine-based solution. It can be seen from the graphs that pressure exerted on the specimen by gas, water and solution No. 1 produced almost no hardening (i.e., strengthening) effect on the rock. This is explained by the complete permeability of the rock to these fluids. Pressure due to solutions No. 2 and No. 3 significantly hardened the rock. Solution No. 3 had the superior effect. However, compared to the monotonously ascending curve A, these relationships show a maximum at $P = P_{cr} = 100$ MPa and later a softening (rock weakening) segment under high pressures P. The presence of such an effect is explained by the fact that at pressures $P > P_{cr}$, the rock becomes more permeable to the fluid because the latter is forced into its pores. Until the material is impermeable to the medium confining it, increase in the pressure P leads to rock hardening; after passing through P_{cr} a reverse relationship is observed. As applied to the problem of maintaining stability of the borehole wall by means of pressure exerted by the drilling mud, the test results can be interpreted as follows. For each of the two objects of interaction—*rock* and *drilling mud*—there exists an optimal level of pressure $P = P_{cr}$ at which the maximum rock hardening effect and maximum stability of the borehole wall is achieved. If the depth of the borehole and value of P_{cr} are known, the optimal specific gravity of drilling mud required to obtain maximum effectiveness in stabilising the borehole wall can be determined. Changes in the specific gravity of the drilling mud above or below this optimum value, result in less effective rock strengthening (or hardening).

Curves obtained for drilling muds No. 2 and No. 3 at different levels of temperature are shown in Figs. 4.27 (b) and (c). Increase in temperature resulted in reduction of the hardening effect and in the case of solution No. 3 the maximum value of hardening shifted towards lower pressures. At temperatures above 100°C, the hardening effect of all drilling muds under study became essentially non-existent. Results of these tests are shown in Fig. 4.27 (d). In addition to the foregoing fluid solutions, curves pertaining to three other solutions are shown in the Figure. The main reason for this temperature effect on the process under study is that the viscosity of the solutions tested drops with increase in temperature, and the possibility of fluid percolation into the rock pores then increases. For example, the viscosity of one of the most viscous liquids used, glycerine, decreased more than 100 times when the temperature was varied from 20°C to 100°C. To exclude the effect of thermally induced cracking of the specimens, which could have led to increased permeability, a special sequence of operations was adopted during the tests. The specimen was first subjected to a given hydrostatic confining pressure before the temperature was raised. The conclusion that temperature-related crack formation processes do not occur in rock under these confined loading conditons can be drawn from the fact that under conditions of complete waterproofing of the specimen (testing regime 1), raising the temperature to 100°C had only a negligible effect on (reduction of) the strength characteristic.

In testing rock specimens under regime 2, it is important to take into account the scale effect. When a pressurised fluid acts on a specimen of low permeability, fluid percolates into the surface layer of the material to a particular depth. The volume of material in which the percolated fluid is absorbed is then not subjected to the confining effect of the fluid pressure nór to (the same level of) triaxial compression. The strength of this part of the specimen remains at the level of the uniaxial compression strength. The middle part of the specimen, into which the fluid cannot percolate, is subject to triaxial compression and develops an increased strength. The strength characteristic of the specimen as a whole thus depends on the proportion of the first and second zones in a cross-section of the specimen. This ratio depends on the initial cross-sectional dimension of the specimen. For example, for an initial diameter of 90 mm and at a depth of contact with fluid

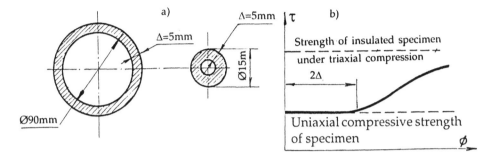

Fig. 4.28: (a) Cross-section of specimens of different diameter, indicating the zone of contact of fluid exerting pressure in the zone; (b) influence of scale effect on rock strength during specimen testing in regime 2.

$\Delta = 5$ mm (see Fig. 4.28), the cross-sectional area of the specimen coming under volumetric compression is 80% of the total area; for an initial diameter of 15 mm, it is only 10%. In line with these observations, for the same confining pressure of the fluid, the strength characteristic obtained for these specimens would differ sharply. Tests showed that the relationship between the strength of a (rock) material and the dimension (diameter) of the specimen during testing under regime 2 at the same value of P, is as shown in Fig. 4.28 (b). The value Δ of the zone in which the fluid penetrates into the rock at a given pressure P can be determined based on this relationship. The value of Δ corresponds to the radius of the specimen, at which the strength τ becomes greater than the uniaxial compression strength.

The scale relationship of rock strength when loaded under regime 2 explains why the disking process of a rock core usually is surceased at great depths as the diameter of the core is increased. While drilling, the column of the core dislodged from the rock mass undergoes elastic unloading deformation in both the axial and lateral directions. Here, in the plane defining the contact between the core and the rock mass still subject to confining pressure, spalling (shearing) stresses arise. The higher the compressive stresses in the rock mass, the higher the elastic unloading strains experienced by the core as it is removed from the massif and the higher the shearing stresses across the (core/rock mass) contact plane. At the particular depth where the confining stresses are sufficiently high, shear stresses reach the shear strength of the rock and the core becomes dislodged from the mass to form a disc.

To prevent core disking, confining pressure must be exerted on the released column of the core to prevent elastic unloading. Development of

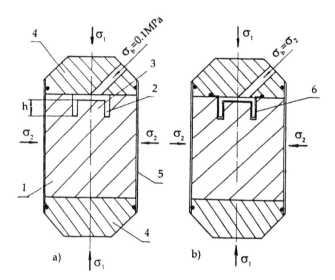

Fig. 4.29: Preparation of specimens to study the core disking process.

such a confining pressure in the borehole face is possible only by use of a drilling mud. However, the effectiveness of confining pressure strengthening (hardening) depends on the permeability of rock to a given drilling mud. Should the confining fluid percolate into the pore space of the drill core, the effect of confinement will be lost and core disking will occur. With increase in diameter of the core, the probability that the central part (assuming that the size of the fluid contact zone Δ is constant) will remain subject to triaxial compression increases, thereby inhibiting the disk-formation process.

Using an experimental relationship (similar to that shown in Fig. 4.28b), reflecting the effect of this scale factor on the rock strength at the fluid pressures of interest to us, it is possible to determine the minimum diameter of a core which may be drilled at a particular depth.

Direct experiments to assess the effect of confining fluid on the core disking process were conducted using the following procedure. The method of specimen preparation is shown in Fig. 4.29. The specimen was a core (1) of large diameter (90 mm) in which a circular cavity (2) was coaxially drilled to a depth h, resulting in the formation of a core (3) of smaller diameter (30 mm). This specimen was placed between metal thrust beatings (4) and encased in an impermeable jacket (5) to protect it from the external pressurising medium. Specimens were tested according to the following schemes.

1) The specimen was subjected to confined pressure tests while maintaining the initial pressure in the cavity (2) equal to atmospheric pressure (Fig. 29a). The confining principal stresses σ_1 and σ_2 were increased while the core remained unloaded. At particular values of σ_1 and σ_2, the core spalled.

2) A specimen similar to that in the first case was subjected to high externally applied confining pressures σ_1 and σ_2 and pressure in the cavity (2) was developed simultaneously. In order to completely exclude fluid penetration into the core material (3) and to provide maximum confining effect, the core (3) and the entire cavity (2), as shown in Fig. 4.29 (b), were covered with a rubber sheet (6). A pressure equal to the confining pressure σ_2 was generated in the cavity (2). Under this loading system spalling did not occur, even though the pressures σ_1 and σ_2 were doubled. When pressure in the cavity (2) was released, the core (3) became detached from its base (i.e., disking occurred).

Now let us analyse the situation when the drilling mud pressure acts on the wall of the borehole. Testing regime 2, described above (i.e., when the pressure of a liquid or gaseous medium acts directly on the rock to strengthen (harden) it), is similar in principle to the conditions that obtain at the borehole wall when we try to change the stress state in the vicinity of the wall by application of drilling mud pressure, thereby hardening the rock. In those cases in which the rock around the borehole is impermeable

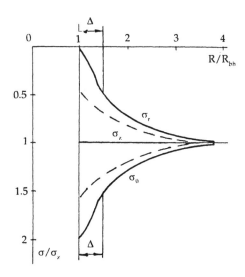

Fig. 4.30: Distribution of stresses in the area around a borehole for permeable rock (continuous lines) and for impermeable rock (dashed lines). Δ—denotes the depth of penetration of fluid pressure.

to the solution, the drilling mud effectively carries out this task, i.e., maintaining stability of the borehole wall. In this case the stress σ_r in the borehole wall is equal to the pressure of the drilling mud column at the given depth and the stress distribution around the borehole is as shown in Fig. 4.26. Should the rock be permeable to the drilling mud, the mud pressure is transmitted via the pores in the rock to a radius of depth Δ, and the presumed effect of confinement, which ought to subject the rock at the borehole wall to a state of triaxial compression resulting in hardening, is either much reduced in magnitude or totally absent. Distribution of stresses in the area around a borehole for permeable rock is shown by the continuous lines in Fig. 4.30. At the borehole wall the stress σ_r becomes equal to 0 and it is only at the boundary of fluid penetration (i.e., at the distance Δ from the wall) that the confining pressure reaches the values equal to the mud pressure for this depth.

The permeability of the rock also influences the level of tangential stress σ_θ. In this case the stress σ_θ computed by formula (4.5) does not occur at the borehole wall, but rather at the boundary of the fluid penetration zone. At the borehole wall, the value of σ_θ then corresponds to the condition under which $P_{bh} = 0$ (see formula 4.5). For purposes of comparison, the stress distribution corresponding to the case of complete impermeability is shown in Fig. 4.30 by dashed lines. These conclusions were confirmed by direct experimental investigations into the stability of the wall of boreholes drilled into rock specimens.

The testing scheme is shown in Fig. 4.31. A rock specimen (externally waterproofed) with a hole drilled axially in the centre was subjected to hydrostatic confining pressure. The ratio between the external and internal diameters of the specimen D/d was taken as 6 to 12 in the tests, which corresponded to the section in which critical pressure P_{ext} (leading to frac-

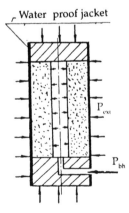

Fig. 4.31: System for testing specimens with a hole drilled in the centre.

Fig. 4.32: Relationship between the stability of a borehole wall and the depth of penetration of drilling fluid into the area around the borehole.

ture of the borehole wall) was almost independent of the D/d ratio. This relationship was computed using Lamé's equation [12].

The objective of these experiments was to establish the level of external confining pressure $P_{ext} = P_{ext}^{er}$ at which the borehole wall would fracture. Tests were conducted under two regimes: (1) with no drilling mud (or other fluid) pressure in the borehole and (2) under a drilling mud pressure P_{bh}. Where the rock was permeable to the fluid, the value of the maximum external confining pressure required to fracture the hole P_{ext}^{er} was almost independent of the pressure P_{bh}. For example, for permeable amphibolite (see curves τ versus ε_1, obtained under different testing regimes, in Fig. 4.25c) the level of critical pressure P_{ext}^{er} remained constant even though P_{bh} was varied from 0.1 to 140 MPa. In tests on marble, when glycerine was used as the fluid, P_{ext}^{er} increased with increase in P_{bh}. The relationship between the stability of the borehole wall and the depth Δ of penetration of fluid into the area around the borehole is illustrated photographically in Fig. 4.32. In this particular case, tests were conducted on limestone and processed oil was used as the fluid. It can be seen from the photograph that fracture of the borehole wall occurred first in that part of the wall where the zone of fluid penetration was the greatest. The zone of fluid penetration is dark on the photograph

It should also be pointed out that the relative stability of a borehole is a function of the scale factor: the higher the ratio Δ/R_{bh}, the lower the relative stability of the borehole. At a fixed value of Δ, the relative stability increases with increases in R_{bh}.

4.4.2 Deformation Processes Occurring in Rocks Undergoing Decompression

Having discussed in the previous section the effects of rock pressure and fluid pressure on cores drilled within the earth's crust and then brought to the surface for various studies, it appears appropriate now to consider the effect of unloading or decompression processes (removal of *in-situ* pressures) on the properties of rocks. The changes induced by decompression may be quite significant, especially if permanent deformations have occurred in the rocks during the loading processes prior to decompression. Cores taken from great depths or specimens withdrawn from a high-pressure cell where they had been subject to high stresses and deformations, may manifest distinct differences in structure and material behaviour compared to their initial state under conditions of triaxial compression. To understand the magnitude of these differences and to establish the effect of decompression processes on rock properties, it was considered necessary to conduct direct experimental studies.

The deformation processes that occur in rocks immediately upon decompression provide a 'first indication' of the deformation after-effects in the rock when it is completely unloaded. The phenomenon of after-effect was discussed in Chapter 2. The fundamental difference between the experiments carried out in the present study and those discussed earlier is that in the first case, forces of elastic unloading were the primary concern. This unloading occurs simultaneously with removal of the external forces. In the second case, the forces due to the residual elastic stresses, i.e., those that remain in the material after it has been completely unloaded, are the key concern.

The effects of hydrostatic confinement—which induces no permanent deformation in rock—and subsequent removal of hydrostatic pressure, have been very thoroughly studied. Hence we shall consider here only those processes of decompression for those cases in which rock has undergone permanent deformation under conditions of non-uniform triaxial compression. Under the conditions that obtain at great depths, tectonic processes can lead to the development of huge differential deformations in rocks. Conventionally, laboratory tests are rarely conducted for relative strains greater than 15%. These restrictions are due to the limitations of standard laboratory test procedures. With such procedures, the cross-sectional dimensions of the specimen change non-uniformly under high axial deformations, causing it to become 'barrel-shaped'. With such distortions of the specimen, the stress and strain distributions are no longer uniform and the processes occurring in the material become very difficult to interpret. However, there is considerable merit in studying large plastic deformation in rock. The investigations described below are probably unique in that they appear to represent the first attempt to develop large permanent deformations in excess of 100% in rock specimens, with reliable recording of the strength and deformation characteristics of the material tested [87].

The method of studying large permanent (irreversible) deformations developed in the laboratory involves repeated testing of the same specimen under identical conditions of triaxial compression. In the first test, after the specimen has undergone 15–20% relative axial deformation, it is first unloaded axially and then the hydrostatic confining pressure is removed, i.e., the specimen is completely decompressed. Next, it is removed from the cell, turned on a lathe along its diameter until its initial dimensions are restored, after which it is again waterproofed and returned to the test cell where it is then subjected to the same level of hydrostatic pressure as in the first deformation, and then deformed axially up to 15% strain. This procedure is repeated several times. It is necessary in these tests to use very long specimens compared to those used for standard test procedures. When the specimen has been shortened considerably due to repeated deformations, it becomes necessary to reduce the diameter also so that the ratio of specimen length to diameter does not fall below 1.5. Sequential repetition of the aforesaid procedure allows relatively huge deformations to be generated in the specimen.

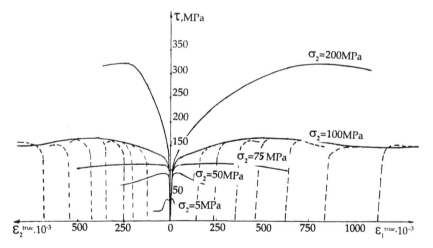

Fig. 4.33: Diagrams of τ versus $\varepsilon_1 - \varepsilon_2$ obtained under several replications of deformation of marble specimens: 8 replications at $\sigma_2 = 100$ MPa, 7 at $\sigma_2 = 200$ MPa and 5 at $\sigma_2 = 75$ MPa.

Figure 4.33 shows diagrams of shear stress (τ) versus axial and lateral strain ($\varepsilon_1 - \varepsilon_2$) obtained through repeated tests on marble specimens under confining pressures $\sigma_2 = 75$, 100 and 200 MPa. In the graph, at $\sigma_2 = 100$ MPa, the diagrams of τ versus $\varepsilon_1 - \varepsilon_2$ are indicated by dashed lines for each stage of testing. The general curve is the envelope of the series of individual loadings, plotted sequentially. The general curve (solid line) at $\sigma_2 = 100$ MPa was obtained from tests that were replicated 8 times. Diagrams at $\sigma_2 = 75$ and 200 MPa were obtained from tests replicated 5 and 7 times respectively. In these tests, the total deformations amounted to more than 100%. It should be noted that graphs were plotted in the so-called 'true strain' units used for computing large plastic deformations. True relative deformations (or strains) were determined by dividing the magnitude of current deformation to the varying current dimensions of the specimen (and not the initial dimensions, as is typically done). The usual (small strain) and true values of deformation are related as follows [26, 42]:

$$\varepsilon_1^{true} = \ln(1 + \varepsilon_1); \ \varepsilon_2^{true} = \ln(1 + \varepsilon_2).$$

These tests, besides giving results associated with structural variations in material when subjected to such large deformations and repeated cycles of compression and decompression (detailed later), are also very interesting in certain other aspects. They show that the limiting (maximum) values of strength in rocks at high levels of σ_2 are attained only after the development of large deformations which, in principle, cannot be achieved in a single conventional laboratory test on a specimen. For example, the limit strength of marble at $\sigma_2 = 200$ MPa is attained only after an axial deformation of the specimen equal to 75%. It is also interesting to note that the modulus of elasticity E, obtained on the first loading of the original specimen, remained almost constant during the subsequent loading cycles of specimens that had been pre-strained in previous cycles. For example, the value $E = 0.6 \times 10^5$ MPa, observed for the first loading of the specimen under conditions $\sigma_2 = 100$ MPa, differs from the value obtained in the eighth loading by just 1%—and that, too, on the lower side. This seems strange since the properties of a specimen pre-strained under high confining pressures in atmospheric conditions, differ sharply from the properties of the initial specimen. For example, Fig. 1.38 (Chapter 1) presents diagrams of τ versus ε_1, obtained in uniaxial compression of (1) an original (previously unloaded) marble specimen and (2) one that had been pre-strained at a confining pressure $\sigma_2 = 100$ MPa. It is evident that under these conditions the mechanical properties of the two specimens are quite different.

Consider the effects induced by hydrostatic confining pressure, resulting in very significant changes in the mechanical properties of rock. First, we shall examine the testing procedure. The action of the hydrostatic pressure, subsequent axial loading of the specimen followed by unloading, changes the dimensions of the specimen and the volume of crack-pore space. To obtain more comprehensive information about the processes occurring in the specimen during testing, several different methods of recording these processes were used. The volumetric changes in the specimen were measured by three methods as listed below:

1) computation based on the results of measurement of axial and lateral deformation in the specimen, using the formula ($\theta = \varepsilon_1 + 2\varepsilon_2$);

2) through direct measurements of the volume of open crack-pore space in the specimen using a U-shaped manometer, working on the principle of maintaining the gas pore-pressure in the specimen constant at 1 atm (see Fig. 4.4 and description in Section 4.2);

3) by hydrostatic weighing to determine the density of the specimen before and after testing.

Some insight into structural changes in the specimen can also be gained based on results of tests conducted using standard geophysical techniques, i.e., by measurement of wave velocities and electrical resistivity.

The method of hydrostatic weighing allows the initial and final volumes of the specimen to be determined accurately (within 1–2%). Moreover, the final volume is determined under atmospheric conditions, i.e., after the confining pressure acting on it has been released. In this case, the difference in volumes of the specimen, prior to and after testing, reflects the combined change in volume, i.e., the permanent deformation produced under loading and the changes resulting from the subsequent unloading. As shown by experience, the complete history of the variation in volume of the rock, both during deformation in the high-pressure cell and during unloading, can be recorded with a high degree of reliability using a U-type manometer. Figure 4.34 shows the volumetric strains measured in marble specimens at (a) $\sigma_2 = 25$ MPa and (b) $\sigma_2 = 50$ MPa by a U-shaped manometer (curves 2) and hydrostatic weighing (points 3). Point A on curves 2, which depict

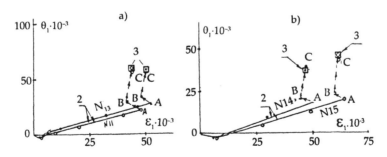

Fig. 4.34: Volumetric changes observed in marble specimens (Nos. 11, 13, 14, 15) during axial deformation at two levels of confining pressure: (a) $\sigma_2 = 25$ MPa; (b) $\sigma_2 = 50$ MPa; followed by unloading.

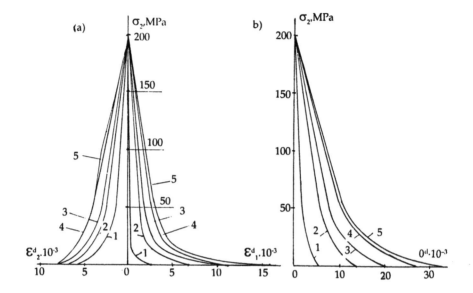

Fig. 4.35: Relationships between axial ε_1^d, lateral ε_2^d and volumetric θ^d strains in a specimen during the release of hydrostatic pressure (i.e., decompression) after subjecting the specimen to different levels of initial permanent axial strain at $\sigma_2 = 200$ MPa.

the entire history of deformation during the test, corresponds to the end of the deformation produced by axial loading of the specimen. Point B corresponds to the state after the specimen is relieved of differential axial load. Point C characterises the volume of the specimen after release of hydrostatic pressure, i.e., under total decompression. As can be seen, the magnitudes of total volumetric strain of the specimens as measured by the two different methods (U-shaped mamometer and hydrostatic weighting) almost coincide.

With regard to the effect of decompression on the state of a rock previously subjected to permanent deformation, test results indicate that the volumetric changes caused by direct irreversible deformation and subsequent decompression are comparable in magnitude. For the test conditions shown in Fig. 4.34, the magnitude of volumetric strain produced during decompression is even greater than the volumetric strain of the specimen in a high-pressure cell with superimposed axial loading. For the confining pressure $\sigma_2 = 25$ MPa, the decompression volumetric strain was 1.4 times larger than the volumetric strain obtained during axial deformation in the high-pressure cell. For $\sigma_2 = 50$ MPa, the ratio was 1.25.

Now let us consider how the magnitude of the preliminary permanent strain affects the deformation caused by decompression. Tests were conducted as described above. A single specimen of marble was subjected to

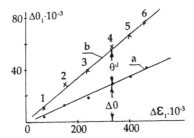

Fig. 4.36: Relationships between the permanent change in volume (θ) of a specimen and magnitude of permanent axial strain $\Delta\varepsilon_1$, measured under (a) conditions prevailing in a high-pressure cell at $\sigma_2 = 200$ MPa and (b) atmospheric pressure after unloading (decompression).

repetitive sequential deformation after it had been machined to its original diameter between each loading. Curves shown in Fig. 4.35 (a) indicate the development of axial ε_1^d and lateral ε_2^d strains in the specimen during release of the hydrostatic pressure σ_2 (during decompression) after each stage of loading and permanent deformation. The serial number of the loading cycle preceding the release of confining pressure is indicated by the numbers on the curves. Permanent axial strain in each test was about 8–10%. The confining pressure was equal to 200 MPa. The graphs in Fig. 4.35 (b) show the variation in volume θ^d of the specimen, caused by decompression after each cycle of loading. It can be seen that the strains produced upon unloading increased after each sequential cycle of loading. The relationships between the permanent changes in specimen volume $\Delta\theta$ and the value of permanent axial strain $\Delta\varepsilon_1$, measured under conditions of a high-pressure cell at (a) $\sigma_2 = 200$ MPa and (b) in atmospheric conditions after decompression, are shown in Fig. 4.36. These curves were plotted from the results of six test replications on the same specimen. As in the previous Figure, the numbers near the experimental points indicate the number of the test. Both relationships are shown by straight lines. The difference in ordinate between the two curves ($\Delta\theta^d$) is the volumetric strain caused by the process of decompression. This quantity is proportional to the quantity of initial irreversible strain. For the given test conditions, the values of the two volumetric strains ($\Delta\theta$ and θ^d) are quite close.

The inverse relationships of those shown in Fig. 4.35 are shown in Fig. 4.37. These curves show the relationships of axial ε_1^k, lateral ε_2^k, and volumetric θ^k strains as a function of the hydrostatic confining pressure σ_2 (i.e., under conditions of compression). The curve indicated by 0 was obtained by loading of the unstrained specimen; curve 1 was obtained by (compression) loading after the first cycle of testing (including restoration of the specimen to its original diameter by machining on a lathe); curve 2

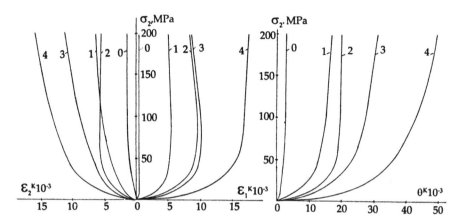

Fig. 4.37: The variation of the axial ε_1^k, lateral ε_2^k and volumetric θ^k strains in a specimen when the hydrostatic pressure was increased to $\sigma_2 = 200$ MPa (compression) after each (sequential) turning of the specimen in a lathe (to restore it to its original diameter).

after the second cycle etc. The large deformations, compared to those developed during decompression, are particularly evident. This difference is due to the fact that while turning the specimen in a lathe, the surface is damaged mechanically, such that when hydrostatic confining pressure is (re)applied, the specimen yield is increased. Overall, the curves shown in Figs. 4.35 and 4.37 reflect similar patterns of variation in deformation for different pressure levels. Maximum contration of the specimen was observed at low confining pressure, while the extent of yield decreased sharply with increase in pressure.

Figure 4.38 shows the results of tests in which the specimen was subjected to cyclical loading and unloading. Each cycle consisted of initial confinement of the specimen to a given hydrostatic pressure, ranging from 0 to $\sigma_2 = 150$ MPa, followed by axial loading and deformation to a given level, after which the specimen was completely decompressed in the reverse sequence. These cycles of compression and decompression differed from those discussed earlier in that here the specimen was not removed from the test cell and was not turned in a lathe to restore its diameter. The velocity (V) of elastic wave propagation in the specimen was measured during the test. Fig. 4.38 (b) shows the change in velocity V for each cycle. The lower portions of the curves, below the horizontal axis, indicate the effect of hydrostatic pressure σ_2 ; the upper portions indicate the effect of axial loading $\Delta\sigma_1$. It can be seen that: 1) permanent strain causes a reduction in the value of V; 2) as a result of permanent strain, the velocity V decreases more rapidly during release of confining pressure σ_2; moreover, the higher

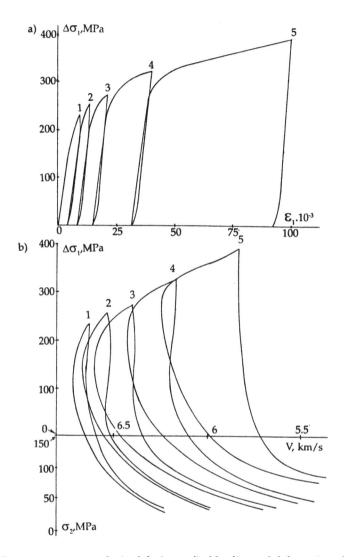

Fig. 4.38: Curves $\Delta\sigma_1$ versus ε_1 obtained during cyclical loading and deformation of a marble specimen at (a) $\sigma_2 = 150$ MPa and (b) curves indicating the variation in elastic wave velocity through the specimen during compression and decompression.

the initial permanent strain, the more rapid the fall in velocity; 3) return of the specimen in each subsequent cycle to the stress state from which the process of unloading (decompression) started in the previous cycle (points 1–4) practically restores the value of the wave velocity V, which signifies a return of the material structure to its initial form. An almost identical

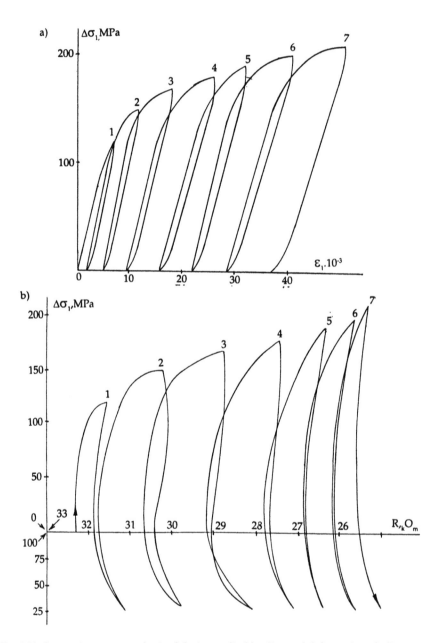

Fig. 4.39: Curves $\Delta\sigma_1$ versus ε_1 obtained during cyclical loading and deformation of a limestone specimen at (a) $\sigma_2 = 100$ MPa and (b) curves showing the variation in electrical resistivity in the specimen during compression and decompression.

picture was observed for similar tests in which measurements were recorded by electrical resistivity (Fig. 4.39). These tests were conducted on limestone at a hydrostatic pressure $\sigma_2 = 100$ MPa.

The important conclusion to be drawn from the results of these tests is that rocks suffering permanent strain exhibit radically different structure and properties upon unloading, from those of the same rocks before decompression. The degree of these changes depends on the magnitude of the permanent strain. Reloading the rock to restore the conditions at the depth from which it was extracted, tends to restore its original (prior to decompression) characteristics.

4.5 FLUID FLOW PROCESSES DURING HYDRAULIC FRACTURING

In this section we discuss the methods used to investigate the mechanics of hydraulic fracture (including fluid flow at the borehole wall and in the extending fracture) under laboratory conditions. The effectiveness of rupturing rock by hydrofracture is determined by the hydraulic losses which accompany the fracturing process. These losses are associated with: 1) leakage of the working fluid into the borehole wall when it is adequately pressurised; 2) filling the volume of the open crack with the working fluid; 3) fluid flow into the rock through the crack surfaces; and 4) elastic deformation of the borehole per se due to the pressure of the working fluid. The proportion of each of these losses depends on several factors, in particular the magnitude of the pressure of the hydrofracture, on the fluid flow properties of the rock, on the ability of the working fluid to penetrate into the rock, on the rate of pressure development in the borehole etc. The goal of the studies was to simulate, in the laboratory, the various conditions of hydrofracture for different rocks and to determine the overall balance of hydraulic losses. To assess the effectiveness of variations in the hydrofracturing process, the balance of hydraulic losses is written as follows:

$$V_{\text{total}} = V_{\text{bh}}^{\text{ft}} + V_{\text{ck}}^{\text{ft}} + V_{\text{ck}} + V_{\text{el}},$$

where $V_{\text{bh}}^{\text{ft}}$ is the volume of flow leak-off through the borehole wall; $V_{\text{ck}}^{\text{ft}}$ is the volume of flow leak-off through the crack surfaces; V_{ck} is the volume of fluid flowing (filling) in the crack; V_{el} is the fluid volume associated with elastic deformation of the borehole.

Since the aim of hydraulic fracturing is to create a crack, losses associated with the crack becoming filled with the working fluid V_{ck} are unavoidable. Therefore, we shall relate all other losses to V_{ck}, taking the latter to be equal to unity.

Experimental studies were conducted according to the program developed by J. D. Diadkin of St. Petersburg Mining Institute in collaboration with Cho Huan Shin (China). The testing machine and specimen are shown schematically in Fig. 4.3. The specimen was a thick-walled cylinder with an inner co-axial hole through it (simulating a borehole). Steel 'packers' were attached with epoxy resin to seal the upper and lower ends of the 'borehole'. These packers were connected to the rod of the cell and the thrust bearings. The specimen was externally waterproofed to protect it from the confining pressure fluid in the cell. A stress state of the type $\sigma_1 = \sigma_2 = \sigma_3$ or $\sigma_1 > \sigma_2 = \sigma_3$ was developed in the specimen. A device in the form of an injector was used to create pressure in the packed-off interval of the inner hole and to study flow and deformation processes in the borehole.

The test procedure was as follows. The pump (12) fills the entire hydraulic system with the (working) fracturing fluid, while also evacuating any air contained in the borehole and in the system through an open valve (15). Valves (15) and (14) are then closed and the pressure of the fluid increased in the borehole by the injector. The pressure developed is recorded by the gauge (23). The injector can be programmed to develop various pressure histories in the hole, such as a smooth or stepped pressure increase and, if need be, pressure reduction and release, etc.

The entire hydraulic pressurisation system is designed to be as stiff as realistically possible. All parasitic volumes in which the pressurised working fluid is contained (in addition to that in the borehole) are reduced to a minimum. The length of the pressure tubing is also kept to a minimum.

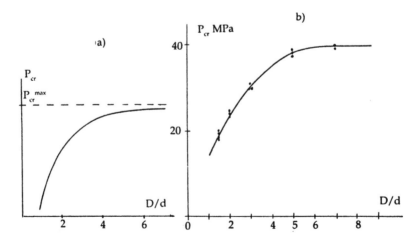

Fig. 4.40: (a) Variation of the critical internal pressure P_{cr} in a thick-walled cylinder versus the ratio of external D to internal d diameters computed by the Lamé equation; (b) Experimental relationship obtained for hydraulic fracturing tests on thick-walled cylinders of acrylic plastic.

The volume of the injector is likewise held to the minimum necessary to create pressure for hydrofracturing and to compensate for flow leakages (leak-offs). The force required to move the piston (7) is generated mechanically, rather than hydraulically. To accomplish this, a screw-feed is used instead of a piston (9) in the injector for fracturing tests. This specific feature was introduced in order to minimise elastic energy stored in the volume of compressed fluid, as a means of controlling growth of the crack developed by hydrofracturing.

The dimensions and shape of specimens have an effect on the test results in almost every type of experiment. Attention was paid to this issue in the present studies, in particular by selecting the minimum possible ratio between specimen diameter D and borehole diameter d, and the minimum distance between the packers—which determine the open pore region of the borehole.

Since the test specimens were thick-walled cylinders, loaded by internal and external pressure, selection of an optimal ratio D/d was based on the Lamé equation [12]. Figure 4.40 (a) shows a graph based on the Lamé equation, of the change in the critical (fracturing) value of internal pressure P_{cr} in a thick-walled cylinder versus the ratio D/d. The maximum value of the fracturing pressure is indicated by the dashed line. A similar experimental graph, obtained by the authors in fracturing thick-walled cylinders of acrylic plastic, is shown in Fig. 4.40 (b). Based on these relationships, all tests were conducted with specimens with $D/d > 6$; thus the fracturing pressure would correspond essentially to the field situation where, as a matter of fact, the 'outer diameter' is infinite.

The choice of ratio l/d (where l is the spacing between the packers; d is the borehole diameter) was based on the experimental relationship P_{cr} versus l/d (see Fig. 4.41) obtained from tests on marble specimens. The spacing

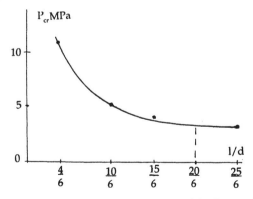

Fig. 4.41: Critical hydrofracturing pressure P_{cr} versus ratio of the distance l between the packers and the diameter d of the borehole, obtained for thick walled marble cylinders of $D = 90$ mm and $d = 6$ mm.

between packers should be not less than 20 mm when the diameter of the borehole is 6 mm (all test specimens discussed below had an internal diameter of 6 mm). Experience showed that the distance between the (internal) end of the packer and the end of the specimen should be not less than the distance between the borehole wall and the external cylindrical surface of the specimen, if hydrofracture turning towards the end surface of the specimen during its propagation is to be prevented.

In the study of fluid flow processes occurring during hydrofracturing, the injector is the main measuring device in the test apparatus. To avoid errors associated with the compressibility of the fluid when measuring volume of flow into the borehole wall and the extending hydraulic fracture, an initial calibration of the entire hydraulic system is carried out with respect to compressibility of the fracturing (working) fluid. This calibration is made using a steel specimen with an inner co-axial hole through it which is an exact replics of the rock specimen. This calibration exercise is designed to obtain a relationship between the pressure P, developed in the hydraulic system during movement of the piston (7), and the change in volume ΔV_{compr} in this system (volume measured by gauge 17). An example of the calibration graphs obtained for (1) water and (2) glycerine is shown in Fig. 4.42. The total volume of the hydraulic system in the extreme (initial) position of the piston, i.e., at the start of pressurisation, was equal in this case to 4.2 cm^3. These curves in Fig. 4.42 contain a non-linear section at low pressures. This is due to the small volume of air that remains in the system when it is being filled with fluid. As the pressure rises, the air is compressed and ceases to have any significant effect on the characteristic ΔV_{compr} versus P pertaining to the working fluid. In order to ensure that air is completely eliminated from the system, it should be put under vacuum in advance of a test. This air evacuation operation was not conducted in the experiments for which the curves shown in Fig. 4.42 were obtained. Since the initial volume of the system varied, depending on the

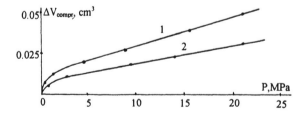

Fig. 4.42: Typical calibration graphs of the compressibility of the fracturing fluid in the test system obtained with (1) water and (2) glycerine.

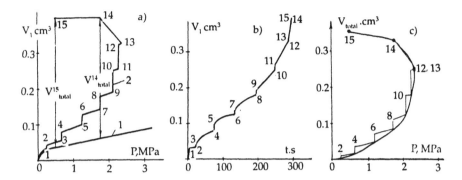

Fig. 4.43: Typical curves recorded in hydraulic fracturing experiments to reflect the relationships between: (a) pressure P in the system and volume V of fluid being fed into the specimen; (b) volume V and duration t of the process. The relationship between V_{total} and P, obtained as a result of processing data from diagram (a), is illustrated in diagram (c).

type of experiment and volume of fluid in the injector, individual calibration relationships ΔV_{compr} versus P were obtained for each test.]

The most informative regime in the study of fluid flow processes during hydrofracturing is that obtained with a stepped increase of the pressure. Typical curves recorded during testing are shown in Fig. 4.43. The compressibility calibration curve, similar to that shown earlier, is indicated by the number 1 in Fig. 4.43 (a); curve 2 indicates the relationship between the volume V of fluid fed into the system containing a real specimen and the magnitude of the pressure P. Fig. 4.43 (b) shows the relationship between volume and time. The pressure in the system was kept constant at each stage until fluid flow stabilised. 'Stabilisation' may occur when flow ceases, as during the initial stages of pressure increase in the system, or when the rate of fluid flow becomes constant. Constancy of pressure in the system is maintained automatically by the injector. Here, the rate of fluid flow into the borehole/fracture system is controlled by the rate of movement of the piston.

The difference in readings between relationships 1 and 2 is the total volume of fluid V_{total}, i.e., the sum of the quantity of fluid filtering into the borehole wall, filling the developed hydraulic fracture (crack), flowing into the crack walls, and the elastic deformation of the borehole as the pressure in the system is increased. It should be emphasised that volume V_{total}, determined as the difference between the readings of curves 2 and 1 when the pressure was increased, is obtained by taking into account the effect of fluid compression; in the case of a pressure decrease (as curve 2 goes through a maximum), the difference is obtained taking into account the effect of elastic expansion. An example of the relationship V_{total} versus P,

Table 4.5

Rock	K_{ft}, m^2	E, MPa	σ_{compr}, MPa	σ_p, MPa
Marble	10^{-18}	4.0×10^4	76	5
Granite	10^{-20}	5.5×10^4	175	11
Acrylic Plastic	0	4.9×10^3	140	40

obtained after processing data from the curves shown in Fig. 4.43 (a), is shown in Fig. 4.43 (c). The method used to determine the fraction of each of the individual components in the total volume of hydraulic losses during hydrofracturing is discussed below.

The magnitude of fluid flow during the process of hydrofracture and, consequently, the associated hydraulic losses, depends both on the properties of the rock and the properties of the working fluid. It is therefore necessary, in order to obtain a more complete insight into the development of the fluid flow processes during hydrofracture, to vary the magnitude of the fluid flow losses. To accomplish this, rocks of different permeabilities and fluid flow properties were tested. Tests were conducted on specimens of marble and granite together with tests on acrylic plastic. The acrylic tests were intended to simulate the behaviour using a medium with zero permeability. The values of permeability K_{ft}, modulus of elasticity E, uniaxial compression strength σ_{compr} and uniaxial tensile strength σ_p for marble, granite and acrylic plastic are given in Table 4.5.

Water, mineral oil and glycerine were used as the working fluids. The viscosity indices of these liquids were 0.001, 0.027 and 0.38 (μPa·s) respectively.

Different approaches were followed in order to identify the various components of fluid flow losses at various stages of the hydrofracturing process:

1) Observation of the entire process of fluid flow through the borehole wall and (hydrofracture) crack surfaces throughout the entire hydrofracturing process.

2) The following method was used to separate that portion of the fluid which flowed through the surfaces of a well-developed crack from the total volume of flow. The borehole wall was waterproofed with epoxy resin to eliminate flow through it. Flow losses were thus restricted to the immediate region of the crack opening and were totally associated with the process of crack (i.e., hydrofracture) formation and growth.

3) Since fluid losses during growth of the hydrofracture are associated not only with flow into the pore spaces of the crack surfaces, but also with the process of crack opening, and the associated increase in its volume, the following method was used to separate these two components. Specimens made from acrylic plastic were subjected to hydrofracturing. Fluid flow losses into the crack surfaces were thereby completely excluded. The

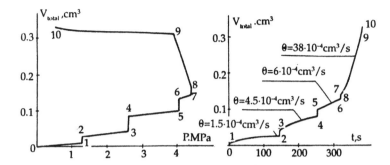

Fig. 4.44: Diagrams of V_{total} versus P, and V_{total} versus t, obtained during hydrofracturing of a marble specimen of $D/d = 90/6$ mm. Water was used as the working fluid.

volume of fluid injected due to crack growth and associated crack volume increase could then be determined by recording the change in pressure in the system as a function of the movement of the injector piston in response to the crack growth. Because the specimen was transparent, it was possible to measure the crack area and compute the width of opening of the crack surfaces. Since the test specimens of rock and acrylic plastic were of the same dimensions, it could be presumed with a certain degree of confidence that the losses associated with filling of the crack were the same in all specimens.

4) The hydraulic losses induced by elastic deformation of the borehole wall were measured in the tests where the borehole wall was waterproofed and were then calculated using the analytical solution for elastic deformation of an internally pressurised circular cylindrical hole.

Individual results of experimental determinations of the hydrofracture parameters are shown below, together with details of the method of testing, in order to provide some idea of the capabilities of the test apparatus.

Figure 4.44 shows diagrams of V_{total} versus P and V_{total} versus t, obtained during hydrofracturing of a marble specimen of $D/d = 90/6$ mm. Water was used as the working fluid. Flow quantities at different stages of pressurisation are indicated on the curve V_{total} versus t. Since marble is fairly permeable to water, increase in pressure P in the borehole produces an increase in flow. Although some decrease in flow rate was observed over time at low pressures, the flow rates became constant at pressures exceeding 2 MPa. A photograph of a cross-section of the test specimen showing the zone of fluid flow as dark area (dyed water was used in the test) is presented in Fig. 4.45 (a). Use of the dye aided analysis of the flow process during hydrofracture.

Flow occurred through the borehole wall only up to a pressure of 4 MPa. In this case, the total volume of fluid absorbed by the specimen was 0.17 cm^3. At $P = 4$ MPa, a hydrofracture crack was initiated, as evidenced in

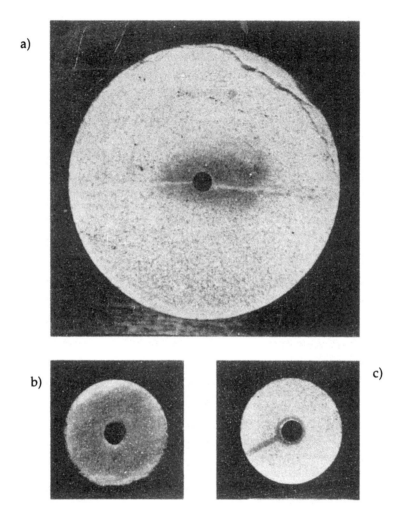

Fig. 4.45: Cross-sections of marble specimens tested under different regimes with zone of fluid flow shown as dark area.

the test by the onset of acoustic emission and a sharp increase in the slope of the relationship V_{total} versus P. Here the rate of fluid flow reached 6×10^{-4} cm^3 s^{-1}. At the limit level of pressure P_{cr}, the flow rate was 15×10^{-4} cm^3 s^{-1} (section 7–8 on curve V versus t). Crack growth was later followed by a decrease in pressure in the test system and a concomitant increase in flow rate to $Q = 38 \times 10^{-4}$ cm^3 s^{-1}. The crack growth could be controlled up to point 9 shown on the graph. The type of flow stain in Fig. 4.45 (a) attests to the uniform occurrence of the flow process through the crack

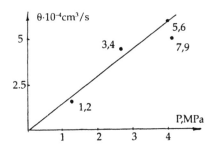

Fig. 4.46: Distribution of the volumes of fluid involved in flow into the borehole wall and in crack growth as a function of pressure P.

walls, which were constant in size. Based on the geometry of the staining produced by the fluid flow and the curves shown in Fig. 4.44, the variation in fluid outflow rate as a function of the pressure of working fluid in the borehole could be assessed. Figure 4.46 shows a graph of the relationship between Q and P plotted from these data. Numbers marked along the graph refer to corresponding sections on the curves in Fig. 4.44. In computations, the area of flow up to the pressure $P = 4$ MPa was assumed to be equal to the borehole cross-sectional area. Points 7–9 were obained taking into account the changed area of flow due to crack formation over the length of the coloured zone. This graph demonstrates the linear nature of variation in flow current with change in borehole pressure, which provided the basis on which to apply Darcy's equations in analysing the fluid flow processes during hydrofracture in the given rock.

Crack growth was unstable and dynamic in nature from point 9 to point 10. The growth acceleration process occurred in a fraction of a second. Assuming that fluid flow into the borehole walls and the crack is negligible over such a short time interval, then the volume increase observed in the graphs shown in Fig. 4.44 should conform to the volume of the opened crack. In this case, the volume was about 0.02 cm^3. Based on these tests, the total balance of flow losses (for the condition $V_{ck} = 1$) can be represented in the following manner, neglecting any loss due to elastic deformation of the borehole wall (tests for determining these losses will be described later):

$$V_{total} = V_{bh}^{ft} + V_{ck}^{ft} + V_{ck} = 8.5 + 7.2 + 1 = 16.7.$$

The distribution of the components involved in the balance of flow losses depends strongly on the rate of pressure increase in the borehole. This aspect is of special significance for permeable materials. Diagrams of V_{total}

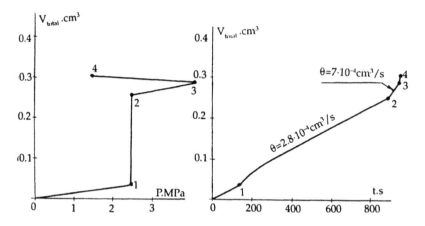

Fig. 4.47: Relationships V_{total} versus P and V_{total} versus t for marble at different (water) pumping rates.

versus P and V_{total} versus t for marble, reflecting the process of flow of water for various increases in the rate of pressurisation of the borehole at different stages of the hydrofracturing process, are shown in Fig. 4.47. When the pressure $P = 2.5$ MPa was maintained constant in the borehole, a constant flow rate of $Q = 2.8 \times 10^{-4}$ cm^3 s^{-1} was attained in section 1–2. In this case, the volume of hydraulic losses was proportional to time. The volume of fluid used in the test over $t = 900$ s was 0.25 cm^3. Increase in the rate of fluid flow into the borehole to $Q = 7 \times 10^{-4}$ cm^3 s^{-1}, led to a sharp rise in pressure and initiation of the hydrofracture. The total volume of absorbed fluid from point 2 to point 4 reached 0.05 cm^3, of which 0.012 cm^3 was spent on crack opening. In this case, the balance of flow losses became

$$V_{total} = 19 + 2.9 + 1 = 22.9.$$

Test experience indicated that the critical quantity of water flow rate into the borehole at which the pressure required for hydrofracture (P_{cr}) was obtained for marble, was approximately 5×10^{-4} cm^3 s^{-1}. When the rate of fluid flow into the system was maintained constant at $Q < 5 \times 10^{-4}$ cm^3 s^{-1}, the entire volume of the specimen became completely saturated. In this case, there was no hydrofracture. The photograph of a specimen completely saturated with coloured water is shown in Fig. 4.45 (b). Maintaining a high fluid pumping rate into the borehole from the moment of initiation of the hydrofracture yielded minimal losses. The zone of fluid flow formed when the fracturing process was extended (at high pumping rate) for about 5 s,

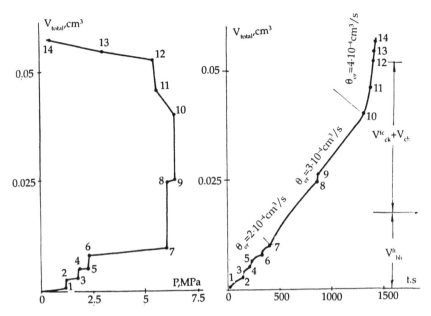

Fig. 4.48: Relationships V_{total} versus P and V_{total} versus t for a marble specimen of $D/d = 90/6$, tested with glycerine.

is illustrated in Fig. 4.45 (c). The balance of hydraulic losses in this case was: $V_{total} = 1 + 0.9 + 1 = 2.9$.

The balance of hydraulic losses also depends to a large extent on the ability of a fluid to flow through the specific rock. For example, glycerine compared to water has high viscosity and low flow capability. The relationship V_{total} versus P and V_{total} versus t, obtained during hydrofracturing of a marble specimen with glycerine, is illustrated in Fig. 4.48. Up to point 7, increase in pressure is followed by a small increase in flow losses since at each pressure (points 1–6), flow tends to exhibit a decaying trend. Point 7 corresponds to initiation of a hydrofracture crack. From this moment, the total hydraulic losses include those associated with continued flow through the borehole, flow through the changing surface area of the crack, and flow to fill the volume of the open crack. To ensure a constant level of flow (which depends on the fluid pressure) during the test, the pressure in the borehole was held constant until the hydrofracture crack approached the outer surface of the specimen. The relationships V_{total} versus P and V_{total} versus t shown in Fig. 4.49 allow the hydraulic losses associated with flow through the borehole to be identified over the entire duration of the test. These relationships were obtained over the same range of variation in pressure and for the same test duration as used in the previous test. The present

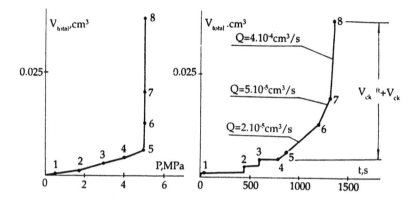

Fig. 4.49: Relationships V_{total} versus P and V_{total} versus t for a marble specimen of $D/d = 90/6$, tested with glycerine in a waterproofed borehole.

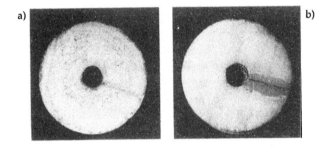

Fig. 4.50: Zone of flow around a crack for two rates of pressurised water in a waterproofed borehole.

test differed from the previous one in one aspect only—the borehole was waterproofed and flow through it thereby eliminated. The entire consumption of fluid consists of two components only—$V_{total} = V_{ck}^{ft} + V_{ck}$. By substituting this value into the curve V_{total} versus t of Fig. 4.48, we can determine the value of V_{ck}^{ft} during the entire process of hydrofracture. The distribution of fluid around hydrofracture cracks extending from a waterproofed borehole is photographically shown in Fig. 4.50. Here the cracks were formed at two rates of water pumping: (a) at high speed and (b) at low speed.

In hydrofracture tests conducted on specimens of acrylic plastic, in which no flow leakage into the specimen takes place, the volume of the crack formed V_{ck} in specimens of $D/d = 90/6$ was in the range of 0.008–0.03 cm^3.

Choosing the average V_{ck} = 0.02, the hydraulic losses for the cases depicted in Figs. 4.48 and 4.49 would be respectively:

$$V_{ck} = 0.8 + 1.1 + 1 = 2.9 \text{ and } V_{ck} = 0 + 1.1 + 1 = 2.1.$$

In all the cases discussed above, the loss V_{el} associated with elastic deformation of the borehole wall under the effect of pressure was neglected. Experimental evaluation of volumetric strains in the borehole under fluid pressure was carried out as follows. Calibrated relationships of compressibility of the working fluid ΔV versus P, illustrated in Fig. 4.42, were obtained when pressure was developed in a hole drilled into a steel specimen. Actually, data of these curves reflect compressibility of volume of fluid and also the elastic deformation in the entire hydraulic system (including the hole) subjected to pressure. In the tests conducted to determine elastic deformation in the borehole, parasitic volumes (pressure tubing and the fluid volume of the injector) were reduced to the minimum and amounted to less than 10% of the volume of the borehole. This error was taken into account in the interpretation of the experimental results. Since the modulus of elasticity of rock specimens is considerably different from the modulus for steel, this leads to a difference between the relationship ΔV versus P and the calibrated relationship for such specimens. The magnitude of volumetric strain in a borehole drilled in rock can be obtained by adding to this difference the calculated volumetric changes of the hole in the steel specimen. During experimental evaluation of the relationships ΔV versus P the borehole wall was thoroughly waterproofed and glycerine was used to generate the fluid pressure in order to avoid leakage of the working fluid into the rock. The relationships V_{el} versus P obtained in this way for granite

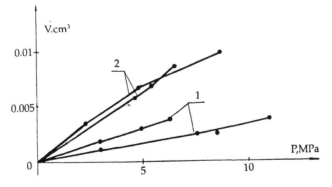

Fig. 4.51: (1) Graphs showing the increase in volume of a borehole in granite specimens as a function of the fluid pressure P; (2) Relationships taking into account flow of glycerine into the borehole wall and increase in volume of the borehole with increase in fluid pressure P.

are shown in Fig. 4.51 (graphs 1). For purposes of comparison, the relationships obtained when glycerine was used as the working fluid but when the borehole was not waterproofed (curve 2) are also shown.

Finally, it should be noted that these test results show wide variation in the hydraulic loss parameters during hydrofracture. Correlation between the components of the total hydraulic losses was determined by a set of factors, including flow properties of rock; ability of the fluid to penetrate into the rock (dependent, in particular, on the fluid viscosity); rate of pressure increase in the borehole and presence/absence of waterproofing on the borehole wall.

The following results are worthy of note with respect to determination of strength parameters in hydraulic fracturing.

1) A tendency towards decrease in the hydraulic fracturing pressure P_{cr} with increase in extent of the zone of fluid flow around the borehole. Under high rates of pumping water into the borehole or when glycerine was used as the working fluid, the level of P_{cr} varied over the range 5 to 9.8 MPa. With water, and a low rate of pressure increase in the borehole, hydrofracture occurred at P_{cr} = 3.2–5 MPa. The causes of this effect are twofold: (i) influence of fluid saturation on the strength of the rocks and (ii) increase in pore pressure in the zone of flow around the borehole, which leads essentially to a reduction in the ratio D/d and, consequently, to a decrease in the strength of a thick-walled cylinder.

2) Results of hydrofracturing under different levels of external confining pressure P_{ext} acting on the specimen showed that the hydrofracture pressure P_{cr} for acrylic plastic specimens rose proportional with pressure P_{ext}, while for rock specimens P_{cr} rose more rapidly than P_{ext}. At higher values of P_{ext}, rock hardened (i.e., appeared to increase in strength) by tenths percents. This result indicates that increase in triaxial stress in heterogeneous solids such as rock, leads to an increase in fracture resistance, while homogeneous materials (acrylic plastic) do not undergo hardening.

4.6 EXPERIMENTAL DETERMINATION OF CONNECTED POROSITY IN ROCKS UNDER IN-SITU CONDITIONS AND PARAMETERS OF FLUID RECOVERY FROM PORES

The porosity of rocks is affected both by the applied stress state and the level of pore pressure. It is important, therefore, in calculating reserves of liquid or gaseous fuels in any given deposit in the earth's crust, to take into account the state of the reservoir under conditions of deposition. Such information can be obtained through evaluation of the essential parameters of specimens taken from the reservoir in question, upon subjecting them to laboratory test conditions that simulate the reservoir *in-situ* conditions. Tests for this purpose were conducted on the machine illustrated in Fig. 4.3, using the following procedure.

The specimen (3), enclosed in a waterproof jacket, is placed in a high-pressure cell. The appropriate stress state $\sigma_1 = \sigma_2 = \sigma_3$ or $\sigma_1 > \sigma_2 = \sigma_3$ is generated by hydrostatic pressure in the cell and an additional axial load imposed by means of the press. Later, with a closed cock (13) and an open cock (15), pore space in the specimen is evacuated using a vacuum pump (16). Cock (15) is then closed. The volume of open porosity in the specimen is measured by means of an injector. To prepare the injector for testing, fluid at the required *in-situ* pressure is fed into the cavity of the cylinder (6) by pump (12) and cock (14) is then closed. In the low-pressure cavity, the corresponding hydrostatic head is created by pump (11). When cock (13) is opened, fluid enters the specimen, filling all available pore space, while the fluid pressure in the injector is maintained constant by the thrust from the low-pressure chamber of piston (9). This pressure can be conveniently developed by compressed air accumulated in the receiver (18). The pore pressure is controlled throughout the test by readings of gauge (23).

The quantities measured during the test are the volume of fluid filling the specimen and the duration of filling. The volume is determined from readings of gauge (17), recording the movement of piston (9). After the pore space in the specimen is completely saturated with fluid, piston (9) stops moving. The magnitude of the piston movement determines the volume of fluid percolating into the pores in the specimen. This method of determining volume of open pore space has the advantage that it is independent of the pressure or compressibility of the fluid being used in the tests; the volume of pores under study can be established directly through movement of the piston of the injector (9) with no additional computations necessary for the complicated question of change in compressibility of various fluids with change in fluid pressure.

The volume of fluid displaced from the pore space in the specimen at the moment of pressure release, is determined as follows: Cock (13) is closed, thereby maintaining the pore pressure in the specimen. Cock (14) is then opened and all fluid is drained from the high-pressure cavity (6) by moving piston (7) to the left until it stops. Cock (14) is then closed and cock (13) opened. When the forward thrust pressure in the low-pressure cavity is dropped to zero, the excess fluid contained in the specimen is drained into the cavity (6) and displaces piston (7) until the pore pressure drops to zero. The volume of fluid taken up by the specimen (or drained from it) is determined by the movement of the piston. This method helps in determining the fluid yield from the specimen since, in this case, the fluid volumes drained from the specimen are very small (due to the low compressibility of the liquid). To record these small values requires very

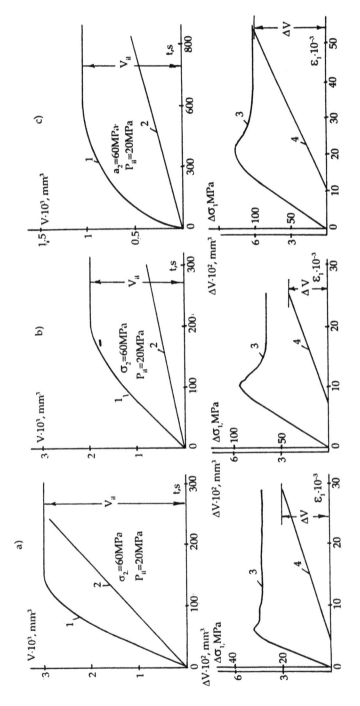

Fig. 4.52: Typical relationships obtained in tests conducted to determine the flow properties and open porosity under *in-situ* condtions, and also the recoverability of liquid from the pore space when the pore pressure was released.

sensitive instrumentation. In the case of gaseous pore fluids, the drained volume is measured in another manner. Gas from the closed cavity of the specimen (when cock (13) is closed) is released through the upper main by cock (15). The volume of released gas is then measured by displacement of the liquid from a measuring vessel fitted at the outlet of the high-pressure cell.

We shall now consider several specific examples of determination of the open porosity of rock specimens under *in-situ* conditions and recoverability of fluid from the specimens when pore pressure is released. Fig. 4.52 shows a series of curves obtained by testing specimens of (a) limestone and two varieties of sandstone (b) and (c). A mixture of organic oil and kerosene was used as the pore fluid. Prior to pumping the fluid into the pores of the specimens, a non-uniform triaxial compression (σ_1 = 85 MPa; $\sigma_2 = \sigma_3$ = 60 MPa) was applied to them. Fluid pressurised to 20 MPa was injected by the injector into the pore space in the specimens without prior evacuation of the pores, since the specimens had been thoroughly dried at high temperatures prior to testing. Curves showing the rate of filling of the pore space in the specimens by liquid are indicated by the number 1. When the pore space was completely filled, the curves became horizontal. Volume V_{il} of fluid entering the specimen corresponds to the volume of open pore space in the specimen for the given stress state and pore pressure. Porosity values η determined by dividing V_{il} by the initial volumes of the specimens are given in Table 4.6. For specimens (a), (b) and (c), the calculated porosities are 14.9%, 9.65% and 5% respectively. The time required to reach complete saturation of the specimens depends on the porosity of the rock. For the specimens under study, the time was found to be 150, 200 and 600 s (see curves 1).

The fluid flow properties of the specimens were determined after they had been saturated. The volume of liquid passing through the specimen was also measured by means of an injector. To develop fluid flow in the specimens, cock (15) was opened, while maintaining constant pressure (P_{il} = 20 MPa) at the inlet. The volume of liquid passing through the specimen in a specified period was measured. The relationships between the flow volume and time are indicated in Fig. 4.52 by the number 2. The relationships are seen to be linear. It should be noted that the fluid flow process taking place in the specimen during the time that fluid is being absorbed occurs more rapidly than in a saturated specimen. Curves 1 are steeper than curves 2. This is probably due to the action of capillary forces that accelerate the flow of liquid through the specimen during the saturation process. The values of permeability K_{ft} calculated from curves 2 are given in Table 4.6.

The volumes of liquid drained from the specimen upon release of the pore pressure were determined under two conditions of applied load: (i) in unstrained specimens and (ii) after development of permanent deforma-

Table 4.6

Rock	(a) Limestone	(b) Sandstone	(c) Sandstone
Initial porosity, η, %	14.9	9.65	5
Coefficient of permeability, $K_{ft} \times 10^{-2}$, md	15	3	0.6
(Initial) volume of pores, $V_{il} \times 10^3$, mm^3	3	1.98	1.1
Post-strain volume of pores, $V_{df} \times 10^3$, mm^3	3.3	2.21	1.7
Increase in volume, $\Delta V \times 10^3$, mm^3	0.3	0.23	0.6
Pre-strain discharge, $V_0 \times 10^3$, mm^3	0.032	0.023	0.014
Post-strain discharge, $V_0 \times 10^3$, mm^3	0.060	0.051	0.065
Increase in actual post-strain discharge	0.028	0.028	0.051
Calculated value of $\Delta V_0 \times 10^3$, mm^3	0.0026	0.002	0.0062

tions in the specimens. Values obtained under both conditions are given in Table 4.6 under 'Discharge'. These experimental results should be noted. The change in void space in the specimens was recorded during the course of deformation. Graphs showing the relationships between the change in load $\Delta\sigma_1$ and the irreversible increase in volume ΔV with deformation ε_1 are indicated in the diagrams by numbers 3 and 4 respectively. If it is assumed that liquid is drained from the specimen due solely to the elastic expansion that occurs as a result of the reduction in pressure, then volume of liquid discharged by the specimen upon deformation should increase by the compressibility of the volume ΔV. For specimens (a), (b) and (c) these values were 2.6 mm^3, 2 mm^3 and 5.2 mm^3 respectively. The values actually obtained from the tests were higher by one order of magnitude, i.e., 28 mm^3, 28 mm^3 and 51 mm^3 respectively. This indicates that when pore pressure in a rock specimen is released, the specimen is deformed by the external confining pressure, which results in expulsion of liquid from the material. Since the liquid has a low compressibility, the fraction of fluid drained from the material by external pressure is comparable to the fraction

Table 4.7

Rock	(d) Sandstone	(e) Sandstone
Initial porosity, η, %	9.65	5.5
Permeability coefficient $K_{ft} \times 10^{-2}$, md	3.5	0.017
(Initial) volume of pores, $V_{il} \times 10^3$, mm^3	2.8	1.6
Post-strain volume of pores, $V_{df} \times 10^3$, mm^3	3.1	2.2
Increase in volume, $\Delta V \times 10^3$, mm^3	0.3	0.6
Pre-strain discharge, $V_0 \times 10^3$, mm^3	875	431
Post-strain discharge, $V_0 \times 10^3$, mm^3	970	595
Actual increase in post-strain discharge	95	164
Calculated values of $\Delta V_0 \times 10^3$, mm^3	94	162

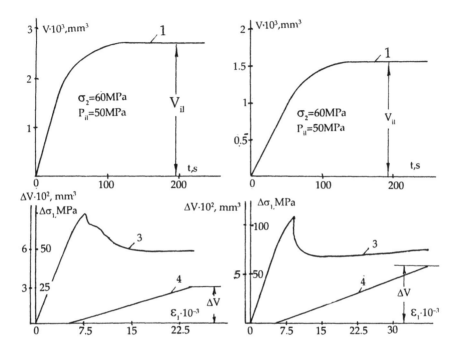

Fig. 4.53: Typical relationshp obtained in tests conducted to determine (i) the flow properties and open porosity in rocks under in-situ conditions, and (ii) the recoverability of gas from pore space during the release of pore pressure.

released due to the expansion of liquid in the total balance expression. The first fraction, with rise in prior permanent rock deformation, becomes one order of magnitude greater than the latter. Therefore, it is advisable to subject rock to permanent deformation in order to increase the effectiveness of draining liquid from pore space by release of the pore pressure, thereby developing a greater fluid yield.

The situation is different in the case of gas drained from rock. Results of tests conducted using gas are given in Fig. 4.53. Tests were conducted on two types of sandstones. The initial stress state in the specimens was $\sigma_1 = 85$ MPa and $\sigma_2 = \sigma_3 = 60$ MPa. The pressure of the gaseous fluid (nitrogen) fed into the specimen from the injector was maintained at 50 MPa. Curves 1 show the progressive saturation of the pore space in specimens with compressed gas as a function of time. As in the previous cases, volume V_{il} of pore space was determined based on the volume of fluid pumped into the specimen under pressure. Values of porosity η, permeability K_{ft}, pore volume in the specimens prior to deformation V_{il} and after deformation V_{df}, increase in volume of pores after deformation ΔV, fluid discharge V_0 from the specimen during release of pressure in pre-strain and post-strain stages, and increase in discharge ΔV_0 as a result of deformation, are given in Table 4.7. In contrast to the previous tests using liquid, the volume of gas drained from the specimen as a result of the reduction in pore space due to confinement of the specimen under external pressure σ_2, is negligible compared to the volume released due to elastic expansion of the gas. This is because of the high compressibility of gas. Values of ΔV_0 determined by experiments and by calculation are very similar. In the case of gas, permanent deformation demonstrates the effectiveness of the recoverability of fluid from pore space. In this case, the effectiveness is associated directly with increase in volume of the void space in which the compressed gas is contained. In the case of gas, increase in compressibility of the rock with increase in permanent deformation is of no consequence.

Literature Cited

1. Aragon, A. S. and Orowan, E. 1961. *Nature*, 192: 447
2. Atomic Mechanism of Failure. 1959. Proc. Intl. Conf. on Failure-related Problems. Swampscotte, USA. [Gosudarstvennoe nauchno-teknicheskoe izdatel'stvo literaturi po chyornai i tsvetnoi metallurgii. Moscow (1963) 660 pp.] (translated from English)
3. Barton, N. 1976 The shear strength of rock and rock joints. *Rock Mechanics Review*. Pergamon Press, 13: 255–279.
4. Bessonov, M. I. 1964. Mekhanicheskoe razrusheniye tverdykh polimerov [Mechanical fracturing of hard polymers]. *Uspekhi fizicheskhik nauk*. Izd. AN SSSR, 53 (1): 107–135.
5. Bieniawski, Z. T. 1967. Mechanism of brittle fracture of rocks, Parts I, II and III. *Int. J. Rock Mech., Mining Sci.*, 4: 395–430.
6. Bieniawski, Z. T. 1970. Time-dependent behaviour of fractured rock. *Rock Mech.*, 2 (3): 123–137.
7. Bich, Ya. A. and Muratov, N. A. 1990. *Profilaktika gornykh udarov* [Prevention of Rockbursts]. Izd-vo Dal'nevost University, Vladivostok, 248 pp.
8. Brace, W. F. 1964. Brittle fracture of rocks. In: *State of Stress in the Earth's Crust*. Elsevier, Amsterdam, pp. 111–180.
9. Brace, W. F. 1964. *International Conference on State of Stress in the Earth's Crust* (W. Judd, ed.). Elsevier, New York, pp. 110–178.
10. Bridgeman, P. 1955. *Issledovaniya bol'shikh plasticheskikh diformatsii i razryva* [Investigations into Large Plastic Deformations and Fracture]. IL, Moscow, 350 pp.
11. Chirkov, S. E. 1976. Prochnost gornykh porod pri tryokhosnom neravnokomponentnom szhatti [Strength of rocks under volumetric inequal component compression]. FTPRPI, 1: 10–21.
12. Tsiklis, D. S. 1976. *Tekhnika fiziko-khimicheskikh issledovanii pri vysokikh i sverkh vysokikh davieniyakh* [Technique for Conducting Physicochemical Studies under High and Very High Pressures]. Khimiya, Moscow, 290 pp.
13. Coffin, L. F. 1950. Fracture of grey cast iron. *Intl. Appl. Mech.*, 17 (3): 233–242.
14. Cook, N. G. W. 1965. The failure of rock. *Intl. J. Rock Mech. Min. Sci.*, 2: 389–403.
15. Cook, N. G. W. 1965. A note on rockbursts considered as a problem of stability. *J. South Afr. Intl. Min. Metal.*, 65: 437–446.
16. Crouch, S. L. and Fairhurst, C. 1974. Mechanics of coal mine bumps. *Trans. Soc. Min. Eng., AIME*, 256: 317–324.
17. Davidenkov, N. N. 1936. *Mekhanicheskiye ispytaniya metallov* [Mechanical Testing of Metals]. ONTI.
18. Davidenkov, N. N. and Stavrogin, A. N. 1954. O kriterii prochnost pri khrupkom razrushenii i ploskom napryzhennom sostoyanii [Strength criteria under brittle failure and plane stress state]. *Izv. AN SSSR. Otdeleniye tekhnicheskikh nauk*, 8: 101–109.
19. Joystone, B. and Gilmon, D. J. 1960. Skorost peredvizheniya, plotnost dislokatsii i plastichnost kristallov ftoristogo litiya [Speed of movements, density of dislocations and plasticity of lithium fluoride crystals]. In: *Uspekhi fizicheskhik nauk*. Izd. AN SSSR, 20.
20. Eily, R. E. 1965. Strength of graphite tube specimens under combined stresses. *J. Amer. Soc. Eng.* 48 (10): 505–508.

21. Fairhurst, C. and Cook, N. J. W. 1966. *Proc. First Cong. Intl. Soc. Rock Mech.* Lisbon, 1: 687–692.
22. Libovits, T. (ed.). 1976. Razrusheniye [Failure]. In: *Neorganicheskiye materiyaly* (Translation). Mir, Moscow, vol. 7, pt. I, 630 pp.
23. *Fizicheksaya priroda khrupkogo razrusheiya metallov* [Physical Nature of Brittle Failure of Metals]. 1965. Respublikanskiy mezhvedomstvenniy sbornik: Seriya Metallofizika. Naukova Dumka, Kiev.
24. Filatov, N. A. and Belyakov, V. D. 1977. Rekomendatsii po izucheniyu modelei deskretno-narushennikh cred metodami fotomekhaniki [Recommendations for Studying Models of Discretely Fractured Media Using Methods of Photomechanics]. VNIMI, Leningrad, 48 pp.
25. Freedman, Ya. B. 1943. *Edinaya teoriya prochnosti materialov* [Unified Theory of Strength of Materials]. Oborongiz, Moscow.
26. Freedman, Ya. B. 1952. *Mekhanicheskiye svoistva metallov* [Mechanical Properties of Metals]. Oborongiz, Moscow, pp. 94–95.
27. Glebov, V. D. 1965. Rezul'taty issledovanii prochnosti avtoklavnykh silikatnykh betonov pri ploskom napryashyonnom sostoyanii [Results of investigations into strength aspects of steam-cured concrete under plane stress state]. *Izv. Vuzov. Stroitel'stvo i arkhitektura*, 12: 60–62.
28. Goncharov, N. G. 1960. *Prochnost kamennykh materialov v usloviyakh razlichnykh napryazhyonnykh sostoyanii* [Strength of Brick Materials under Different Stress States]. Gosud. izd. liter. po stroitel'stvu, arkhitekture i stroitel'nom materialam. Moscow-Leningrad, 75 pp.
29. Grassi, R. C. and Cornet, J. 1949. Fracture of grey cast iron. *J. Appl. Mech.*, 16 (2): 178–192.
30. Haupert Rene. 1974. Le rate du temps dans le comportment à la rupture des roches. *Adv. in Roch Mech.*, 2 (Part A): 325–329.
31. Heard, C. 1963. Effect of large changes in strain rate in the experimental deformation of marble. *J. Geology*, 71 (2): 90–99.
32. Hudson, J. A., Crouch, S. L. and Fairhurst, C. 1972. Soft, stiff and servo-controlled testing machines: A review with reference to rock failure. *Eng. Geol.*, 6: 155–189.
33. Jaeger, J. C. 1967. *Transactions of the 8th Rock Mechanics Symposium, Univ. Minnesota.* AIME, New York, pp. 3–57.
34. Griggs, J. 1936. Deformation of rocks under high confining pressures. *J. Geol.*, 44 (5): 541–577.
35. Karmanskiy, A. T. 1981. Metodika issledovaniya gornykh porod pri slozhnykkh napryazhyonnykh sostoyaniyakh s uchyotom gazovogo faktora [Methodology for studying rocks under complex stress states, considering the gas factor]. In: *Boryba s gornymi udarami.* Trudy VNIMI, Leningrad, coll. no. 119: 27–31.
36. Kuz'min, E. A. and Puch, V. P. 1959. Skorost rosta khrupkoi treshyny v stekle ikanifoli. Nekotorye problemy prochnosti tvyordogo tela [Rate of growth of a brittle crack in glass and resin. Certain problems related to strength of a solid body]. In: *Sbornikstateyi po svyashyonnykh 80-letiyu N. N. Davidenkova*, 367 pp.
37. Kumar, A. 1968. The effect of stress rate and temperature on the strength of basalt and granite. *Geophysics*, 33 (3): 501–510.
38. Linkov, A. M. 1994. *The Problem of Stability and Dynamic Phenomena in Mines.* Intl. Soc. Rock Mechanics, P-1799 Lisboa Cedex, 132 pp.
39. Lodus, E. V. 1983. Vliyaniye skorosti deformirovaniya i vidov napryashynogo sostoyaniya na zapredel'nye kharakteristiki udaroopasnykh i vybrosoopasnykh gornykh porod [Effect of rate of stress and types of stress state on post-peak characteristics of burst-prone and bump-prone rocks]. In: *Regional'nye mery predotvarasheniya gornykh udarov.* VNIMI, Leningrad, pp. 35–39.
40. Lundberg, B. 1976. A split Hopkinson bar study of energy absorption in dynamic rock fragmentation. *Intl. J. Rock Mech. Min. Sci.*, 13 (6): 187–197.

41. Malyshev, M. V. 1963. O vliyanii srednogo glavnogo napryazheniya na prochnost grunta i o poverkhnostyakh skol'zheniya [Effect of average principal stress on strength of soil and planes of slippage]. *Osnovaniye, fundamenty i mekhanika gruntov*, 1: 28–37. Moscow.

42. Nadai, A. 1954. *Plastichnost i razrusheniye tvyordykh tel* [Plasticity and Fracture of Solid Bodies]. (translated from English, IL, Moscow (1963), vol. 11, 565 pp.)

43. Peng, S. S. and Podnieks, B. R. 1972. Relaxation and behaviour of failed rock. *Intl. J. Rock Mech. Min. Sci.*, 9 (6): 699–712.

44. Peng, S. S. 1973. Time-dependent aspects of rock behaviour as measured by a servo-controlled hydraulic testing machine. *Intl. J. Rock Mech. Min. Sci.*, 10 (3): 235–246.

45. Perkins [sic]. 1970. *Intl. J. Rock Mech. Min. Sci.*, 7 (5): 527–535.

46. Petukhov, I. M. and Linkov, A. M. 1979. The theory of post-failure deformations and the problem of stability in rock mechanics. *Intl. J. Rock Mech., Min. Sci. and Geomech. Abstr.* 16: 79–87.

47. Petukhov, I. M. and Linkov, A. M. 1983. *Mekhanika gornykh udarov i vybrosov* [Mechanics of Rockbursts and Bumps]. Nedra, Moscow, 280 pp.

48. Robertson, E. C. 1955. Experimental study of the strength of rocks. *Bull. Geol. Soc. Amer.*, 66: 1275–1314.

49. Rummel, F. and Fairhurst, C. 1970. Determination of the post-failure behaviour of brittle rock using servo-controlled testing machine. *Intl. J. Rock Mech. Min. Sci.*, 2 (4): 189–204.

50. Rusch, H. 1960. Researches towards a general flexural theory for structural concrete. *Amer. Concr. Inst. Proc.*, 57: 1–28.

51. Salamon, M. D. G. 1970. Stability, instability and design of pillar working. *Intl. J. Rock Mech. Min. Sci.*, 7: 613–631.

52. Spivak, A. I. and Popov, A. M. 1986. *Razrusheniye gornykh porod pri burenii skvazhin* [Fracture of Rocks during Drilling of Boreholes]. Nedra, Moscow, 190 pp.

53. Shemyakin, E. I. 1986. O pasporte prochnosti gornykh porod [Strength Chart for Rocks]. IGD SO AN SSSR (1974), Novosibirsk, pp. 3–12.

54. Shestopalov, L. M. 1958. *Deformirovaniye metallov i volny plastichnosti v nikh* [Deformation of Metals and Waves of Plasticity in Them]. Izd. AN SSSR, Moscow, 300 pp.

55. Shannikov, V. M. and Kan, K. N. 1964. Issledovaniye staticheskoi prochnosti zhyostkikh plastmass pri ploskom napryazhyonnom sostoyanii [Study of static strength of rigid plastmass under plane stress state]. *Plasticheskiye massy.* AN SSSR, 1: 25–29.

56. Stepanov, V. A. and Kurov, I. E. 1962. Dolgovechnost materialov pri kruchenii [Longevity of materials subjected to torsion]. *Fizika tvyordogo tela.* Izd. AN SSSR, Moscow, pp. 137–142.

57. Stokes, R. J., Johnston, T. L. and Li, C. H. 1958. *Phil. Mag.*, 6: 9.

58. Stokes, R. J., Johnston, T. L. and Li, C. H. 1960. *Trans. AIME*, 218: 655.

59. Stokes, R. J. 1976. Mikroscopicheskiye aspekty razrusheniya keramiki [Microscopic aspects of failure of ceramic materials]. In: *Sb. Razrusheniye* (M. Libovits, ed.). Mir, Moscow, vol. 7, 182 pp.

60. Wawersik, W. R. and Fairhurst, C. 1970. A study of brittle rock fracture in laboratory compression experiments, pt. I. *Intl. J. Rock Mech. Min. Sci.*, 7: 561–575.

61. Wawersik, W. R. 1968. Detailed analysis of rock failure in laboratory compression experiments. *Intl. J. Rock Mech. Min. Sci.*, 7: 561–575 [sic].

62. Wawersik, W. R. and Fairhurst, C. 1970. A study of brittle rock fracture in laboratory compression experiments, pt. II. *Intl. J. Rock Mech. Min. Sci.*, 7: 613–631.

63. Von-Burgen [sic]. 1962. *Deffekty v kristallakh* [Defects in Crystals]. Izd. IL, Moscow (translation).

64. Vitman, F. F., Zlatin, N. A. and Shestopalov, L. M. 1950. O svyazi mezhdu energiyei aktivatsii metallov i ikh soprotivleniem deformirovaniyu [Relation between energy activation of metals and their resistance to undergo strain]. *Sb. Posvyashenny semidesyatiletiyu A. F. Ioffe.* Izd. AN SSSR, Moscow, pp. 331–340.

65. Stavrogin, A. N. 1966. *Ob usloviyakh predel'nykh sostoyanii gornykh porod* [Conditions of Ultimate States of Rocks]. Tezisy dokladov II soveshaniya po fizicheskim svoistvam

gornykh porod pri vysokikh davleniyakh. Institut Fiziki Zemli im. O. Yu. Shmidta, AN SSSR, 118 pp.

66. Stavrogin, A. N. and Georgievskiy, V. S. 1967. Vliyaniye vida nagruzheniya na protsess deformirovaniya gornykh porod [Effect of Type of Loading on the Process of Deformation in Rocks]. Tezisy dokladov vsesoyuznoi mezhvuzovskoi nauchnoi konferentsii s uchastiem nauchno-issledovatel'skikh institutov po fizike gornykh porod i protsessov. Moscow, 23 pp.

67. Stavrogin, A. N. 1968. Prochnost i deformatsiya gornykh porod [Strength and strain in rocks]. Diss. na soiskaniye uchyonoi stepeni doktora tekhnicheskikh nauk, VNIMI, Leningrad, 234 pp.

68. Stavrogin, A. N. 1968. Eksperimental'nye issledovaniya polzuchesti i dolgovechnosti gornykh porod [Experimental Studies of Creep and Longevity of Rocks]. Trudy koordinatsionnykh soveshanii po gidrotekhnike. VNIIG im. Vedenenva. Izd. Energiya, 38 pp.

69. Stavrogin, A. N. 1969. Issledovaniye predel'nykh sostoyanii i deformatsii gornykh porod [Study of limit states and strains in rocks]. *Izv. AN SSSR, Fizika Zemli*, 12: 54–69.

70. Stavrogin, A. N. and Karmanskiy, A. T. 1971. Metodika a rezultaty issledovaniy mekhanicheskikh svoistv gornykh porod pri raznoi vlazhnosti v usloviyakh slozhnogo napryazhynnogo sostoyaniya [Methodology and results of studies on mechanical properties of rocks at different moisture levels under conditions of complex stress state]. *Trudy VNIMI Sbornik*, Leningrad, 82: 155–159.

71. Stavrogin, A. N. and Georgievskiy, V. S. 1972. *Katalog mekhanicheskikh svoistv gornykh porod* [Handbook of Mechanical Properties of Rocks]. Izd. VNIMI, Leningrad, 267 pp.

72. Stavrogin, A. N., Georgievskiy, V. S. and Lodus, E. V. 1973. *Katalog mekhanicheskikh svoistv gornykh porod pri dlitel'nykh ispitaniyakh v usloviyakh odnoosnogo szhatiya* [Handbook of Mechanical Properties of Rocks in Long-term Tests under Conditions of Uniaxial Compression]. VNIMI, Leningrad, 110 pp.

73. Stavrogin, A. N. and Lodus, E. V. 1974. Polzuchest i vremennaya zavisimost prochnosti gornykh porod [Creep and time-dependent relationship of strength of rocks] FTPRPI, 6: 3–10.

74. Stavrogin, A. N. 1974. Statisticheskiye osnovy prochnosti i deformatsii gornykh porod pri slozhnykh napryazhonnykh sostoyaniyakh [Statistical fundamentals of strength and strain of rocks under complex stress states]. FTPRPI, 4: 24–31.

75. Stavrogin, A. N. 1975. Rock strength and plasticity under high pressure and strain velocity variation in the range of twelve orders. In: *High Temperature-High Pressure*. Moscow.

76. Stavrogin, A. N. et al. 1976. *Katalog mekhanicheskikh svoistv gornykh porod pri shirokoi variatsii vidov napryazhyonnogo sostoyaniya i skorosti deformirovaniya* [Handbook of Mechanical Properties of Rocks under Widely Varying Types of Stress State and Rates of Deformation]. VNIMI, Leningrad, 170 pp.

77. Stavrogin, A. N. and Protosenya, A. G. 1979. *Plastichnost gornykh porod* [Plasticity of Rocks]. Nedra, Moscow, 300 pp.

78. Stavrogin, A. N., Pevzner, E. D. and Tarasov, B. G. 1981. Zapredel'nye kharakteristiki khrupkikh gornykh poro [Post-peak characteristics of brittle rocks]. FTPRPI, 4: 8–15.

79. Stavrogin, A. N., Tarasov, B. G. and Shirkes, O. A. 1981. Prochnost i deformatsiya gornykh porod v dopredel'noi i zapredel'noi oblastyakh [Strength and strain in rocks in the pre-peak and post-peak zones]. FTPRPI, 6: 2–11.

80. Stavrogin, A. N., Tarasov, B. G. and Pevzner, E. D. 1982. Vliyaniye skorosti deformatsii na zapredel'nye kharakteristiki gornykh porod [Effect of deformation rate on the post-peak characteristics of rocks]. FTPRPI, 5: 8–15.

81. Stavrogin, A. N. and Lodus, E. V. 1982. Vliyaniye strukturnykh faktorov na mekhanicheskie svoistva gornykh porod [Effect of structural factors on the mechanical properties of rocks]. In: *Failure of Rocks during Drilling of Boreholes*. Ufa, pp. 49–52.

82. Stavrogin, A. N., Pevzner, E. D., Tarasov, B. G. and Shirkes, O. A. 1983. Universal'naya laboratornaya ustanovka s vysokoy zhyostkostyu dlya issledovaniya gornykh porod pri

slozhnykh napryazhyonnykh sostoyaniyakh [Universal highly stiff laboratory testing machine for studying rocks under complex stress states]. In: *Fizika i Mekhanika Razrusheniya Gornykh Porod.* Ilim, Frunze, pp. 44–49.

83. Stavrogin, A. N. and Protosenya, A. G. 1985. *Prochnost gornykh porod i ustoichivost gornykh vyrabotok na bol'shykh glubinnakh* [Strength of Rocks and Stability of Mine Workings at Great Depths]. Nedra, Moscow, 271 pp.

84. Stavrogin, A. N. and Tarasov, B. G. 1985. Balans energii pri khrupkom razrushenii gornykh porod [Energy balance during brittle failure]. FTPRPI, 1: 18–27.

85. Stavrogin, A. N., Yurel, G. N. and Tarasov, B. G. 1986. Mekhanicheskiye, filtratsionnye i petrograficheskiye svoistva vybrosoopasnykh i nevybrosoopasnykh peschannikov Donbassa [Mechanical, filtration and petrographic properties of burst-prone and non-burst-prone sandstones of Donbass]. FTPRPI, 2: 11–18.

86. Stavrogin, A. N. and Shirkes, O. A. 1986. Yavleniye posledeistviya v gornykh porodakh, vyzvannoe preshestvuyushei neobratimoi deformatsiei [Phenomenon of after-effects in rocks, caused by preceding permanent deformation]. FTPRPI, 4: 16–27.

87. Stavrogin, A. N., Tarasov, B. G. and Shirkes, O. A. 1990. Statisticheskaya model deformatsii neodnorodnykh tvyordykh tel (gornykh porod) v usloviyakh vysokikh davlenii i bol'shykh deformatsii [Statistical model of deformation in heterogeneous solids (rocks) under conditions of high pressures and large deformations]. FTPRPI, 1: 10–17.

88. Stavrogin, A. N. and Karmanskiy, A. T. 1992. Vliyaniye vlazhnosti, vida napryazhyonnogo sostoyaniya i skorosti nagruzheniya na fiziko-mekhanicheskiye svoistva gornykh porod [Effect of moisture, type of stress state and rate of loading on the physicomechanical properties of rocks]. FTPRPI, 4: 3–10.

89. Stavrogin, A. N., Tarasov, B. G., Shirkes, O. A. et al. 1992. Pronitsaemost gornykh porod i effektivnost podderzhaniya ustoichivosti stvola glubokikh i sverkhglubokikh skvazhin davleniyem burovogo rastvora [Permeability of rocks and effectiveness of maintaining stability of well of deep and ultradeep boreholes by means of drilling mud pressure]. FTPRPI, 5: 7–17.

90. Tarasov, B. G. 1981. Izucheniye mekhanizma ostatochnoi deformatsii khrupkikh gornykh porod [Study of mechanism of residual deformation in brittle rocks]. *Sb. Rasrabotka i obogasheniya tvyordykh polezhnykh iskopaemykh.* IPKON AN SSSR, Moscow, pp. 93–98.

91. Tarasov, B. G. 1982. Vliyaniye skorosti nagruzheniya i vida napryazhyonnogo sostoyaniya na povedeniye gornykh porod v dopredel'noi i zapredel'noioblastyakh [Effect of rate of loading and type of stress state on behaviour of rocks in pre-peak and post-peak strength zones]. *Sb. Fiziko- tekhnicheskiye i tekhnologicheskiye problemy razrabotki i obogasheniya tvyordykh polezhnykh iskopaemykh.* IPKON AN SSSR, Moscow, pp. 111–117.

92. Tarasov, B. G. 1983. Energoyomkost protsessov khrupkogo razrusheniya gornykh porod [Energy consumption in processes of brittle failure of rocks]. Diss. na soiskaniye uchyonoi stepeni kandidata tekhnicheskikh nauk. VNIMI, Leningrad, 233 pp.

93. Tarasov, B. G. 1987. Sposob opredeleniya koeffitsienta dinamicheskogo uprochneniya materialya [Method for Determination of Coefficient of Dynamic Hardening of Material]. A S. No. 1293547, B.I. No. 8.

94. Tarasov, B. G. 1989. Balans energii khrupkogo razrusheniye v usloviyakh obyomnogo naprayazhyonnogo sostoyaniya [Energy balance during brittle failure under conditions of volumetric stress state]. FTPRPI, 1: 2–28.

95. Tarasov, B. G. 1990. Uproshyonny metod opredeleniya stepeni vliyaniya skorosti deformatsii na prochnost i energoyomkost razrusheniya gornykh porod [Simplified method for determining degree of effect of deformation rate on strength and energy consumption for fracturing rocks]. FTPRPI, 4: 29–35.

96. Tarasov, B. G. 1991. Universal'naya diagramma mekhanicheskogo sostoyaniya gornykh porod i statisticheskaya priroda razvitiya deformatsiyonnykh protsessov v nikh [Universal diagram of mechanical state of rocks and statistical nature of growth of deformation processes in them]. Tez. Doklada 4: Vsesoyuznogo seminara *Fizicheskiye Osnovy prognozirovaniya razrusheniya gornykh porod.* Fiztech, Leningrad.

97. Tarasov, B. G. 1991. O statisticheskoi prirode prochnost gornykh porod [Statistical nature of strength of rocks]. FTPRPI, 4: 30–41.
98. Tarasov, B. G. 1991. O statistichekoi prirode deformatsionnykh protsessov v gornykh porodakh [Statistical nature of deformation processes occurring in rocks]. FTPRPI, 6: 36–44.
99. Tarasov, B. G. 1992. Prochnostnye, uprugiye i deformatsionnye svoistva gornykh porod kak funktsiya strukturnykh osobennostei materialya [Strength, elastic and deformation properties of rocks as a function of structural features of the material]. FTPRPI, 2: 26–35.
100. Tarasov, B. G. 1992. Vliyaniye vida nagruzheniya na protsess deformatsii gornykh porod [Effect of type of loading on the process of deformation in rocks]. FTPRPI, 1: 12–21.
101. Tarasov, B. G. 1992. Zakonomernosti deformirovaniya i razrusheniya gornykh porod pri vysokikh davleniyakh [Laws of deformation and fracture of rocks under high pressures]. Diss. na soiskaniye uchyonoi stepeni doktora tekhnicheskikh nauk. Gornyi Institut, Leningrad, 378 pp.
102. Tarasov, B. G., Stavrogin, A. N. and Shirkes, O. A. 1994. Mekhanizm formirovaniya porovogo prostranstva v gornykh porodakh v usloviyakh deformirovaniya pri vysokikh davleniyakh [Mechanism of formation of pore space in rocks under conditions of deformation at high pressures]. FTPRPI, 3: 23–36.

List of Inventions Used in Designing the Testing Machines Described in this Monograph

103. Stavrogin, A. N., Tarasov, B. G. and Pevzner, E. D. 1980. Sposob ispytaniya materiyalov na dinamicheskoe szatiye [Method of testing materials under dynamic compression]. Avtorskoe svidetel'stvo No. 736742, B.I. No. 45.
104. Stavrogin, A. N., Tarasov, B. G. and Pevzner, E. D. 1980. Gidroprivod k pressu dlya ispytaniya obraztsov na prochnost [Hydraulic drive attached to the press for testing strength of specimens]. Avt. svid. No. 785680, B.I. No. 45.
105. Stavrogin, A. N., Tarasov, B. G. and Pevzner, E. D. 1980. Ustanovka dlya ispytaniya obraztsov na szatiya c kontrolem skorosti deformatsii obrazta [Machine for testing compressive strength of specimens while controlling rate of deformation in the specimen]. Avt. svid. No. 911208, B.I. No. 9.
106. Stavrogin, A. N., Tarasov, B. G. and Pevzner, E. D. 1981. Ustanovka dlya ispytaniya obraztsov pri tryokhosnom szatii tipa $\sigma_1 > \sigma_2 = \sigma_3$ [Machine for testing specimens under volumetric compression of the type $\sigma_1 > \sigma_2 = \sigma_3$]. Avt. svid. No. 815583, B.I. No. 11.
107. Stavrogin, A. N. and Tarasov, B. G. 1981. Ustanovka dlya ispytaniya obraztsov na szatiye [Machine for testing specimens under compression]. Avt. svid. No. 911239, B.I. No. 9
108. Stavrogin, A. N. and Tarasov, B. G. 1983. Ustanovka dlya izucheniya balansa energii v sisteme *nagruzhayushee ustroistva—obrazets* pri razrushenii obraztsa [Machine for studying energy balance in the system *loading complex—specimen* during fracture of rock]. Avt. svid. No. 1024796, B.I. No. 23.
109. Stavrogin, A. N. and Tarasov, B. G. 1983. Ustanovka dlya dinamichekikh ispytanii materiyalov [Machine for dynamic testing of materials]. Avt. svid. No. 1016728, B.I. No. 4.
110. Stavrogin, A. N., Tarasov, B. G. and Pevzner, E. D. 1984. Ustanovka dlya udarnogo szatiya [Machine for shock (impact) compression]. Avt. svid. No. 1067403, B.I. No. 2.
111. Stavrogin, A. N., Tarasov, B.G. and Shirkes, O. A. 1985. Gidroprivod k pressu dlya ispytaniya obraztsov na prochnost [Hydraulic drive attached to the press for testing specimen strength]. Avt. svid. No. 1180745, B.I. No. 35.
112. Stavrogin, A.N. and Tarasov, B. G. 1985. Gidroprivod k pressu dlya ispytaniya obraztsov na prochnos [Hydraulic drive for the press for testing specimen strength]. Avt. svid. No. 1155902, B.I. No. 18.

113. Stavrogin, A. N. and Tarasov, B. G. 1985. Ustroistvo dlya dinamicheskikh ispytanii [Dynamic testing machine]. Avt. svid. No. 1174826, B.I. No. 31.
114. Stavrogin, A. N. and Tarasov, B. G. 1985. Bhystrodeistvuyushii klapan [A fast-acting cock]. Avt. svid. No. 1171629, B.I. No. 29.
115. Stavrogin, A. N. and Tarasov, B.G. 1985. Ustroistvo dlya dinamicheskikh ispytanii obraztsov materiyalov [Machine for dynamic testing of materials]. Avt. svid. No. 1180750, B.I. No. 35.
116. Stavrogin, A. N. and Tarasov, B. G. 1986. Gidroprivod k pressu dlya ispytaniya obraztsov na prochnost [Hydraulic drive attached to the press for testing rock strength]. Avt. svid. No. 1241090, B.I. No. 24.

APPENDICES

Appendix I

Specimen no.	+C = σ_2/σ_1	No. of specimens tested	Coeff. variation, %	Peak: comp. strength τ_{us}, MPa	Elas. limit τ_{el}, MPa	Coeff. perm. lateral strain, μ	Young's modulus, E	Poisson ratio, ν	Brief petrographic description
(1)	(2)	(3)	(4)	(5)	(6)	(7)	(8)	(9)	(10)
1	0.00	18	6.7	47.0	39.0	2.15	4.8×10^4	0.3	Talc chlorite (from Seg Lake deposit, Karelia). Dark green, consists of scaly type aggregate of talc, with fractions of chlorite and carbonates of Ca, Mg and Fe up to 25%. Specific gravity 2.916, bulk weight 2.91 g/cm^3, true porosity 0.21%.
	0.069	13	4.0	61.5	44.0	1.8			
	0.116	6	3.6	63.0	45.0	1.11			
	0.178	6	4.5	71.0	48.0	1.10			
	0.233	12	5.0	90.0	50.0	0.82			
	0.322	12	4.3	95.0	68.0	0.61			
	0.407	19	4.0	113.0	73.0	0.519			
	0.510	12	6.9	123.0	73.0	0.50			
2	0.00	37	3.9	38.2	30.0	2.3	6.2×10^4	0.28	White marble (Koelga, the Urals). Texture of marble dense, uniformly granular. Grains 0.15–1.40 mm in size with smooth, even edges, having direct contact. Specific gravity 2.713, bulk weight 2.71 g/cm^3, true porosity 0.11%.
	0.069	14	8.2	52.0	40.0	1.29			
	0.116	11	5.9	56.0	44.0	1.26			
	0.178	10	5.8	67.0	47.5	1.03			
	0.232	14	13.2	102.5	55.0	1.2			
	0.321	9	12.1	174.0	61.0	0.70			
	0.408	9	13.0	208.0	59.0	0.555			
	0.508	6	15.0	209.0	65.0	0.48			

Appendix I: (Contd)

(1)	(2)	(3)	(4)	(5)	(6)	(7)	(8)	(9)	(10)
3	0.00	9	7.4	58.0	42.5	1.834	4.0×10^4	0.3	White marble (presumably from Kararsk, Italy). Crystalline grain texture. Grains of calcite isometric, 0.1 to 0.35 mm in size. Major portion of grains 0.1–0.25 mm in size. Grain edges smooth, contact between grains direct. Bulk weight 2.71 g/cm^3, effective porosity 0.92%.
	0.069	7	8.3	79.5	58.0	1.463			
	0.116	6	5.6	83.5	61.0	1.47			
	0.178	8	12.3	149.5	74.0	0.795			
	0.232	8	10.2	266.5	99.0	1.22			
	0.313	3	6.4	382.5	137.0	0.80			
	0.405	3	0.6	386.0	149.0	0.60			
	0.515	3	2.9	430.00	145.0	0.482			
4	0.00	9	6.6	101.0	85.0	1.59	8.7×10^4	0.24	Diabase (from Bratsk Hydroelectric Station, Siberia). Darkish grey, poikilitic-diabasic (ophitic) texture. Plagioclase labrador content about 30–40%, of long prismatic shape, 0.25 to 1.3 mm in size. Monoclinal pyroxene-augite (40%), 0.8–12 mm in size. Specific gravity 2.99, bulk weight 2.97 g/cm^3, true porosity 0.98%.
	0.069	5	8.3	203.0	158.0	1.37			
	0.116	6	9.2	268.5	159.0	1.03			
	0.182	4	7.4	280.0	205.0	0.70			
	0.227	9	12.3	351.5	272.5	0.63			

									Description
5	0.00	17	11.6	61.0	5.0	2.32	1.8×10^4	0.1	Burst-prone sandstone (Donbass). Coarse-grained feldspar-quartzite. Fragments 0.2 to 1 mm in size (85–90%) (fractions 0.5 to 0.7 mm predominant) containing quartz, feldspar and sedimentary rocks. Cement of porous basal conglomerate type (10–15%), clayey with small quantity of sericite. Porosity 5.8–6.1%.
	0.069	6	6.9	133.0	99.0	2.29			
	0.116	6	6.8	194.0	109.0	1.5			
	0.178	6	11.5	250.5	165.0	1.76			
	0.227	6	19.4	382.0	252.0	0.667			
6	0.00	21	21.1	72.0	60.0	1.8	1.2×10^4	0.1	Non-burst-prone sandstone (NBP) (Donbass). Medium granular quartzite-feldspar. Fragments (65–70%) 0.1–0.5 mm in size (those of 0.25–0.3 mm predominant), containing quartz feldspar and sedimentary rocks. Cementing material of basal type, thin scaly texture; micaceous (15–20%) with portions of clay material. Calcite also encountered. Porosity 5.9–6.0%.
	0.069	8	23.0	122.0	99.0	1.7			
	0.116	7	19.3	179.0	133.0	1.14			
	0.178	7	23.4	206.5	152.0	0.95			
	0.227	6	16.5	402.0	250.0	0.76			

Appendix I: (*Contd*)

(1)	(2)	(3)	(4)	(5)	(6)	(7)	(8)	(9)	(10)
7	0.00	4	11.4	67.0	52.5	5.7	2.1×10^4	0.18	Sandstone D-8 (Donbass, from Oktyabr'sk mine). Fine grained, grey, with thin interlayers of mica and carbonised plant detritus. Fragments (65–70%) 0.04 to 0.3 mm in size, contain quartz feldspar. Cementing material basal type with porous sections. Effective porosity 7.4%.
	0.069	3	3.3	104.5	74.5	2.29			
	0.116	3	2.9	129.0	75.5	3.15			
	0.176	3	7.9	195.0	128.5	1.32			
	0.232	3	4.0	280.5	147.0	1.3			
	0.321	3	—	—	203.5	0.8			
8	0.00	3	6.3	117.5	9.3	2.4	6.5×10^4	0.28	Sandstone P-O (Donbass, Nesvetaiantratsit group). Grey, quartzitic feldspar, containing 60–70% silicate fragments and 30–40% clayey. Sericitic quartzitic and calcitic cement of binding type with more inclusions of basal type. Fragments 0.04–0.5 mm in size, quartz, plagioclase and sedimentary rocks. Bulk weight 2.72 g/cm³. Effective porosity 0.36%.
	0.069	5	12.0	166.0	109.0	2.16			
	0.116	2	7.0	225.0	141.5	1.34			
	0.178	3	21.0	158.5	181.0	1.30			
	0.223	4	9.4	383.0	—	—			

									Description
9	0.00	8	7.4	37.5	22.5	10.7	2.2×10^4	0.17	Sandstone P-04. Highly porous, quartzitic rock. Deposit from which collected, not known. Brownish-red, medium size, ferruginous grains. Consists of ferruginous cement of porous type 5–7% and silicate fragments 93–95%. Fragments of 0.06–0.6 mm in size contain quartz. Bulk weight 2.1 g/cm^3. Effective porosity 18.6%.
	0.069	4	3.0	62.5	39.5	4.9			
	0.116	4	5.3	92.0	53.0	3.9			
	0.178	5	4.2	116.0	71.3	1.1			
	0.232	6	2.9	252.0	96.5	0.95			
	0.322	6	3.7	144.0	—	—			
10	0.00	4	4.2	79.0	56.5	1.5	2.6×10^4	0.29	Sandstone D-12 (Donbass). Medium-size grains. Fractions (75%) 0.08–0.5 mm in size (those of 0.2 mm predominant). Cementing material (25%) of porous type, clayey, mixed with sericite 3%, muscovite 5–7%, secondary quartz 7%, sericite 1–3%.
	0.116	3	6.1	153.0	115.0	1.7			
	0.233	3	6.2	276.0	161.0	0.90			

Appendix I: (Contd)

(1)	(2)	(3)	(4)	(5)	(6)	(7)	(8)	(9)	(10)
11	0.00	6	7.3	92.0	77.5	1.26	7.2×10^4	0.29	Limestone D-6 (Donbass mine 1–2). Detritous greyish limestone consists of shell fragments of 0.02 to 0.4 mm, cemented by pelitic calcite with one inclusion of carbonaceous materials (5–7%). Porosity 1%.
	0.069	3	5.9	125.0	88.0	1.25			
	0.116	3	6.2	179.0	110.5	1.02			
	0.185	3	1.8	213.5	150.0	0.81			
	0.233	3	1.5	310.0	165.0	0.7			
12	0.00	5	15	9.05	65.0	1.35	3.7×10^4	0.25	Aleurolite (siltstone) D-19 (Donbass). Coarse-grained, transformed into fine-grained limestone. Micaceous argillite (mudstone) or pyrite layers (10–12%) present. Fragments (75–80%) of 0.02–0.25 mm in size. Fragments 0.0–0.15 mm in size mainly quartzitic, predominant. Cementing material (20–25%) of porous type, clayey with one inclusion of sericite.
	0.116	3	12	167.0	115.0	0.94			
	0.227	3	16	399.5	163.5	0.67			

13	0.00	5	4.3	56.5	48.5	1.0	2.3×10^4	0.36	Argillite (Donbass, Obukhovsk-Zapadnaya mine). Dark grey quartzitic. Major portion clayey mixed with sericite, quartzitic with rare fragments of silicates and thinly scattered carbonised material. Bulk weight 2.8 g/cm³. Effective porosity 1.4–1.6%.
	0.190	5	9.1	94.0	60.8	0.95			
	0.316	5	10	154.5	82.5	0.74			
14	0.00	2	5.1	68.5	57.0	1.65	2.8×10^4	0.31	Sandstone P-026 (Kizelovsk basin, the Urals). Bright greyish. Fragments (70–80%) of quartz feldspars 0.04–0.35 mm in size, having direct contact. Cement (20–30%) of porous type, clayey-sericitic with calcite up to 10–15%. Bulk weight 2.5 g/cm³. Specific gravity 2.64. Effective porosity 5.3%, true porosity 5.5%.
	0.069	5	22.0	138.5	98.7	1.8			
	0.116	5	11.0	160.0	109.5	1.44			
	0.178	5	7.8	225.0	161.0	1.2			
	0.232	4	3.2	316.5	207.5	0.96			

Appendix I: (Contd)

(1)	(2)	(3)	(4)	(5)	(6)	(7)	(8)	(9)	(10)
15	0.00	3	8.0	153.0	125.0	2.2	6.1×10^4	0.31	Sandstone P-02 (Kizelovsk basin, the Urals). Grey, medium-size grains. Fragments (85–95%) 0.1–0.5 mm in size Fragments of 0.3 size predominant. Cementing material from 5–15%, of porous type, quartzitic clayey-sericitic with calcite 7–10%. Specific gravity 2.64. Bulk weight 2.62 g/cm^3. Effective porosity 2%, true porosity 0.84%.
	0.068	4	3.2	262.0	204.0	2.04			
	0.116	4	5.5	311.5	228.0	1.8			
16	0.00	5	7.1	140.5	105.0	2.44	5.8×10^4	0.27	Sandstone P-03 (Kizelovsk basin, the Urals). Grey, unevenly fragmented (Fine- and medium-size grains). Fragments up to 0.55 mm in size (80%); those of 0.15–0.3 mm predominant. Cementing material (20%) of porous type, quartzitic-clayey-sericitic. Bulk weight 2.66 g/cm^3. Effective porosity 1.49%.
	0.070	5	3.7	220.5	136.0	1.3			
	0.116	5	10.5	254.5	190.0	1.07			
	0.178	5	8.0	326.0	225.0	0.92			
	0.227	6	12.1	510.0	250.0	0.68			

								Description
17	0.00	6.9	47.5	33.5	3.8	2.4×10^4	0.22	Sandstone P-021 (Kizelovsk basin, the Urals). Bright grey, metamorphic, calcinated, fine-grained. Quartz fragments (60–70%) 0.04–0.25 mm in size. Cement (30–40%) of basal type, clayey-micaceous, with calcite up to 7–10%. Bulk weight 2.5 g/cm^3. Effective porosity 5.7%.
	0.070	11.7	68.5	39.0	1.27			
	0.116	8.9	101.0	46.0	1.03			
	0.189	6.9	144.0	—	—			
	0.233	25.0	249.0	—	—			
18	0.00	6.3	39.5	28.3	0.75	2.1×10^4	0.24	Limestone (Estanoslanets, Estonia). Clayey limestone, slimy detritus. Major portion very fine-grained (0.02–0.04 mm) with fine-grained (0.10–0.15 mm) calcite mixed with celite based clayey material; silicate fragments 0.02–0.04 mm in size, up to 3%; residues of detritus shells 0.04–1.0 mm in size, up to 25–30%.
	0.070	4.6	56.5	37.0	0.60			
	0.116	6.6	58.5	44.0	0.55			
	0.192	5.7	69.0	48.5	0.50			
	0.245	8.3	112.5	65.0	0.44			
	0.322	8.3	126.0	66.0	0.45			

Appendix I: (*Contd*)

(1)	(2)	(3)	(4)	(5)	(6)	(7)	(8)	(9)	(10)
19	0.00	3	11.8	105.5	90.0	3.3	6.9×10^4	0.22	Urtite (Kola peninsula). Free from feldspar, highly alkaline, effusive nepheline rock. Texture pseudomorphic grains. Consists of 75% nepheline particles 0.25–2.5 mm in size, 15% alkaline non-ferrous minerals (pyroxene and hornblende). Bulk weight 2.7 g/cm^3. Effective porosity 1%.
	0.069	6	13.0	182.5	162.0	2.0			
	0.116	4	12.1	266.0	190.0	1.25			
	0.175	6	14.5	308.5	231.0	1.20			
	0.227	3	18.0	430.0	—	—			
20	0.00	4	17.0	65.0	53.0	1.14		0.4	Foyaite (Kola peninsula). Nepheline-albitite-orthoclase rock. Pseudomorphic trachytic texture. Contains nepheline particles 1.0–1.5 mm in size; orthoclase-albitite particles 4.25 mm in size in non-ferrous minerals. Bulk weight 2.6 g/cm^3. Effective porosity 1.7%.
	0.069	4	20.0	121.0	80.0	0.79			
	0.116	3	10.7	227.0	170.0	1.1			
	0.175	3	10.0	298.0	210.0	0.98			
	0.227	3	19.4	447.5	334.0	0.87			

21	0.00	5	13.3	90.0	71.5	3.3	7.2×10^4	0.23	Diabase (Kola peninsula). Black. Texture ophitic with irregular tabular augite in interstices between prisms of plagioclase. Plagioclase-labrador up to 30–40%, idiomorphic crystals, highly fissured 0.75–3.0 mm in size. Monoclinic pyroxene of augite type (30–35%) 0.65 to 2–3 mm in size, olivine (5–7%) 0.25–0.45 mm in size. Effective porosity 0.58%.
	0.069	5	11.2	125.0	88.5	0.88			
	0.116	5	14.1	173.0	132.5	1.20			
	0.175	3	6.4	236.0	165.0	1.25			
	0.227	5	4.2	300.0	201.0	0.67			
22	0.00	3	5.6	184.0	178.5	1.2	7.9×10^4	0.27	Diorite No. II (Kola peninsula). Dioritic porphyrite, dark grey, weathered. Hypomorphic fine crystalline structue. Consists of turbid, composite plagioclase (30–40%), tabular and lengthy prismatic (0.8 mm) quartz (5–10%), biotite (7–10%) up to 0.6 mm in size. Effective porosity 0.15%.
	0.069	3	17.1	283.5	198.0	1.5			
	0.116	3	5.2	321.0	265.0	1.08			
	0.175	3	3.0	375.0	291.0				
	0.227	3	30.4	430.0	330.0	0.55			

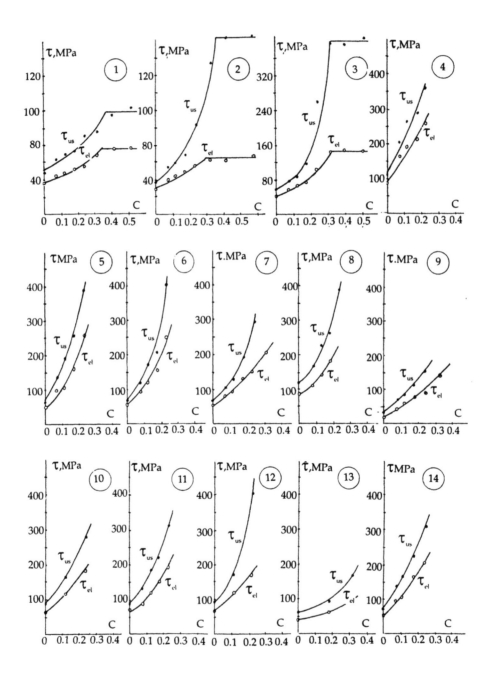

Fig. 1. Appendix I: (Contd.)

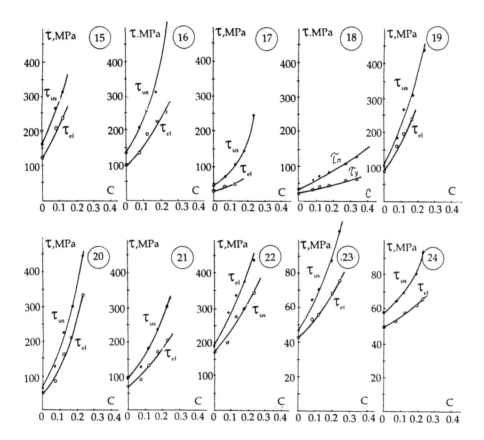

Fig. 1. Appendix I: Relationships τ versus C plotted for a series of rocks according to the Table given above.

Appendix I: (Contd)

(1)	(2)	(3)	(4)	(5)	(6)	(7)	(8)	(9)	(10)
23	0.00	6	8.0	46.5	42.5	0.6	2.9×10^4	0.21	Concrete (specimens cored from blocks). Prepared from portland cement-500; sand to cement ratio 2; water to cement ratio 0.4. Quartzitic sand grains 0.2–0.3 mm in size. Cement 40–50% of basal type, pelitomorphic, sometimes grains 0.01–0.02 mm in size. Voids (up to 10%) of up to 0.5 mm size seen. Bulk weight 2.35 g/cm^3. Effective porosity 18–20%.
	0.070	3	10.4	64.0	52.5	0.6			
	0.116	2	2.3	70.0	55.6	0.5			
	0.20	2	4.2	86.5	68.0	0.4			
	0.238	5	14.2	103.5	74.0	0.25			
24	0.00	4	10.0	58.0	49.5	0.7	2.0×10^4	0.20	Cement (cored specimens) with no added sand. Water to cement ratio 0.34. Results of microsection analysis: major portion pelitomorphic, some grains 0.02–0.1 mm in size and sharp edged fragments 0.1–0.35 mm found; sometimes hollow voids up to 5 mm in size and fissures up to 10–15% also encountered. Bulk weight 1.85 g/cm^3. Effective porosity 25–30%.
	0.070	6	7.5	64.7	52.5	0.39			
	0.116	3	4.8	68.5	57.0	0.35			
	0.20	6	11.0	80.5	62.5	0.32			
	0.238	3	4.9	93.0	64.0	0.20			

Appendix II
Brief Petrographic Description of Rocks

DIABASE (from a deposit in the region of Bratsk Hydroelectric Station)

Poikilitic-ophitic texture. Consists of coarse plate-like, allotriomorphic grains of monoclinal pyroxene, in which idiomorphous crystals grow— basaltic plagioclase. Plagioclase-labrador (55–60%) in the form of elongated prismatic crystals from 0.1×0.4 to 1.5×0.25 mm in size, polysynthetically twinned, differently oriented, forming poikilitic intergrowths in pyroxene and sometimes completely locked into large pyroxene crystals. Monoclinal pyroxene (35–40%) of augite type in the form of large plate-like crystals (up to 3.5 mm), slightly brownish, with low double refraction, having oblique extinction, angle of extinction 42–44°, with frequent poikilitic intergrowths of plagioclase. Ore mineral 3–4%, black, not transparent, irregular outlines, ranging in size from 0.02×0.9 to 0.6×1.2 mm.

Portions of chlorite (up to 1%) and biotite (1%) are also encountered. Bulk weight 2.92–2.93 g/cm^3. Total open porosity 0.48–0.49%.

LIMESTONE (Leningrad shales)

Clayey, highly detrital, microfine grain texture. Major portion—microfine granular (0.01–0.1 mm) calcite with portions of fine grains (0.15 mm) mixed with clayey-sericitic substance, with aleurolite fragments of silicate minerals in varying quantity from 5 to 20%, with detrital shells (15–25% of 0.1–3 mm size) and shale substance from 7 to 25%. Dolomite, colourless mica and ore-bearing minerals (0.01–0.04 mm) in quantities ranging from 5–10 to 15%. Bulk weight 2.48–2.49 g/cm^3. Effective porosity 6.52–8.24%. Water saturation limit 2.62–3.33%. Pore size 0.01–0.09 mm.

MARBLE I (Koelga deposit)

Granular structure. Massive texture. Rock consists of carbonate grains, more or less isometric in form (from 0.05–0.85 mm in diameter, with those of

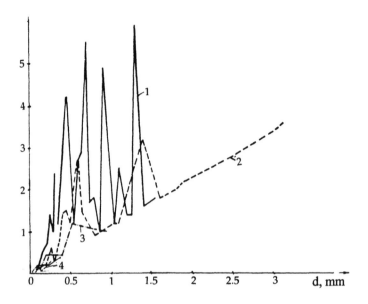

Fig. 1. Appendix II: Grain size distribution function for diabase. 1—plagioclase (57.3%), 2—pyroxene (36.9%), 3—ore grains (3.55%), 4—biotite (0.9%)

0.3–0.4 mm predominant). Round and smooth granular edges. Direct contact between grains. Specific gravity 2.713 g/cm^3.

ROCK SALT (Starobinsk deposit, Byelorussia)

Structure, crystalline-granular, coarse granular and highly coarse granular. Rock salt (halite) grains cubic, rarely parallelepiped or irregular form (0.24–16 mm in size, with those 1–12 mm in size predominant, about 84%). Finely dispersed clayey substance with microgranular structure and microaggregate polarisation encountered in interstices between grains of halite and sometimes in accumulations on grains and at granular edges. Crystals of microgranular (0.01–00.05 mm) carbonate and anhydride (0.1–0.15 up to 0.4 mm) of prismatic and zigzag fibrous structure can be distinguished in composition of clayey substance. Bulk weight 2.08–2.09 g/cm^3. Total open porosity 0.78–1.06%.

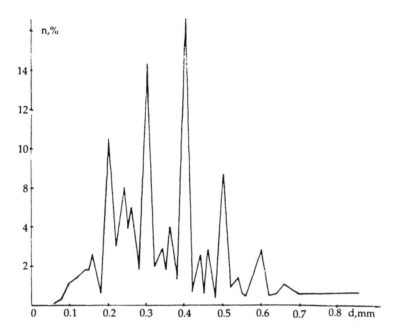

Fig. 2. Appendix II: Grain size distribution function for marble.

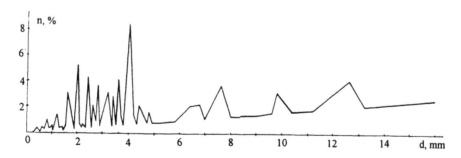

Fig. 3. Appendix II: Grain size distribution function for rock salt from the Starobinsk deposit.

SYLVINITE (Starobinsk deposit, Byelorussia)

Crystalline-granular, fine- and medium-granular structure. Finely striped microtexture. Stripes differ in composition, form and quantity of grains, and amount of impurities in halite. Thickness of stripes from 3 to 7–10 mm. Halopelite bands 1–3 mm thick encountered.

Average composition of sylvinite: sylvine 68–70% and rock salt 30–32%; predominant size of sylvine grains 0.2–0.5 up to 1 mm, 47%; 1–3 mm, 15%; larger than 3 mm, 4%; predominant size of rock salt grains: 0.2–0.5 up to 1 mm, 23%; 1–6 mm, 9%. Specific gravity 1.97–1.99 g/cm^3. Bulk weight 1.94–1.95 g/cm^3. Open porosity 0.32–1.18%. Total porosity 1.52–2.02%.

ROCK SALT (Tadzhikistan)

Crystalline-granular, coarse granular, highly coarse granular and gigantic granular structure. Rock salt consists of halite grains (from 0.1–7.5 mm), cubic and rarely parallelepiped in shape. Predominant sizes of halite grains 0.5–1.5 mm total 90–92%. Pilitomorphic dark brown clayey substance (3–5%) with microgranular carbonate (1–2%) and crystals of fine granular (0.25 mm) calcite (1%) encountered in intergranular gaps of halite and at grain edges enveloping these grains. Bulk weight 2.06–2.09 g/cm^3. Open porosity 2.13–3.21%.

BURST-PRONE SANDSTONE (Donbass, Gorlovsky Region, *Kochegarka mine*)

Fine granular, light greyish. Psammitic structure, with fine grains. Fragments of silicate (70–75%) 0.04–0.4 mm in size, not well sorted: grains with irregular edges (0.2–0.25 mm) consisting of quartz, feldspars, quartzites, effusive rocks etc. predominant. Cementing substance (25–30%) of porous type, clayey-sericitic with microfine granular carbonate 7%, calcite 5–6%, siderite 1–2%. The following fragments are also encountered: muscovite (3%) of film type (up to 0.5 mm), thin scaly colourless chlorite (2%), secondary quartz (5–7%) and charred plant detritus. Microsection of rock, fissured and porous. Pores (10–15%) 0.004–0.25 mm in size. Bulk weight 2.5 g/cm^3. Effective porosity 6.5–7.5%. Water saturation limit 2.8–3%.

NON-BURST-PRONE SANDSTONE (Donbass, Gorlovsky Region, *Kochegarka mine*)

Coarse aleurolite mixed with sideritic sandstone, dark grey. Aleurolitic structure coarse, mixed with sandstone. Fragments (50–60%) 0.02–0.2 mm in size, semi-rounded grains (0.06–0.08 mm) abundant; mainly quartz,

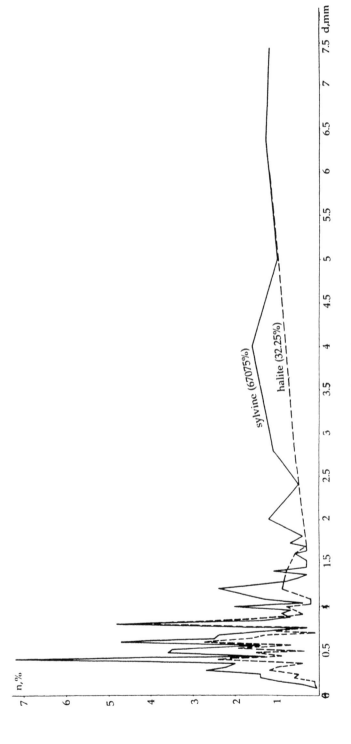

Fig. 4. Appendix II: Grain size distribution function for sylvine and rock salt contained in sylvinite from the Starobinsk deposit. 1—sylvine (67.7%), 2—rock salt (32.2%)

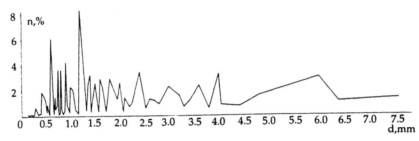

Fig. 5. Appendix II: Grain size distribution function for rock salt from Tadzhikistan.

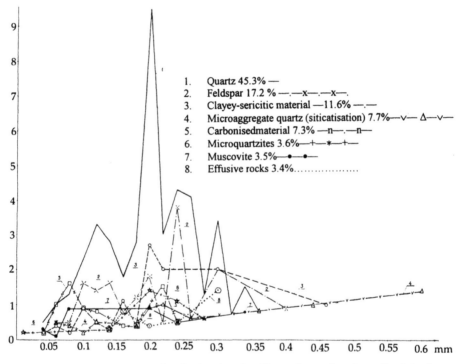

Fig. 6. Appendix II: Grain size distribution function (for various minerals as listed in the legend) for burst-prone sandstone from Kochegarka mine (Donbass).

feldspar, quartzite, effusive rocks etc. Cement (40–50%) of basaltic type, pelitomorphic-fine granular siderite (25%) with insignificant inclusion of pelitomorphic-clayey substance, muscovite films (3%) (up to 0.15 mm) and charred plant-based detritus (5%). Detritus along with spots (0.25 mm) of pelitomorphic siderite, sometimes with muscovite, forms thin (fractions of mm) radiating interlayers. Bulk weight 2.46 g/cm^3. Effective porosity 6.7–7.1%. Water saturation limit 2.9%.

QUARTZITIC SANDSTONE

Coarse granular, quartzitic feldspar, brownish-red, highly porous. Structure psammitic, coarse and medium granular. Semi-rounded fragments (90%) 0.2 to 1 mm in size and those of 0.5 mm predominant, mainly of quartz, feldspars, rarely microquartzite. Film-like cement (5%) clayey-sericitic, streaked with chlorite (colourless chlorite, radially disseminated). Voids (up to 20%) of 0.05 to 0.6 mm and a few up to 1 mm also encountered in this rock. Bulk weight 2.15 g/cm^3. Effective porosity 18.72%. Water saturation limit 8.75%.

BROWN COAL (LIGNITE) (Shurabsk deposit. Mine named after Lenin, Coal seam B)

Lignite-humus, fusain-semi-fusain, dull, black with pale tarred lustre, relatively soft. Unevenly fractured coal. Fractured surface rough.

Massive texture. Homogeneous structure. Attrition—fragmented microstructure. Major portion of coal of fusain-semi-fusain composition: represented by alternate fragmented and aggregate residues of stem and leaf parts of plants. Mineral substance (up to 10%) present. Moisture content in coal 20.4–26.2%. Specific gravity 1.33–1.35 g/cm^3. Bulk weight 0.87–0.9 g/cm^3. True porosity 33.5–35.5%.

BITUMINOUS COAL (Kizelovsk basin)

Micro-macrosporinite (microsporinitic durain), black humus coal, matte, viscous, dense with tarred lustre, rough surface.

Massive texture. Scratched structure. Major portion of coal consists of fusain-sporinite. Coal composed of: lipoidic components (group of leucoptinite) up to 90%, represented by sporinite—fragments of exine microspar (60%) 0.02–0.16 mm in size and macrospar (25%) 1.4 mm in length and 0.2 mm in thickness, and rarely by cutinite (3–5%), fusainite (1–3%)—thin, elongated lenses (0.88 × 0.12 mm, 1.6 × 0.16 mm) with well-preserved cage-like structure; semi-fusainite (3–5%) lenses (1.2 × 0.2 mm); vitrinite (1–2%) bands 0.6 mm in thickness and small lenses (0.1 mm); fragmentary accumulations (up to 1%) of broken fusainite (0.02–0.18 mm). Major cementing substance consists of small fragments—lumps of plant tissues 0.007–0.04 mm in size and mixture of finely disseminated clayey material and grains of pyrite (1–2% of 0.007–0.25 mm size). Singular, thin (0.02–0.1 mm) cracks filled with clay. Bulk weight 1.4 g/cm^3. Open porosity 0.96–1.22%.

ANTHRACITE (Donbass)

Anthracite: highly metamorphosed humus coal, black, hard, containing 1–4% volatile components, with bright metallic lustre and golden tone, irregular fractures, sections of small voids in the plane; terraced surface of fractures with smooth sections.

Massive texture. Microstructure often thinly striped, scratched. Major microcomponent, vitrinite (90%).

Banding (striation) of coal from stripes and layer-wise lenses of fusain (1–3%) and semi-fusain (3–5%) vary in length from 0.1 to 23 mm and thickness from 0.03 to 3 mm; spaced at 0.03–0.24 mm to 2–3 mm. Fractured coal. Empty and mineralised cracks. Thickness of cracks 0.03–0.18 mm, spaced 5–10 mm apart. Spot grains of pyrite (1–2%) encountered.

Bulk weight 1.53–1.55 g/cm^3. Open porosity 1.15–1.55%.